Complex Systems Design & Management

Omar Hammami, Daniel Krob,
and Jean-Luc Voirin (Eds.)

Complex Systems Design & Management

Proceedings of the Second International Conference on Complex Systems Design & Management CSDM 2011

Springer

Editors

Prof . Omar Hammami
ENSTA ParisTech
32 Bvd Victor
75739 Paris Cedex 15
France
E-mail: hammami@ensta.fr

Prof. Daniel Krob
Ecole Polytechnique
DIX/LIX
91128 Palaiseau Cedex
France
E-mail: dk@lix.polytechnique.fr

Jean-Luc Voirin
Thales Systèmes Aéroportés
10, avenue de la 1ère DFL / CS 93801
29238 BREST Cedex 3
France
E-mail: jean-luc.voirin@fr.thalesgroup.com

ISBN 978-3-642-25202-0 e-ISBN 978-3-642-25203-7

DOI 10.1007/978-3-642-25203-7

Library of Congress Control Number: 2011941489

Typeset & Cover Design: Scientific Publishing Services Pvt. Ltd., Chennai, India.

Printed on acid-free paper

9 8 7 6 5 4 3 2 1

springer.com

Preface

Introduction

This volume contains the proceedings of the Second International Conference on "Complex System Design & Management" (CSDM 2011; see the conference website: `http://www.csdm2011.csdm.fr` for more details).

The CSDM 2011 conference was jointly organized by the research & training Ecole Polytechnique – Thales chair "Engineering of Complex Systems" and by the non profit organization C.E.S.A.M.E.S. (Center of Excellence on Systems Architecture, Management, Economy and Strategy) from December 7 to December 9 at the Cité Internationale Universitaire of Paris (France).

The conference benefited of the permanent support of many academic organizations such as Ecole Centrale de Paris, Ecole Nationale Supérieure des Techniques Avancées (ENSTA), Ecole Polytechnique, Ecole Supérieure d'Electricité (Supélec), Université Paris Sud 11 and Télécom Paristech which were deeply involved in its organization. A special thank is also due to MEGA and Thales companies which were the main industrial sponsors of the conference. All these institutions helped us a lot through their constant participation to the organizing committee during the one-year preparation of CSDM 2011. Last, but least, we would also like to point out the assistance of Alstom Transport & EADS in the same matter.

Why a CSDM Conference?

Mastering complex systems requires an integrated understanding of industrial practices as well as sophisticated theoretical techniques and tools. This explains the creation of an annual *go-between* forum at European level (which did not existed yet) dedicated it both to academic researchers and industrial actors working on complex industrial systems architecture and engineering in order to facilitate their *meeting*. It was actually for us a *sine qua non* condition in order to nurture and develop in Europe this complex industrial systems science which is now emergent.

The purpose of the "Complex Systems Design & Management" (CSDM) conference is exactly to be such a forum, in order to become, in time, *the* European academic-industrial conference of reference in the field of complex industrial systems architecture and engineering, which is a quite ambitious objective. The first CSDM 2010 conference – which was held in end of October 2010 in Paris – was the first step in this direction with more than 200 participants coming from 20 different countries with an almost perfect balance between academia and industry.

The CSDM Academic–Industrial Integrated Dimension

To make the CSDM conference this convergence point of the academic and industrial communities in complex industrial systems, we based our organization on a principle of *complete parity* between academics and industrialists (see the conference organization sections in the next pages). This principle was first implemented as follows:

- the Programme Committee consisted of 50 % academics and 50 % industrialists,
- the Invited Speakers are coming equally from academic and industrial environments.

The set of activities of the conference followed the same principle. They indeed consist of a mixture of research seminars and experience sharing, academic articles and industrial presentations, software and training offers presentations, etc. The conference topics covers in the same way the most recent trends in the emerging field of complex systems sciences and practices from an industrial and academic perspective, including the main industrial domains (transport, defense & security, electronics & robotics, energy & environment, health & welfare services, media & communications, e-services), scientific and technical topics (systems fundamentals, systems architecture & engineering, systems metrics & quality, systemic tools) and system types (transportation systems, embedded systems, software & information systems, systems of systems, artificial ecosystems).

The CSDM 2011 Edition

The CSDM 2011 edition received 71 submitted papers (representing a 15 % progression with respect to 2010), out of which the program committee selected 21 regular papers to be published in these proceedings and 3 complementary industrial full presentations at the conference, which corresponds to a 33 % acceptance ratio which is fundamental for us to guarantee the high quality of the presentations. The program committee also selected 18 papers for a collective presentation in the poster session of the conference.

Each submission was assigned to at least two program committee members, who carefully reviewed the papers, in many cases with the help of external referees. These reviews were discussed by the program committee during a physical meeting held in Ecole Nationale Supérieure des Techniques Avancées (ENSTA, Paris) by June 1, 2011 and via the EasyChair conference management system.

We also chose 16 outstanding speakers with various industrial and scientific expertise who gave a series of invited talks covering all the spectrum of the conference during the two first days of CSDM 2011, the last day being dedicated to the presentations of all accepted papers. The first day of the conference was especially organized around a common topic – Sustainable Design – that gave a coherence to all the initial invited talks.

Futhermore, we had a poster session, for encouraging presentation and discussion on interesting but "not-yet-polished" ideas, and a software tools presentation session, in order to provide to each participant a good vision on the present status of the engineering tools market offer.

Acknowledgements

We would like finally to thank all members of the program and organizing committees for their time, effort, and contributions to make CSDM 2011 a top quality conference. A special thank is addressed to the CESAMES nonprofit organization team which managed permanently with an huge efficiency all the administration, logistics and communication of the CSDM 2011 conference (see `http://www.cesames.net`).

The organizers of the conference are also greatly grateful to the following sponsors and partners without whom the CSDM 2011 would not exist:

- **Academic Sponsors**
  - École Centrale de Paris,
  - École Polytechnique,
  - École Supérieure d'Electricité (Supélec),
  - ENSTA Paritech,
  - Télécom Paristech,
  - Commissariat à l'Energie Atomique (CEA-LIST)
- **Industrial Sponsors**
  - Thales,
  - Mega International,
  - EADS,
  - Bouygues Telecom,
  - Veolia Environnement,
  - EDF,
  - BelleAventure

- **Institutional Sponsors**
  - Digiteo labs,
  - Région Ile-de-France,
  - Ministère de l'Enseignement Supérieur et de la Recherche
- **Supporting Partners**
  - Association Française d'Ingénierie Système (AFIS),
  - International Council on Systems Engineering (INCOSE),
  - Société de l'Electricité, de l'Electronique et des Technologies de l'Information et de la Communication (SEE)
- **Participating Partners**
  - Atego,
  - Dassault Systèmes,
  - Enbiz,
  - IBM Rational Software,
  - Lascom,
  - Knowledge Inside,
  - Maplesoft,
  - Mathworks,
  - Obeo,
  - Pôle de compétivité System@tic,
  - Project Performance International.

Paris, July 31, 2011

Omar Hammami – ENSTA Paristech
Daniel Krob – Ecole Polytechnique & CESAMES
Jean-Luc Voirin – Thales

Conference Organization

Conference Chairs

- *General & Organizing Committee Chair:*
  - Daniel Krob, Institute Professor, Ecole Polytechnique, France
- *Program Committee Chairs:*
  - Omar Hammami, Associate Professor, ENSTA ParisTech, France (academic co-chair of the Program Committee)
  - Jean-Luc Voirin, Thales, France (industrial co-chair of the Program Committee)

Program Committee

The PC consists of 30 members (15 academic and 15 industrial): all are personalities of high international visibility. Their expertise spectrum covers all of the conference topics.

Academic Members

- *Co-chair:*
  - Omar Hammami, Associate Professor, ENSTA ParisTech, France
- *Other Members:*
  - Farhad Arbab (Leiden University, Netherlands)
  - Jonas Andersson (National Defence College, Sweden)
  - Daniel Bienstock (Columbia University, USA)
  - Manfred Broy (TUM, Germany)
  - Daniel D.Frey (MIT, USA)

- Leon Kappleman (University of North Texas, USA)
- Kim Larsen (Aalborg University, Danemark)
- Jon Lee (University of Michigan, USA)
- Timothy Lindquist (Arizona State University, USA)
- Gerrit Jan Muller (Buskerud University College, Norway)
- Jacques Printz (CNAM, France)
- Bernhard Rumpe (RWTH Aachen, Germany)
- Jan Rutten (CWI, Netherlands)
- Ricardo Valerdi (MIT, USA)

Industrial Members

- *Co-chair:*
  - Jean-Luc Voirin, Thales, France
- *Other Members:*
  - Erik Aslaksen (Sinclair Knight, Australia)
  - Ian Bailey (Model Futures, Great-Britain)
  - Yves Caseau (Bouygues Telecom, France)
  - Hans-Georg Frischkorn (Verband der Automobilindustrie, Germany)
  - Yushi Fujita (Technova, Japon)
  - Jean-Luc Garnier (Thales, France)
  - Gerhard Griessnig (AVL LIST, Austria)
  - Andreas Mitschke (EADS, Germany)
  - Jean-Claude Roussel (EADS, France)
  - Bran Selic (Zeligsoft, Canada)
  - Jeffrey D. Taft (CISCO, USA)
  - David Walden (Sysnovation, USA)
  - Ype Wijnia (D-cision, Netherlands)
  - Nancy Wolff (Auxis, USA)

Organizing Committee

- *Chair:*
  - Daniel Krob, Institute Professor, Ecole Polytechnique, France
- *Other Members:*
  - Marc Aiguier (Ecole Centrale de Paris, France)
  - Karim Azoum (System@tic, France)
  - Paul Bourgine (Ecole Polytechnique, France)
  - Isabelle Demeure (Télécom ParisTech, France)
  - Claude Feliot (Alstom Transport, France)
  - Gilles Fleury (Supélec, France)

- Pascal Foix (Thales, France)
- Vassilis Giakoumakis (Université de Picardie, France)
- Antoine Lonjon (MEGA, France)
- Clothilde Marchal (EADS, France)
- Isabelle Perseil (Inserm, France)
- Bertrand Petit (Innocherche, France)
- Sylvain Peyronnet (Université Paris Sud, France)
- Antoine Rauzy (Ecole Polytechnique, France)
- Jacques Ariel Sirat (EADS, France)

Invited Speakers

Societal Challenges

- Yves Bamberger, CEO scientific adviser, EDF - France
- Denise Pumain, professor, Université Paris 1 - France
- Philippe Martin, R&D vice president, Veolia Environnement - France
- Carlos Moreno, SINOVIA chairman and scientific adviser for the business unit FI&SA, GDF Suez - France

Industrial Challenges

- Eric Gebhardt, vice president Energy Services, General Electric - USA
- François Briant, chief architecte, IBM France - France
- Jérôme Perrin, director of CO2- Energy- Environment Advanced Projects, Renault - France
- Renaud de Barbuat, chief information officer, Thales - France

Scientific State-of-the-art

- José Fiadeiro, professor, University of Leicester - UK
- Marc Pouzet, professor, Ecole Normale Supérieure - France
- Dimitri Mavris, professor, Georgia Tech - USA
- Nancy Leveson, professor, MIT - USA

Methodological State-of-the-Art

- Derek Hitchins, professor, Cranfield University - UK
- Michael Hinchey, professor, University of Limerick - Ireland
- Alberto Tobias, head of the Systems, Software & Technology Department, Directorate of Technical & Quality Management, European Spatial Agency - Netherlands
- Michel Riguidel, professor, Télécom ParisTech - France

Contents

Chapter 1
An Overview of Design Challenges and Methods in Aerospace Engineering

Dimitri N. Mavris and Olivia J. Pinon

Abstract. Today's new designs have increased in complexity and need to address more stringent societal, environmental, financial and operational requirements. Hence, a paradigm shift is underway that challenges the way complex systems are being designed. Advances in computing power, computational analysis, and numerical methods have also significantly transformed and impacted the way design is conducted. This paper presents an overview of the challenges and enablers as they pertain to the Conceptual, Preliminary and Detailed design phases. It discusses the benefits of advances in design methods, as well as the importance of visualization and knowledge management in design. Finally, it addresses some of the barriers to the transfer of knowledge between the research community and the industry.

Keywords: Advanced Design Methods Surrogate Modeling Probabilistic Design Multidisciplinary Analysis and Optimization.

1 Introduction

As Keane and Nair [28] noted, "Aerospace engineering design is one of the most complex activities carried by mankind". In a world characterized by fierce competition, high performance expectations, affordability and reduced

Dimitri N. Mavris · Olivia J. Pinon
Georgia Institute of Technology,
School of Aerospace Engineering
270 Ferst Drive, Atlanta, GA 30332-0150, U.S.A.
Tel.: +1-404-894-1557; +1-404-385-2782
Fax: +1-404-894-6596
e-mail: dimitri.mavris@aerospace.gatech.edu, olivia.pinon@asdl.gatech.edu

time-to-market [54], this is an activity on which a company bets its reputation and viability with each new design opportunity [20].

Today's new aerospace designs have increased in complexity and need to address more stringent societal, environmental, financial and operational requirements. Hence, a paradigm shift is underway that challenges the way these complex systems are being designed. Advances in computing power, computational analysis, and numerical methods have also significantly transformed and impacted the way design is conducted, bringing new challenges and opportunities to design efforts.

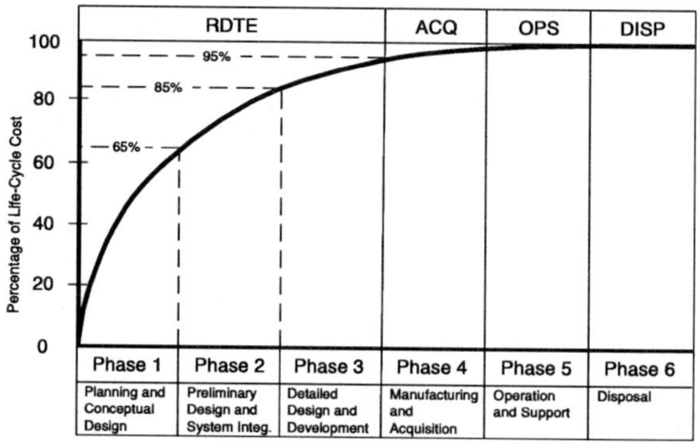

Fig. 1 Percentage of Life Cycle Cost during the Aircraft Life Cycle Phases [55]

Design activities involve an irrevocable allocation of resources (money, personnel, infrastructure) as well as a great amount of analysis to help choose an alternative that satisfies the strategic goals, product objectives, customer needs and technical requirements [19, 24]. In particular, the lock-in of aircraft life cycle costs early on in the design process, as illustrated in Figure 1, emphasizes the need for new methods to help generate data and knowledge, support synthesis and analysis, and facilitate decision making during these phases.

This paper first presents the current challenges and enablers as they pertain to these first three design phases. It further illustrates the importance of visualization and knowledge management in design. Finally it concludes on the benefits and importance of advances in design methods and later addresses some of the barriers to the transfer of knowledge between the research community and the industry.

2 The Design Process – Challenges and Enablers

Design is a problem-solving activity that maps a set of requirements to a set of functions, leading to a set or series of decisions that contribute to the final description of a solution [41, 44] meeting particular needs. Specifically, design involves defining and exploring a vast space of possibilities that requires the building up of knowledge and a familiarity with the constraints and trades involved [45]. Designers typically follow a three-phase design process, namely Conceptual, Preliminary, and Detailed design (Figure 2), during which the level of detail of the representations and analyses increases with each phase. Consequently, the breadth and depth of the analysis and trades considered, along with their level of uncertainty and accuracy, vary significantly between each phase. For example, Preliminary design is characterized by higher fidelity analyses and tools than Conceptual design. Similarly, uncertainty (in particular disciplinary uncertainty) is much more prevalent in Conceptual design (Figure 3) and must be quantified with fewer parameters than in Preliminary design. All these aspects have concrete and significant implications on the way design is conducted. The following sections review the main activities and challenges that characterize Conceptual, Preliminary and Detailed Design.

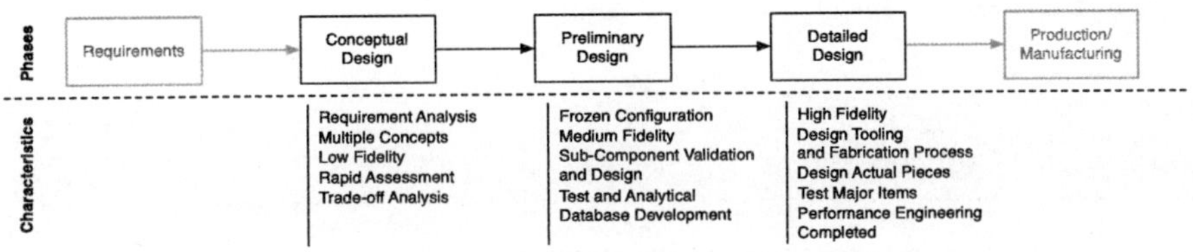

Fig. 2 The Three Phases of Design [43, 53]

2.1 Conceptual Design

The goal of Conceptual Design is to enable the identification and selection of a feasible and viable design concept to be further developed and refined during Preliminary design. This phase is thus the most important one in terms of the number of design concepts to be developed and examined, the feasibility studies to be conducted, the technologies to be considered, and the mappings that need to occur between requirements and configurations. It is also a key element of the design process. Indeed, as changes to the design concept at the later stages come at high expenses, decisions made during Conceptual design have a strong and critical impact on the overall cost, performance and life cycle of the final product. At this stage of the design process, however, the designer is faced with a lack of relevant knowledge and data regarding the problem, its requirements, its constraints, the technologies to be infused, the analytical tools and models to be selected, etc. The following sections

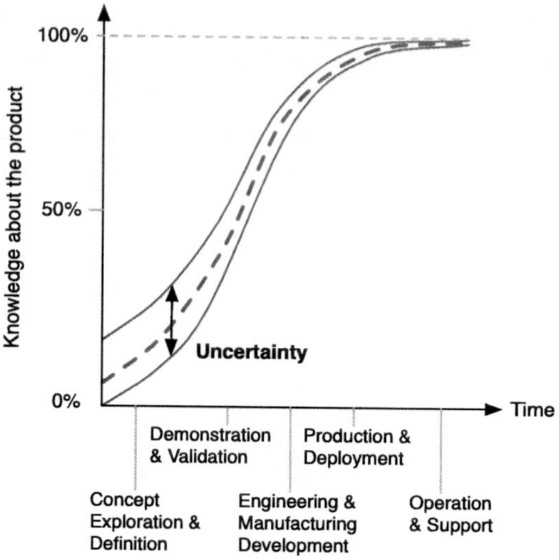

Fig. 3 Uncertainty Variation in Time (notional) [43]

review these challenges in more detail and discuss some of the methods and techniques to address them.

2.1.1 Requirements Definition and Sensitivity

The definition of the design problem and requirements is the starting point of any design exercise and represents a central issue in design. Design requirements are a direct driver of cost and, therefore, affordability. It is thus critical to understand and capture the significance of requirements' impact on the design, as well as the sensitivity of affordability metrics to requirements. It is also important to acknowledge that the formulation of crisp and specific requirements by the customers is a grueling task. This is particularly true in commercial programs, where changes in the market and unknown implications of design specifications prevent customers from expressing more than desired outcomes or perceived needs [9, 20, 73]. Consequently, new programs often start without clearly defined requirements [73], which prompts requirements to be refined, during the initial phase, as design trade studies are being conducted and customers' understanding and knowledge about the design problem increase [20, 73]. This requirement instability has serious implications as variation in requirements may result in the selection of a different design concept. Consequently, unless multiple scenarios are investigated and documented, a static approach to design will rapidly show its limitations. Additionally, decisions regarding the concept to pursue must be made in the presence of multiple, usually conflicting criteria. Such decisions can be facilitated through the use of Multiple Attribute Decision Making (MADM) selection techniques. Finally, decision making implies considerations beyond the technical aspects. Therefore, the final solution may be different from one decision maker's perspective to the next. There is thus a need to move from deterministic, serial, single-point designs to dynamic parametric trade environments. In particular, the decision maker should be provided with a parametric formulation that has the appropriate degrees of freedom to play the what-if games he is interested in. Such enabler is further discussed in the following section.

As discussed at the beginning of the previous section, the decisions made during the Conceptual design phase have a tremendous impact on the life cycle cost of the system. To support informed decision making, analyses need to be conducted, hence requiring the use of mathematical modeling. Indeed, as emphasized by Bandte [2], it is during this phase that mathematical modeling can have the strongest influence on the decisions made. Hence, while the use of mathematical models is traditionally associated with Preliminary design, there is a strong incentive and value in moving modeling efforts upstream in the design process, i.e in the conceptual design phase. Additional challenges to be tackled in Conceptual design arise from the definition and nature of the system itself. As defined by Blanchard and Fabricky [3],

> "The total system, at whatever level in the hierarchy, consists of all components, attributes, and relationships needed to accomplish an objective. Each system has an objective, providing a purpose for which all system components, attributes, and relationships have been organized. Constraints placed on the system limit its operation and define the boundary within which it is intended to operate. Similarly, the system places boundaries and constraints on its subsystems."

A system is thus a collection of multiple complex and heterogeneous subsystems and elements (wing, tail, fuselage, landing gear, engine, etc.) whose defining disciplines (aerodynamics, structures, propulsion, etc.) and their interactions need to be pursued concurrently and integrated (Figure 4) [40]. Hence, the increasing complexity of Aerospace designs has incited a growing interest in Multidisciplinary Analysis and Design Optimization (MDA/MDO) to support the identification and selection of designs that are optimal with regard to the constraints and objectives set by the problem.

As mentioned by Sobieszczanski-Sobieski and Haftka [69], MDO methods, in Aerospace Engineering, were initially developed and implemented to address issues in detailed structural design and the simultaneous optimization of structures and aerodynamics. Today, as discussed by German and Daskilewicz [20], MDO methods have transcended their structural optimization origins to encompass all the phases of design, and it is now recognized that

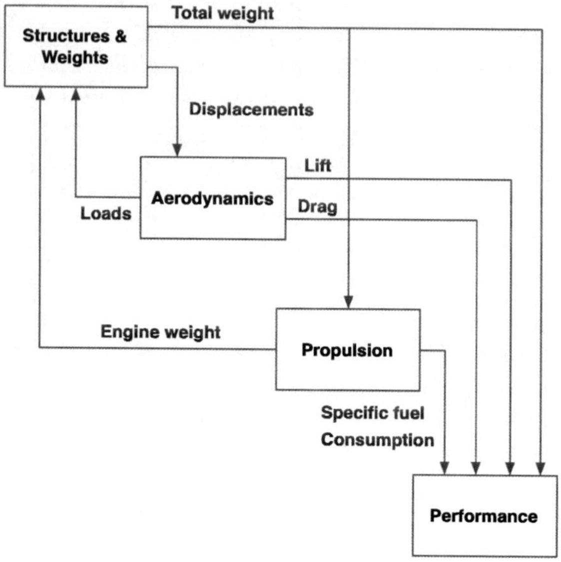

Fig. 4 Potential Interactions in a Multidisciplinary System [28]

the greatest impact and benefits of optimization are experienced during the Conceptual design phase. These aspects also support the ongoing effort, further discussed in Section 2.2, to bring more analysis earlier in the design process.

2.1.2 Integration of Multiple Disciplines

With increasing system complexity comes a growing number of interactions between the subsystems. This added complexity can originate, for example, from the recently recognized need to integrate both aircraft thermal management and propulsion systems, along with their resulting dynamic effects, early on in the design process [8, 39]. Integrating the different disciplines also represents an issue as it raises the number of constraints and inputs involved and increases the complexity of the modeling environments. In addition, the different disciplines may have competing goals, as it is the case between aerodynamic performance and structural efficiency. Hence, trade-off studies, which are at the basis of design decisions, need to be carried out. The codes and models used may also have different levels of fidelity, resolution, and complexity (sometimes within the same discipline [1, 69]), which makes impractical the realization of sensitivity studies and the identification of driving parameters. In particular, high-fidelity models are also known to represent a serious obstacle to the application of optimization algorithms. Finally, the disciplinary interactions are so intricately coupled that a

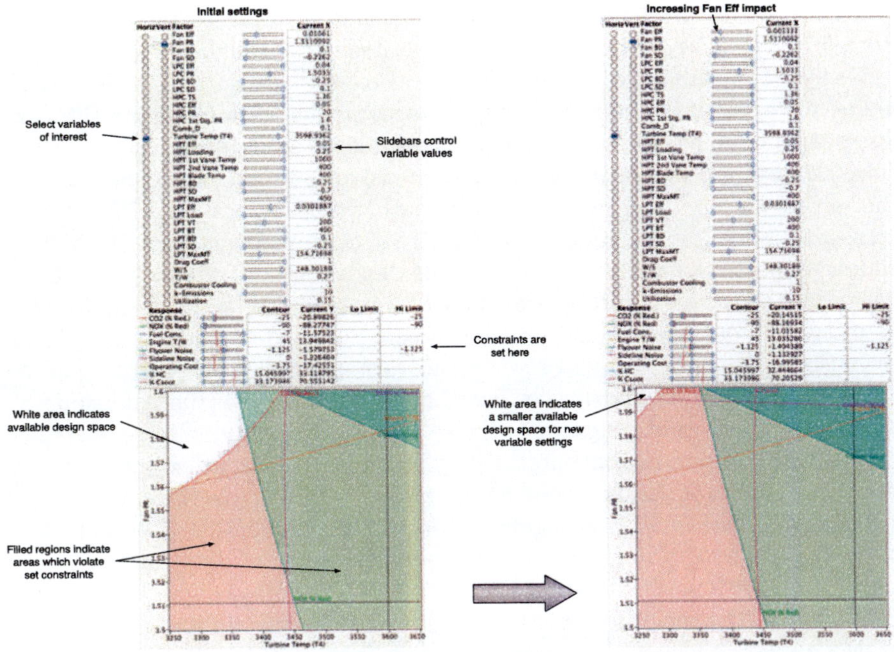

Fig. 5 Design Space Exploration Enabled by Parametric Design Formulation

parametric environment is necessary to avoid re-iterating the design process until all requirements are met.

A parametric environment is thus necessary to provide the user with the power to test a multitude of designs, evaluate the sensibility and feasibility of the chosen concepts and assess, in more detail, their corresponding design variables. In particular, a parametric, dynamic, and interactive environment, such as the one illustrated in Figure 5, allows the designer to rapidly explore hundreds or thousands of potential design points for multiple criteria, while giving him the freedom to change the space by moving both the design point and the constraints. The designer is thus able to visualize the active constraints and identify the ones that most prevent him from obtaining the largest feasible space possible and, consequently, from gaining the full benefits of the design concept [45]. Such a parametric formulation should thus have the appropriate degrees of freedom to allow the decision maker to play the what-if games he is interested in. However, this parametric environment, depending on the level of fidelity of the disciplinary models it is composed of, may require thousands of function evaluations. The resulting computational burden may in turn significantly limit the designer's ability to explore the design space. Hence, as noted by Keane and Nair [28], "the designer rarely has the luxury of being able to call on fully detailed analysis capabilities to study the options in all areas before taking important design decisions."

To address these challenges, full codes are often replaced by approximations [61]. While several approximation methods are available [28, 69], a particularly well-established technique that also provides an all encompassing model for exploring a complex design space, is surrogate modeling [11, 26]. Surrogate modeling enables virtually instantaneous analyses to be run in real-time [37] by approximating computer-intensive functions or models across the entire design space with simpler mathematical ones [64, 76]. Hence, surrogate modeling techniques, by constructing approximations of analysis codes, support the integration of discipline-dependent and often organization-dependent codes, and represent an efficient way to lessen the time required for an integrated parametric environment to run. Surrogate modeling techniques can also be used to "bridge between the various levels of sophistication afforded by varying fidelity physics based simulation codes" [19]. Additionally, these techniques yield insight into the relationships between design variables (inputs) and responses (outputs), hence facilitating concept exploration and providing knowledge about the behavior of the system across the entire design space as opposed to a small region. In particular, as mentioned by Forrester et al. [19], "surrogate models may be used in a form of data mining where the aim is to gain insight into the functional relationships between variables." By doing so, the sensitivity of the variables on the variability of a response as well as their effects can be assessed, hence increasinf the designer's knowledge of the problem. Finally, by enabling virtually instantaneous analyses to be computed in real-time, surrogate modeling supports the use of interactive and integrative visual environments [37]. These environments, further discussed in [45] and in Section 3.1 of this paper, in turn facilitate the designer's understanding of the design problem. Surrogate modeling thus represents a key enabler for decision making.

Surrogate models can be classified into data fit models and multi-fidelity models (also called hierarchical models) [15]. A data fit surrogate, as described by Eldred et al [15], is "a non physics-based approximation typically involving interpolation or regression of a set of data generated from the high-fidelity model," while a multi-fidelity model is a physics-based model with lower fidelity. The most prevalent surrogate modeling techniques used to facilitate concept exploration include Response Surface Methodology (RSM) [5, 4, 48], Artificial Neural Networks (ANN) [7, 67], Kriging (KG) [60, 59], and Inductive Learning [34]. The reader is invited to consult [25, 50, 64, 65, 76] for reviews and comparative studies of these different techniques. The application of one of the most popular of these surrogate modeling techniques, RSM, is illustrated in more detail in Mavris et al [45].

By allowing design teams to consider multiple disciplines simultaneously and facilitating design space exploration, parametric environments thus represent a key enabler to the identification and selection of promising candidate designs [20, 28]. Such parametric environments are also greatly benefit from the significant improvements in speed, size, and accessibility of data storage systems, which allow the reuse of the data and results from previous design evaluations. Finally, as discussed by many [28, 63, 65, 76], surrogate models support optimization efforts by 1) enabling the connection of proprietary tools, 2) supporting parallel computation, 3) supporting the verification of simulation results, 4) reducing the dimensionality of the problem and the search space, 5) assessing the sensitivity of design variables, and 6) handling both continous and discrete variables.

The last challenge to be discussed in the context of Conceptual design stems from the need to reduce uncertainty and doubt to allow reasonable decisions to be made. Uncertainty has been previously defined by DeLaurentis and Mavris [13] as "the incompleteness in knowledge (either in information or context), that causes model-based predictions to differ from reality in a manner described by some distribution function." It is thus a fact that any engineering design presents some degree of uncertainty [28]. However, failing to identify and account for risk and uncertainty at the Conceptual design phase will have serious and costly consequences later on in the design process. The following section addresses the nature of uncertainty in Conceptual design and briefly discusses methods that allow the designer to account for it.

2.1.3 Uncertainty

The design problem at this stage of the process is plagued with many uncertainties. Uncertainty in conceptual design exists at different levels and has many origins: approximations, simplifications, abstractions and estimates, ambiguous design requirements, omitted physics and unaccounted features, lack of knowledge about the problem, incomplete information about the operational environment and the technologies available, unknown boundary conditions or initial conditions, prediction accuracy of the models, etc. [2, 9, 21, 28, 43, 40, 85]. Uncertainty is notably present in the requirements, vehicle attributes, and technologies that define the design concept. The consequences and effects of uncertainty on the overall performance, and eventually, selection of a design concept, can be dramatic. They can depend, as discussed by Daskilewicz et al. [9], on different factors, such as the magnitude of the uncertainties, the performance metrics considered, the proximity of design concepts to active constraint boundaries, etc. Consequently, it is important to characterize, propagate and analyze uncertainty in order to make the design either more robust, more reliable, or both. In particular, efforts should be made to "determine the relative importance of various design options on the robustness of the finished product" [85]. In other words, the sensitivities of the outcomes to the assumptions need to be assessed.

A common practice in the industry is to follow a deterministic approach, either by assuming nominal values or by using safety factors built into the design [28, 85]. While this approach has proven useful for the design of conventional vehicles and their metal airframes [85], it suffers from many shortcomings. First, as discussed by Zang et al., defining safety factors for unconventional configurations is not achievable. In addition, the factor of safety approach assumes a worst-case condition, which, for such configurations, is nearly impossible to identify. Hence, this approach has been shown to be problematic and to eventually result in over designs in some areas [85] and in degraded performance or constraints violations for heavily optimized designs subjected to small perturbations [28].

Many approaches for uncertainty-based design exist and are thoroughly documented in the literature [6, 85]. One particularly established and efficient means to model, propagate and quantify the effects of uncertainty, as well as to support design space exploration, is through the implementation of probability theory and probabilistic design methods [43].

In a probabilistic approach, probability distributions are assigned to the uncertain inputs of analysis codes. This leads to the generation of Probability Density Functions (PDFs) and their corresponding Cumulative Distribution Functions (CDFs) for each design objective or constraint. These distributions, which represent the outcomes of every possible combination of synthesized designs, help assess the feasibility (or likelihood) of meeting specified target values. Indeed, by simultaneously representing specific constraints and their associated CDFs in a single plot, the designer now has the ability to evaluate how feasible the design space is and quickly identify any "show-stoppers", i.e., constraints inhibiting acceptable levels of feasibility. From there he can decide if any targets/constraints need to be relaxed and/or if new technologies need to be infused. Such an approach thus represents a valuable means to form relationships between input and output variables, while accounting for the variability of the inputs [42] and inform the designer regarding the magnitude and direction of the improvements needed to obtain an acceptable feasible space [45].

Many probabilistic analyses are conducted using a random sampling method called Monte Carlo Simulation (MCS). However, while its use does not require the simplifying assumption of normal distributions for the noise variables required by most analytical methods, it often necessitates huge amounts of function calls and is thus too expensive for complex problems characterized by many variables, high-fidelity codes and expensive analysis. This impractical computational burden can be alleviated, however, by building surrogate models for function evaluations, hence allowing MCS to run on the surrogate rather than on the actual code [2, 40]. The type of surrogate model to be used depends on factors such as the number and types of inputs [65], the complexity and dimensionality of the underlying structure of the model (i.e. the computational expense of the simulation) [86], the computational tasks to be performed, the intended purpose of the analyses, etc. Moreover, attention should be paid to ensure consistency between the surrogate model and the model that it approximates. Using probability distributions along with surrogate modeling can thus enable thousands of designs across a user-specified distribution (uniform or other) to be quickly generated and analyzed. This allows the designer to rapidly explore the design space and assess both the technical feasibility and the economic viability of a multitude of synthesized designs.

It is also important to keep in mind that a design is feasible if it meets all requirements concurrently. The requirements associated with two metrics of interest can be evaluated simultaneously using joint probability distributions [2]. Joint distributions can be represented, along with both future target values and Monte Carlo Simulation data, to quickly identify any points that meet the constraints. Technology metric values can then be extracted for any of the points that satisfy these constraints. Finally, these points can be further queried and investigated in other dimensions, through brushing and filtering, as illustrated in [45].

The use of probability theory in conjunction with RSM thus allows the analyst and decision maker to model and quantify the effects of uncertainty and to explore huge combinatorial spaces. It can also support the designer in his selection of technologies by providing him with the capability to continuously and simultaneously trade between requirements, technologies and design concepts. The combination of these methods, as illustrated in [45], can enable the discovery and examination of new trends and solutions in a transparent, visual, and interactive manner.

Identifying and selecting a satisfying design concept is contingent on the designers' ability to rapidly conduct trades, identify active constraints, model and quantify the effects of uncertainty, explore huge combinatorial spaces and evaluate feasible concepts. Once a concept has been chosen that is both technically feasible and financially viable, the design process continues further with the Preliminary design phase. At this stage, the disciplinary uncertainty is reduced [9] and modeling efforts are characterized by higher fidelity tools. This design phase and their attending challenges are discussed in the following section.

2.2 Preliminary Design

The main focus of Preliminary design is to size the various design variables optimally to obtain a design that meets requirements and constraints. To do so requires investigation of the detailed disciplinary interactions between the different subsystems/elements of the selected concept. For example, as described by Sobieszczanski–Sobieski and Haftka [69], the tight coupling between aerodynamics and structural efficiency (which has given rise to aeroelasticity) has always driven aircraft design and has thus been the focus of many studies on simultaneous optimization. Recently, issues such as technological advances and life cycle considerations have also induced a shift of emphasis in preliminary design to include non-conventional disciplines such as maintainability, reliability, and safety, as well as crossover disciplines such as economics and stability and control [68].

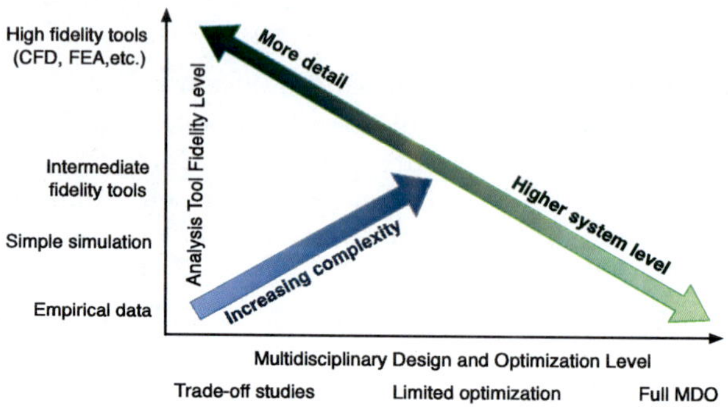

Fig. 6 MDO Taxonomy [32]

The analyses conducted in Preliminary design are much more sophisticated and complex, and require more accurate modeling than in Conceptual design (Figure 6). However, the complexity, accuracy and high-fidelity of the numerical models involved (such as Computational Fluid Dynamics (CFD) and Finite Element Analysis (FEA) codes) lead to unacceptable computational costs. In addition, these models often have different geometrical models and resolution, different equations, and can be of different levels of fidelity. A model can also involve hundreds and hundreds of variables, whose relationship(s) with other models need to be properly identified, mapped and accounted for (certain analyses need to be run in a particular order). Consequently, significant amounts of time are spent defining the structure of the modeling and simulation environment, preparing the data, and running the analyses [28]. This eventually limits the number of options that can be considered and studied. As a result, the designer is forced to trade between the level of modeling and fidelity that he deems appropriate, the level of accuracy required, and the time available to conduct the analyses. This aspect is further illustrated in Figure 7. Moreover, a lot of attention also needs to be put on critical cases, which often necessitates that additional cases be run and that the background, knowledge and expertise of the disciplinarians be captured, synthesized and integrated into the analysis process.

While a parametric environment at this phase of the design process would also be desirable, there are some limitations, given the nature of the models (high-fildelity, nonlinear, coupled, etc.) and the complexity of the problem, as to how much can be formulated, how parametric such an environment can be, how much computational effort is required, etc. However, regardless of the feasibility of a fully parametric and encompassing environment at this stage, it nevertheless remains that optimization efforts using high-fidelity models are computationally impractical and prohibit any assessment of the subsystems contribution to the overall system. Previous work has shown that adoption of approximation techniques could lessen the time required for an integrated parametric environment to run while retaining the required physics and time

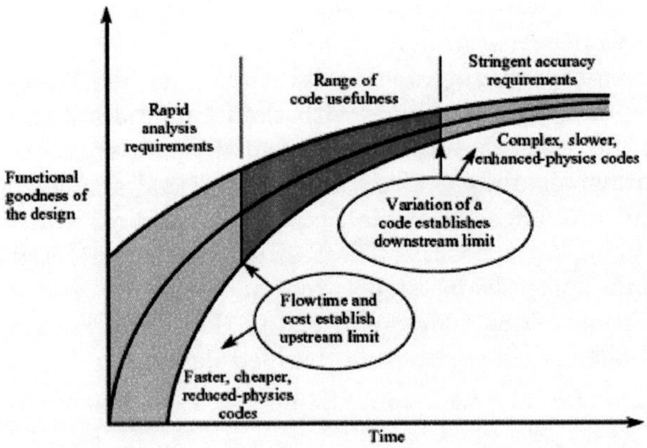

Fig. 7 Range of Usefulness for a CFD Code within the Design Process [57]

behavior of subsystems. In particular, certain surrogates techniques, such as reduced-order models preserve the physical representation of the system [62], while presenting a lower-fidelity than the full order model [15]. As explained by Weickum et al. [80], a reduced-order model (ROM) "mathematically reduces the system modeled, while still capturing the physics of the governing partial differential equations (PDEs), by projecting the original system response using a computed set of basis functions." Hence, these techniques are particularly well-established and frequently used to study fluid structure interactions [22] and aeroelastic simulations [36, 75]. The reader is invited to consult [38, 62] for a detailed description and discussion on the theory and applications of reduced-order models.

Finally, a shift in modeling effort between the Preliminary and Conceptual phases is ongoing. This shift is brought about by the introduction of new complexity in the system and the necessity to account for discipline interactions and dynamic effects earlier in the design. Hence, many in the design community share Moir's opinion [47], which is that "the success or failure in the Aerospace and Defense sector is determined by the approach taken in the development of systems and how well or otherwise the systems or their interactions are modeled, understood and optimized." Indeed, analyses that were once the focus of Preliminary design are now taking place early on in the design process. This trend is exemplified by the approach developed by Maser et al. [39] to account for the integration of both aircraft thermal management and propulsion systems, along with their resulting dynamic effects, as early as in the requirements definition and conceptual design phases. This shift in modeling effort between the Preliminary and Conceptual phases is also illustrated by the recent work presented by De la Garza et al. [12] on the development of a parametric environment enabling the high-fidelity, multi-physics-based aero-structural assessment of configurations during the later stages of Conceptual design.

The detailed phase begins once a favorable decision regarding full-scale development is made [53] and ends with the fabrication of the first product. At this stage, the design is fully defined and the primary focus is on preparing detailed drawings and/or CAD files with actual geometries, topologies, dimensions and tolerances, material, etc. to support part assembly, tooling and manufacturing activities [17, 53]. Testing is also conducted on areas such as aerodynamics, propulsion, structures, and stability and control, and a mockup is eventually constructed and tested. The following section describes some of the challenges in conducting detailed design.

2.3 Detailed Design

Detailed design focuses on the design and fabrication of every single piece of the product based on the design documentation/knowledge, mappings, specifications and geometric data defined or acquired during preliminary design. Design activities at this stage are also supported by a number of commercial Computer-Aided Design (CAD) and Computer-Aided Manufacturing (CAM) tools and systems that need to be integrated with other softwares, databases and knowledge bases. As noted by Szykman2001 [72] and later re-emphasized by Wang and Zhang [77], "interoperability between computer-aided engineering software tools is of significant importance to achieve greatly increased benefits in a new generation of product development systems." Flexible and "plug-and-play" environments are thus required that support the dynamic management of data, knowledge and information across the heterogenous engineering applications and platforms used during the spectrum of design activities. Such efforts, supported by the rapid advancements of information and communication technologies, are currently underway. In addition, design activities may be performed by geographically dispersed and temporally distributed design teams. Consequently, collaborative frameworks and infrastructures are also necessary to allow design teams to remotely access, share, exchange and modify information and files.

2.4 Preliminary Remarks

As mentioned in the previous sections of this paper, Aerospace engineering design is a complex activity whose success is contingent on the ability to:

- Support *requirements exploration, technology infusion trade-offs and concept down selection* during the early design phases (conceptual design) using physics-based methods
- Transition from a reliance on historical data to a *physics-based formulation* (especially when designing unconventional concepts)

- Transition from single-discipline to *multi-disciplinary analysis, design and optimization*
- Move form deterministic, serial single-point designs to *dynamic parametric trade environments*
- Quantify and assess risk by incorporating *probabilistic methods*
- *Speed up computation* to allow for the inclusion of variable fidelity tools so as to improve accuracy
- Automate the integrated design process

The need to address these points is gaining more and more recognition and support from the community. The ongoing research on design methods, as discussed in the previous sections of this paper, is critical to the advancement of the field. However, additional efforts are needed to seamlessly integrate these methods, tools and design systems in order to bring the knowledge forward in the design process, support decision making, and eventually reduce costs and time-to-market. In particular, there is also a need to recognize that the implementation of these methods goes hand-in-hand with the integration of visualization and knowledge management capabilities. These aspects are further discussed in the following section.

3 Integration of Visualization and Knowledge Management into the Design Process

Advances in numerical simulation techniques and computational methods have allowed for significant amounts of data to be generated, collected, and analyzed in order to increase the designer's knowledge about the physics of the problem. However, data by itself has little value if it is not structured and visualized in a way that allows the designer to act upon it. Data also needs to be properly indexed, stored and managed to ensure that it is available in the right place, at the right time, and in the right format for the designers to use. The following sections discuss these aspects.

3.1 Visualization

Humans deal with the world through their senses. The human visual system, in particular, provides us with the capability to quickly identify patterns and structures [81] and supports the transition from cognition, the processing of information, to perception, the obtaining of insight and knowledge. Hence, visual representations are often the preferred form of support to any human cognitive task because they amplify our cognitive ability [58] and reduce the complex cognitive work necessary to perform certain activities [30, 31]. From the early ages, when design was conducted on a piece of paper, up until today with the recent advances in Computer-Aided Design (CAD) models, design has always been conducted and communicated through visual means.

As explained by Wong et al. [83], "visual representations are essential aids to human cognitive tasks and are valued to the extent that they provide stable and external reference points on which dynamic activities and thought processes may be calibrated and on which models and theories can be tested and confirmed." However, it is important to understand that "visualization of information alone does not provide new insights" [46]. In other words, information visualization without interaction between the information and the human cognitive system does little to stimulate human reasoning and enable the generation and synthesis of knowledge or the formulation of appropriate conclusions or actions [46].

A multidisciplinary perspective termed Visual Analytics has originated from the need to address this issue. Visual Analytics, defined by [74] as the "science of analytical reasoning facilitated by interactive visual interfaces" provides visualization and interaction capabilities, allowing the analyst and decision maker to be presented with the appropriate level of depiction and detail to help them make sense of the data and synthesize the knowledge necessary to make decisions. In particular, the authors have previously discussed in [45] how the integration of Visual Analytics in the design process allows analysts and decision makers to:

- Rapidly explore huge combinatorial spaces
- Identify potentially feasible concepts or technology combinations
- Formulate and test hypotheses
- Steer the analysis by requesting additional data as needed (data farming)
- Integrate their background, expertise and cognitive capabilities into the analytical process
- Understand and quantify trade-offs between metrics and concepts
- Study correlations, explore trends and sensitivities
- Provide interactive feedback to the visualization environment
- Synthesize and share information
- Investigate the design space in a highly visual, dynamic, interactive, transparent and collaborative environment
- Document and communicate findings and decisions

Examples of efforts to support visualization-enabled design space exploration are further discussed in the following section.

3.1.1 Visualization-Enabled Design Space Exploration

Design problems are often characterized by a high number of design variables. This "curse of dimensionality" makes it difficult to obtain a good approximation of the response and leads to high computation expenses for surrogate modeling approaches [76] when exploring the design space. Also, the resulting high number of dimensions may encapsulate different kinds of information that are difficult to visualize because "there are more kinds of variation in the information than visual channels to display them" [18]. In addition, because "humans are, by nature and history, dwellers of low-dimensional worlds" [16], they are strongly limited in their ability to build a mental image of a multidimensional model, explore associations and relationships among dimensions [84], extract specific features, and identify patterns and outliers. The importance of visualization-enabled design space exploration in general, and of visualization methods for multidimensional data sets in particular, has thus been widely recognized as a means to support engineering design and decision making [37]. In particular, Wong and Bergeron [82] mention that such techniques have as their objective the synthesis of multidimensional data sets and the identification of key trends and relationships. Companies such as Chrysler, Boeing, Ford, Lockheed Martin or Raytheon, to name a few, have invested significant efforts in the use of visualization to speed and improve product design and development. The research community has also worked on the development and implementation of diverse design space visualization environments. Past efforts to visualize multidimensional data include programs such as XmdvTool [79], Xgobi [71], VisDB[29], and WinViz [35]. More recent work, as reported in [45], discusses the use of an interactive visualization environment to help determine the technologies and engine point designs that meet specific performance and economic targets. In particular, this environment features a scatterplot that allows the designer to display simultaneously both design variables and responses and to filter the discrete designs to determine regions of the design space that are the most promising for further and more detailed exploration. Additional recent interactive and multidimensional design space visualization environments include, among others, the ARL Trade Space Visualizer (ATSV) [49, 70], Cloud Visualization [14], BrickViz [27], the Advanced Systems Design Suite [87, 88], the framework introduced by Ross et al. [56], and the work conducted by Simpson et al. [66]. These environments incorporate diverse visualization techniques (glyphs, parallel coordinates, scatter matrices, 3-D scatter plots, etc.) depending on the nature of the data and the end goal of the environment [70]. Finally, it is important to note that dimensionality reduction techniques also present benefits for visualization purposes as well as data storage and data retrieval. The reader is invited to consult [33] for a survey of dimensionality reduction techniques for classification, data analysis, and interactive visualization.

3.2 Data and Knowledge Management

Being able to summarize, index, store and retrieve information related to previous exploration steps would allow collaborative designers to learn from past design experience and later reuse that information in the development of new design configurations [78]. As noted by Szykman [72], "the industry has an increasing need for tools to capture, represent, exchange and reuse product development knowledge." The management and visualization of the analysis workflow is also necessary in order to visualize the steps that lead to a decision, as well as to quickly make available, in a transparent manner, the assumptions formulated throughout the design process. Such capability would also support the story-telling and facilitate the integration of latecomers [51] or stakeholders that may contribute at different levels of the analysis [23]. Hence, taxonomies and other data models need to be developed to logically structure the data/information. In addition, computer-based tools, such as databases and data storage systems, are needed to capture, articulate, code and integrate the many forms and types of information generated during the design process, and eventually ensure that the data is available in the right place, at the right time, and in the right format [10].

4 Concluding Remarks

This paper has shown that, as the design process progresses, different trades need to be investigated, and different challenges need to be addressed. In particular, the authors illustrated how improvements in design are contingent to the advancements of design methods, as well as the community's ability to leverage the rapid advancement in information, infrastructural and communication technologies. As discussed, the development of advanced methods should be geared towards the:

- *Advances in MDA/MDO* to encompass the holistic nature of the problem, with an emphasis on uncertainty associated with the early design phases
- Creation of *computational architecture frameworks* to allow for easy integration and automation of sometimes organizationally dispersed tools
- Creation of *physics-based approximation models* (surrogate or metamodels) to replace the higher fidelity tools which are usually described as too slow for use in the design process, cryptic in their use of inputs, interfaces and logic, and non-transparent (lack of proper documentation, legacy)
- Use of *probability theory* in conjunction with these metamodels to enable designers to quantify and assess risk; to explore huge combinatorial spaces; to uncover trends and solutions never examined before in a very transparent, visual and interactive manner
- Use of *multi-attribute decision making techniques, Pareto optimality, and genetic algorithms* to account for multiple, conflicting objectives and discrete settings

The rapid advancement in information, infrastructural and communication technologies should also be leveraged to:

- Support the seamless integration of design systems, methods and tools
- Support collaborative design efforts within the design process
- Facilitate data and knowledge creation and sharing across multiple disciplines and geographical locations (through the use of distributed networking and applications, etc.)
- Facilitate information storage, versioning, retrieval and mining

- Support the development of environments that integrate and leverage computational, interaction and visualization techniques to support analytical reasoning and data fusion, and eventually reduce the designers cognitive burden
- Develop an integrated knowledge-based engineering and management framework
- Support the development of immersive visualization environments that leverage advances in computer graphics, large visual displays, etc.

Finally, improvements in Aerospace Design also depend on the industry's ability and desire to leverage the significant amount of work conducted by research and academic institutions. Though the need to consider the aforementioned aspects is gaining more and more recognition from the industry, as illustrated by the recent work from [12, 22, 52], the integration and implementation of the methods discussed in this paper still face some barriers. Indeed, it is well-known that new methods almost by definition go against the grain of established paradigms that are well defined and accepted by the practicing community and are thus always viewed with skepticism, criticism, or in some cases even cynicism. Hence, to foster the transfer of knowledge and facilitate the introduction of new methods, it is important that:

- Designers recognize that, even though traditional design methods have been very successful in the past, their implementation has often resulted in cost and schedule overruns that are not acceptable in today's competitive environment [85]
- The underlying theories, methods, mathematics, logic, algorithms, etc. upon which the new approaches are based be well understood, accepted, scientifically sound and practical.
- Familiarity exists with the underlying theories. Specifically, the material needed for someone to understand the method itself must be readily available
- Training utilizing material written on the overarching method, tutorials, etc. be available and supported with relevant examples
- Proposed methods be grounded on or be complimentary to established practices to have a better chance of succeeding
- Tools automate the proposed method and make it practical for every day use, as without them the method resembles a topic of academic curiosity
- Relevant examples and applications within a given field of study be provided
- The users be familiar with the conditions upon which the method/ techniques can be applied. For example, the types of surrogate model to be used depend on factors such as the number of inputs, the complexity, dimensionality and structure of the underlying models, the intended purposes of the analysis, etc.

To conclude, it is important to remember that there is no "silver bullet" method or technique that can be universally applied when it comes to design. However, appropriate methods and techniques can be leveraged and combined to provide visually interactive support frameworks for design decisions.

References

1. Baker, M., Giesing, J.: A practical approach to mdo and its application to an hsct aircraft. In: Proceedings of the 1st AIAA Aircraft Engineering, Technology, and Operations Congress, AIAA-1995-3885, Los Angeles, CA (1995)
2. Bandte, O.: A probabilistic multi-criteria decision making technique for conceptual and preliminary aerospace systems design. Ph.D. thesis, Georgia Institute of Technology, School of Aerospace Engineering, Atlanta, GA, U.S.A (2000)
3. Blanchard, B.S., Fabrycky, W.J.: Systems Engineering and Analysis, 3rd edn. Prentice Hall International Series in Industrial & Systems Engineering. Prentice Hall (1998)
4. Box, G.E.P., Hunter, W.G., Hunter, J.S.: Statistics for Experimenters. John Wiley & Sons, Inc., NY (1978)
5. Box, G.E.P., Wilson, K.B.: On the experimental attainment of optimum conditions. Journal of the Royal Statistical Society 13(Series B), 1–45 (1951)
6. Cacuci, D.G.: Sensitivity and Uncertainty Analysis: Theory, 1st edn., Boca Raton, FL, vol. I (2003)
7. Cheng, B., Titterington, D.M.: Neural networks: A review from a statistical perspective. Statistical Science 9(1), 2–54 (1994)
8. Dalton, J.S., Miller, R.E., Behbahani, A.R., Lamm, P., VanGriethuysen, V.: Vision of an integrated modeling, simulation, and analysis framework and hardware: test environment for propulsion, thermal management and power for the u.s air force. In: Proceedings of the 43rd AIAA/ASME/SAE/ASEE Joint Propulsion Conference and Exhibit, AIAA 2007-5711, Cincinnati, OH (2007)
9. Daskilewicz, M.J., German, B.J., Takahashi, T., Donovan, S., Shajanian, A.: Effects of disciplinary uncertainty on multi-objective optimization in aircraft conceptual design. Structural and Multidisciplinary Optimization (2011), doi:10.1007/s00158-011-0673-4
10. D'Avino, G., Dondo, P., lo Storto, C., Zezza, V.: Reducing ambiguity and uncertainty during new product development: a system dynamics based approach. Technology Management: A Unifying Discipline for Melting the Boundaries, 538–549 (2005)
11. De Baets, P.W.G., Mavris, D.N.: Methodology for the parametric structural conceptual design of hypersonic vehicles. In: Proceedings of the 2000 World Aviation Conference, 2000-01-5618, San Diego, CA (2000)
12. De La Garza, A.P., McCulley, C.M., Johnson, J.C., Hunten, K.A., Action, J.E., Skillen, M.D., Zink, P.S.: Recent advances in rapid airframe modeling at lockheed martin aeronautics company. In: Proceedings of the AVT-173 - Virtual Prototyping of Affordable Military Vehicles Using Advanced MDO, no. RTO-MP-AVT-173 in NATO Research and Technology Organization, Bugaria (2011)
13. DeLaurentis, D.A., Mavris, D.N.: Uncertainty modeling and management in multidisciplinary analysis and synthesis. In: Proceedings of the 38th AIAA Aerospace Sciences Meeting and Exhibit, AIAA-2000-0422, Reno, NV (2000)

14. Eddy, J., Lewis, K.: Visualization of multi-dimensional design and optimization data using cloud visualization. In: Proceedings of the ASME Design Engineering Technical Conferences - Design Automation Conference, DETC02/DAC-02006, Montreal, Quebec, Canada (2002)
15. Eldred, M.S., Giuntay, A.A., Collis, S.S.: Second-order corrections for surrogate-based optimization with model hierarchies. In: Proceedings of the 10th AIAA/ISSMO Multidisciplinary Analysis and Optimization Conference (2004)
16. Fayyad, U., Grinstein, G.G.: Introduction. In: Information Visualization in Data Mining and Knowledge Discovery, pp. 1–12. Morgan Kaufmann Publishers (2002)
17. Feng, S.C.: Preliminary design and manufacturing planning integration using web-based intelligent agents. Journal of Intelligent Manufacturing 16(4-5), 423–437 (2005)
18. Foltz, M.A., Lee, A.: Infomapper: Coping with the curse of dimensionality in information visualization. Submitted To UIST 2002 (2002)
19. Forrester, A.I.J., Sóbester, A., Keane, A.J.: Engineering Design via Surrogate Modeling: A Practical Guide. In: Progress in Astronautics and Aeronautics, vol. 226. John Wiley & Sons Ltd. (2008)
20. German, B.J., Daskilewicz, M.J.: An mdo-inspired systems engineering perspective for the "wicked" problem of aircraft conceptual design. In: Proceedings of the 9th AIAA Aviation Technology, Integration, and Operations (ATIO) Conference, AIAA-2009-7115, Hilton Head, South Carolina (2009)
21. Green, L.L., Lin, H.Z., Khalessi, M.R.: Probabilistic methods for uncertainty propagation applied to aircraft design. In: Proceedings of the 20th AIAA Applied Aerodynamics Conference, AIAA-2002-3140, St. Louis, Missouri (2002)
22. Grewal, A.K.S., Zimcik, D.G.: Development of reduced order aerodynamic models from an open source cfd code. In: Proceedings of the AVT-173 - Virtual Prototyping of Affordable Military Vehicles Using Advanced MDO, no. RTO-MP-AVT-173 in NATO Research and Technology Organization, Bugaria (2011)
23. Heer, J., Agrawala, M.: Design considerations for collaborative visual analytics. Informtion Visualization 7, 49–62 (2008)
24. Howard, R.A.: An assessment of decision analysis. Operations Research 28(1), 4–27 (1980)
25. Jin, R., Chen, W., Simpson, T.W.: Comparative studies of metamodeling techniques under multiple modeling criteria. Structural and Multidisciplinary Optimization 23(1), 1–13 (2001), doi:10.1007/s00158-001-0160-4
26. Kamdar, N., Smith, M., Thomas, R., Wikler, J., Mavris, D.N.: Response surface utilization in the exploration of a supersonic business jet concept with application of emerging technologies. In: Proceedings of the World Aviation Congress & Exposition, 2003-01-0359, Montreal, QC, Canada (2003)
27. Kanukolanu, D., Lewis, K.E., Winer, E.H.: A multidimensional visualization interface to aid in tradeoff decisions during the solution of coupled subsystems under uncertainty. ASME Journal of Computing and Information Science in Engineering 6(3), 288–299 (2006)
28. Keane, A.J., Nair, P.B.: Computational Approaches for Aerospace Design. John Wiley & Sons, Ltd. (2005)
29. Keim, D., Kriegel, H.P.: Visdb: Database exploration using multidimensional visualization. IEEE Computer Graphics and Applications 14(5), 40–49 (1994)

30. Keim, D.A.: Visual exploration of large data sets. Communications of the ACM 44(8), 38–44 (2001)
31. Keim, D.A., Mansmann, F., Schneidewind, J., Thomas, J., Ziegler, H.: Visual Analytics: Scope and Challenges. In: Simoff, S.J., Böhlen, M.H., Mazeika, A. (eds.) Visual Data Mining. LNCS, vol. 4404, pp. 76–90. Springer, Heidelberg (2008)
32. Kesseler, E.: Advancing the state-of-the-art in the civil aircraft design: A knowledge-based multidisciplinary engineering approach. In: Proceedings of the European Conference on Computational Fluid Dynamics, ECCOMAS CDF 2006 (2006)
33. Konig, A.: Dimensionality reduction techniques for multivariate data classification, interactive visualization, and analysis-systematic feature selection vs. extraction. In: Proceedings of the Fourth International Conference on Knowledge-Based Intelligent Engineering Systems and Allied Technologies, vol. 1, pp. 44–55 (2000), doi:10.1109/KES.2000.885757
34. Langley, P., Simon, H.A.: Applications of machine learning and rule induction. Communications of the ACM 38(11), 55–64 (1995)
35. Lee, H.Y., leng Ong, H., whatt Toh, E., Chan, S.K.: A multi-dimensional data visualization tool for knowledge discovery in databases. In: Proceedings of the 19th Annual International Computer Software & Applications Conference, pp. 7–11 (1995)
36. Lieu, T., Farhat, C.: Adaptation of aeroelastic reduced-order models and application to an f-16 configuration. AIAA Journal 45, 1244–1257 (2007)
37. Ligetti, C., Simpson, T.W., Frecker, M., Barton, R.R., Stump, G.: Assessing the impact of graphical design interfaces on design efficiency and effectiveness. Journal of Computing and Information Science in Engineering 3(2), 144–154 (2003), doi:10.1115/1.1583757
38. Lucia, D.J., Beran, P.S., Silva, W.A.: Reduced-order modeling: New approaches for computational physics. Progress in Aerospace Sciences 40(1-2), 51–117 (2004)
39. Maser, A.C., Garcia, E., Mavris, D.N.: Thermal management modeling for integrated power systems in a transient, multidisciplinary environment. In: Proceedings of the 45th AIAA/ASME/SAE/ASEE Joint Propulsion Conference & Exhibit, AIAA-2009-5505, Denver, CO (2009)
40. Mavris, D.N., Bandte, O., DeLaurentis, D.A.: Robust design simulation: a probabilistic approach to multidisciplinary design. Journal of Aircraft 36(1), 298–307 (1999)
41. Mavris, D.N., DeLaurentis, D.A.: A stochastic design approach for aircraft affordability. In: Proceedings of the 21st Congress of the International Council on the Aeronautical Sciences (ICAS), ICAS-1998-623, Melbourne, Australia (1998)
42. Mavris, D.N., DeLaurentis, D.A.: A probabilistic approach for examining aircraft concept feasibility and viability. Aircraft Design 3, 79–101 (2000)
43. Mavris, D.N., DeLaurentis, D.A., Bandte, O., Hale, M.A.: A stochastic approach to multi-disciplinary aircraft analysis and design. In: Proceedings of the 36th Aerospace Sciences Meeting and Exhibit, AIAA-1998-0912, Reno, NV (1998)
44. Mavris, D.N., Jimenez, H.: Systems Design. In: Architecture and Principles of Systems Engineering. Complex and Enterprise Systems Engineering Series, pp. 301–322. CRC Press (2009)

45. Mavris, D.N., Pinon, O.J., Fullmer, D.: Systems design and modeling: A visual analytics approach. In: Proceedings of the 27th International Congress of the Aeronautical Sciences (ICAS), Nice, France (2010)
46. Meyer, J., Thomas, J., Diehl, S., Fisher, B., Keim, D., Laidlaw, D., Miksch, S., Mueller, K., Ribarsky, W., Preim, B., Ynnerman, A.: From visualization to visually enabled reasoning. Tech. rep., Dagstuhl Seminar 07291 on "Scientific Visualization" (2007)
47. Moir, I., Seabridge, A.: Aircraft Systems: Mechanical, Electrical and Avionics Subsystems Integration, 3rd edn. AIAA Education Series. Professional Engineering Publishing (2008)
48. Myers, R.H., Montgomery, D.C.: Response Surface Methodology: Process and Product Optimization Using Designed Experiments, 2nd edn. Wiley Series in Probability and Statistics. John Wiley & Sons, Inc. (2002)
49. O'Hara, J.J., Stump, G.M., Yukish, M.A., Harris, E.N., Hanowski, G.J., Carty, A.: Advanced visualization techniques for trade space exploration. In: Proceedings of the 48th AIAA/ASME/ASCE/AHS/ASC Structures, Structural Dynamics, and Material Conference, AIAA-2007-1878, Honolulu, HI, USA (2007)
50. Paiva, R.M., Carvalho, A., Crawford, C., Suleman, A.: Comparison of surrogate models in a multidisciplinary optimization framework for wing design. AIAA Journal 48, 995–1006 (2010), doi:10.2514/1.45790
51. Ravachol, M., Bezerianos, A., De-Vuyst, F., Djedidi, R.: Scientific visualization for decision support. Presentation to the Forum Ter@tec (2010)
52. Ravachol, M., Caillot, G.: Practical implementation of a multilevel multidisciplinary design process. In: Proceedings of the AVT-173 - Virtual Prototyping of Affordable Military Vehicles Using Advanced MDO, no. RTO-MP-AVT-173 in NATO Research and Technology Organization, Bugaria (2011)
53. Raymer, D.P.: Aircraft Design: A Conceptual Approach, 4th edn. AIAA Education Series. American Institute of Aeronautics and Astronautics, Inc., Reston (2006)
54. Reed, J.A., Follen, G.J., Afjeh, A.A.: Improving the aircraft design process using web-based modeling and simulation. ACM Transactions on Modeling and Computer Simulation 10(1), 58–83 (2000)
55. Roskam, J.: Airplane Design, Part VIII: Airplane Cost Estimation: Design, Development, Manufacturing and Operating. Darcoporation (1990)
56. Ross, A.M., Hastings, D.E., Warmkessel, J.M., Diller, N.P.: Multi-attribute tradespace exploration as front end for effective space system design. Jourmal of Spacecraft and Rockets 41(1), 20–28 (2004)
57. Rubbert, P.E.: Cfd and the changing world of airplane design. In: Proceedings of the 19th Congress of the International Council of the Aeronautical Sciences (ICAS), ICAS-1994-0.2 (1994)
58. Russell, A.D., Chiu, C.Y., Korde, T.: Visual representation of construction management data. Automation in Construction 18, 1045–1062 (2009)
59. Sacks, J., Schiller, S.B., Welch, W.J.: Design for computer experiments. Technometrics 31(1), 41–47 (1989)
60. Sacks, J., Welch, W.J., Mitchell, T.J., Wynn, H.P.: Design and analysis of computer experiments. Statistical Science 4(4), 409–435 (1989)
61. Schőnning, A., Nayfeh, J., Zarda, R.: An integrated design and optimization environment for industrial large scaled systems. Research in Engineering Design 16, 86–95 (2005)

62. Schilders, W.H., van der Vorst, H.A., Rommes, J.: Model Order Reduction: Theory, Research Aspects, and Applications. Springer, Heidelberg (2008)
63. Shan, S., Wang, G.G.: Survey of modeling and optimization strategies to solve high-dimensional design problems with computationally-expensive black-box functions. Structural and Multidisciplinary Optimization 41, 219–241 (2010)
64. Simpson, T.W., Peplinski, J.D., Koch, P.N., Allen, J.K.: On the use of statistics in design and the implications for deterministic computer experiments. In: Proceedings of the 1997 ASME Design Engineering Technical Conferences (DETC 1997), Sacramento, CA, USA (1997)
65. Simpson, T.W., Peplinski, J.D., Koch, P.N., Allen, J.K.: Metamodels for computer-based engineering design: Survey and recommendations. Engineering with Computers 17, 129–150 (2001)
66. Simpson, T.W., Spencer, D.B., Yukish, M.A., Stump, G.: Visual sterring commands and test problems to support research in trade space exploration. In: Proceedings of the 12th AIAA/ISSMO Multidisciplinary Analysis and Optimization Conference, AIAA-2008-6085, Victoria, British Columbia, Canada (2008)
67. Smith, M.: Neural Networks for Statistical Modeling. Von Nostrand Reinhold, NY (1993)
68. Soban, D.S., Mavris, D.N.: Methodology for assessing survivability tradeoffs in the preliminary design process. In: Proceedings of the 2000 World Aviation Conference, 2000-01-5589, San Diego, CA (2000)
69. Sobieszczanski-Sobieski, J., Haftka, R.: Multidisciplinary aerospace design optimization: Survey of recent developments. Structural Optimization 14, 1–23 (1997)
70. Stump, G., Simpson, T.W., Yukish, M., Bennett, L.: Multidimensional visualization and its application to a design by shopping paradigm. In: Proceedings of the 9th AIAA/ISSMO Symposium on Multidisciplinary Analysis and Optimization, AIAA-2002-5622, Atlanta, GA, USA (2002)
71. Swayne, D.F., Cook, D., Buja, A.: Xgobi: Interactive dynamic data visualization in the x window system. Journal of Computational Graphical Statistics 7(1), 113–130 (1998)
72. Szykman, S., Fenves, S.J., Keirouz, W., Shooter, S.B.: A foundation for interoperability in next-generation product development systems. Computer-Aided Design 33(7), 545–559 (2001)
73. Tam, W.F.: Improvement opportunities for aerospace design process. In: Proceedings of the Space 2004 Conference and Exhibit, AIAA-2004-6126, San Diego, CA (2004)
74. Thomas, J.J., Cook, K.A. (eds.): Illuminating the Path: The Research and Development Agenda for Visual Analytics. IEEE CS Press (2005), `http://nvac.pnl.gov/agenda.stm`
75. Thomas, J., Dowell, E., Hall, K.: Three-dimensional transonic aeroelasticity using proper orthogonal decomposition based reduced order models. Journal 40(3), 544–551 (2003), doi:10.2514/2.3128
76. Wang, G.G., Shan, S.: Review of metamodeling techniques in support of the engineering design optimization. Journal of Mechanical Design 129(4), 370–380 (2007)
77. Wang, H., Zhang, H.: A distributed and interactive system to integrated design and simulation for collaborative product development. Robotics and Computer-Integrated Manufacturing 26, 778–789 (2010)

78. Wang, L., Shen, W., Xie, H., Neelamkavill, J., Pardasani, A.: Collaborative conceptual design - state of the art and future trends. Computer-Aided Design 34, 981–996 (2002)
79. Ward, M.: Xmdvtool: Integrating multiple methods for visualizing multivariate data. In: Proceedings of Visualization, Wahsington, D.C., USA, pp. 326–333 (1994)
80. Weickum, G., Eldred, M.S., Maute, K.: Multi-point extended reduced order modeling for design optimization and uncertainty analysis. In: Proceedings of the 47th AIAA/ASME/ASCE/AHS/ASC Structures, Structural Dynamics, and Materials Conference, AIAA-2006-2145, Newport, Rhode Island (2006)
81. van Wijk, J.J.: The value of visualization. In: Proceedings of IEEE Visualization, pp. 79–86 (2005), doi:10.1109/VISUAL.2005.1532781
82. Wong, P.C., Bergeron, R.D.: 30 years of multidimensional multivariate visualization. In: Proceedings of the Workshop on Scientific Visualization. IEEE Computer Society Press (1995)
83. Wong, P.C., Rose, S.J., Chin Jr., G., Frincke, D.A., May, R., Posse, C., Sanfilippo, A., Thomas, J.: Walking the path: A new journey to explore and discover through visual analytics. Information Visualization 5, 237–249 (2006)
84. Yang, J., Patro, A., Huang, S., Mehta, N., Ward, M.O., Rundensteiner, E.A.: Value and relation display for interactive exploration of high dimensional datasets. In: Proceedings of the Symposium on Information Visualization, Austin, TX (2004)
85. Zang, T.A., Hemsch, M.J., Hilburger, M.W., Kenny, S.P., Luckring, J.M., Maghami, P., Padula, S.L., Stroud, W.J.: Needs and opportunities for uncertainty-based multidisciplinary design methods for aerospace vehicles. Tech. Rep. NASA/TM-2002-211462, NASA Langley Research Center (2002)
86. Zentner, J., Volovoi, V., Mavris, D.N.: Overview of metamodeling techniques for problems with a large number of input parameters. In: Proceedings of the AIAA 3rd Annual Aviation Technology, Integration, and Operations (ATIO) Conference, AIAA-2003-6762, Denver, CO (2003)
87. Zhang, R., Noon, C., Oliver, J., Winer, E., Gilmore, B., Duncan, J.: Development of a software framework for conceptual design of complex systems. In: Proceedings of the 3rd AIAA Multidisciplinary Design Optimization Specialists Conference, AIAA-2007-1931, Honolulu, HI, USA (2007)
88. Zhang, R., Noon, C., Oliver, J., Winer, E., Gilmore, B., Duncan, J.: Immersive product configurator for conceptual design. In: Proceedings of the ASME Design Engineering Technical Conferences - Design Automation Conference, DETC 2007-35390, Las Vegas, NV, USA (2007)

Chapter 2
Complexity and Safety

Nancy G. Leveson

Abstract. Complexity is overwhelming the traditional approaches to preventing accidents in engineered systems and new approaches are necessary. This paper identifies the most important types of complexity related to safety and discusses what is necessary to prevent accidents in our increasingly complex engineered systems.

1 The Problem

Traditional safety engineering approaches were developed for relatively simple electro-mechanical systems. The problem is that new technology, especially software, is allowing almost unlimited complexity in the systems we are building. This complexity is creating new causes of accidents and changing the relative importance of traditional causes. While we have developed engineering techniques to deal with the older, well-understood causes, we do not have equivalent techniques to handle accident causes involving new technology and the increasing complexity of the systems we are building. A potential solution, of course, is to build the simpler systems, but usually we are unwilling to make the necessary compromises.

Complexity can be separated into complexity related to the problem itself and complexity introduced in the design of the solution of the problem. For complexity that arises from the problem being solved, reducing complexity requires reducing the goals of the system, which is something humans are often unwilling to do. Complexity can also be introduced in the design of the solution of the problem and often this "accidental complexity" (in the words of Brooks [cite]) can and should be eliminated or reduced without compromises on the basic system

Nancy G. Leveson
Aeronautics and Astronautics
Massachusetts Institute of Technology

goals. In either case, we need new, more powerful safety engineering approaches to dealing with complexity and the new causes of accidents arising from it.

2 What Is Complexity?

Complexity is subjective; it is not in the system itself but in the minds of observers or users of the system. What is complex to one person or at one point in time may not be to another. Consider the introduction of the high-pressure steam engine in the first half of the nineteenth century. While engineers quickly amassed information about thermodynamics, they did not fully understand what went on in steam boilers, resulting in frequent and disastrous explosions. Once the dynamics of steam were fully understood, more effective safety devices could be designed and explosions prevented. While steam engines may have seemed complex in the nineteenth century, they no longer would be considered complex by engineers. Complexity is relative and changes with time.

With respect to safety, the basic problem is that the behavior of complex systems cannot be thoroughly planned, understood, anticipated, and guarded against, that is there are "unknowns" in predicting system behavior. The critical factor that differentiates complex systems from other systems is *intellectual manageability*.

We can either not build and operate intellectually unmanageable systems until we have amassed the knowledge to fully understand their behavior or we can use tools to stretch our intellectual limits and to deal with the new causes of accidents arising from increased complexity.

Treating complexity as one indivisible property is not very useful in creating tools to deal with it. Some have tried to define complexity in terms of one or two properties of a system (for example, network interconnections). While useful for some problems, it is not for others. I have found the following types of complexity of greatest importance when managing safety:

- *Interactive complexity* arises in the interactions among system components.
- *Non-linear complexity* exists when cause and effect are not related in any obvious (or at least known) way.
- *Dynamic complexity* is related to understanding changes over time.
- *Decompositional complexity* is related to how we decompose or modularize our systems.

Other types of complexity can certainly be defined, but these seem to have the greatest impact on safety.

The rest of the paper discusses each of these types of complexity, their relationship to safety, and how they can be managed to increase safety in the complex systems we build.

2.1 Interactive Complexity

The simpler systems of the past could be thoroughly tested before use and any design errors identified and removed. That left only random component failure as the cause of accidents during operational use. The use of software is undermining this assumption in two ways: software is allowing us to build systems that cannot be thoroughly tested and the software itself cannot be exhaustively tested to eliminate design errors. Note that design errors are the only type of error in software: Because software is pure design, it cannot "fail" in the way that hardware does (including the hardware on which the software is executed). Basically, software is an abstraction.

One criterion for labeling a system as interactively complex, then, is that the level of interactions between the parts of the problem has reached the point where they no longer can be anticipated or thoroughly tested. An important cause of interactive complexity is coupling. Coupling leads to interdependence between parts of the problem solution by increasing the number of interfaces and thus interactions. Software has allowed us to build much more highly coupled and interactively complex systems than was feasible for pure electro-mechanical systems.

Traditionally, accidents have been considered to be caused by system component failures. There may be single or multiple failures involved, and they may not be independent. Usually some type of randomness in assumed in the failure behavior.

In interactively complex systems, in contrast, accidents may arise in the interactions among components, where none of the individual components may have failed. These *component interaction accidents* result from system design errors that are not caught before the system is fielded. Often they involve requirements errors, particularly software requirements errors. In fact, because software does not "wear out," the only types of errors that can occur are requirements errors or errors in the implementation of the requirements. In practice, the vast majority of accidents related to software have been caused by software requirements errors, i.e., the software has not "failed" but did exactly what the software implementers wanted it to do but the implications of the behavior from a system standpoint were not understood and led to unsafe system behavior.

Component interaction accidents were noticed as a growing problem starting in the Intercontinental Ballistic Missile Systems of the 1950's when interactive complexity in these systems had gotten ahead of our tools to deal with it. System engineering and System Safety were created to deal with these types of problems [Leveson, 1995]. Unfortunately, the most widely used hazard analysis techniques stem from the early 1960s and do not handle today's very different types of technology and system design.

An important implication of the distinction between component failure and component interaction accidents is that safety and reliability, particularly in complex systems, are *not* the same although they are often incorrectly equated.

Making all the components highly reliable will not prevent component interaction accidents or those arising from system design errors. In fact, sometimes they conflict and increasing one will even decrease the other, that is, increasing safety may decrease reliability and increasing component reliability may decrease system safety.

The distinction between safety and reliability is particularly important for software-intensive systems. Unsafe software behavior is usually caused by flaws in the software requirements. Either there are incomplete or wrong assumptions about the operation of the controlled system or required operation of the computer or there are unhandled controlled-system states and environmental conditions. Simply trying to get the software "correct" or to make it reliable (however one might define that attribute for a pure abstraction like software), will not make it safer if the problems stem from inadequate requirements specifications.

2.2 Non-linear Complexity

Informally, non-linear complexity occurs when cause and effect are not related in an obvious or direct way. Sometimes non-linear causal factors are called "systemic factors" in accidents, i.e., characteristics of the system or its environment that indirectly impact all or many of the system components. Examples of systemic factors are management pressure to increase productivity or reduced expenses. Another common systemic cause is the *safety culture*, which can be defined as the set of values upon which members of the organization make decisions about safety. The relationship between these systemic factors and the events preceding the accident (the "chain of events" leading to the accident) are usually indirect and non-linear and often omitted from accident reports or from proactive hazard analyses. Our accident models and techniques assume linearity, as discussed below.

Along with interactive complexity, non-linear complexity makes system behavior very difficult to predict. This lack of predictability affects not only system development but also operations. The role of operators in our systems is changing. Human operators previously were directly controlling processes and usually following predefined procedures. With the increase in automation, operators are now commonly supervising automation that controls the process rather than directly controlling the process itself. Operators have to make complex, real-time decisions, particularly during emergencies, and non-linear complexity makes it harder for the operators to successfully make such real-time decisions [Perrow, 1999]. Complexity is stretching the limits of comprehensibility and predictability of our systems.

Newer views of human factors reject the idea that operator errors are random failures [Dekker, 2005; Dekker, 2006]. All behavior is affected by the context (system) in which it occurs. Human error, therefore, is a symptom, not a cause; it is a symptom of a problem in the environment, such as the design of the equipment and human-automation interface, the design of the work procedures, management pressures, safety culture, etc. If we want to change operator behavior, we need to change the system in which it occurs. We are designing systems in which operator error is inevitable and then blaming accidents on operators rather than designers. Operator errors stemming from complexity in the system design will not be eliminated by more training or telling the operators to be more careful.

2.3 Dynamic Complexity

Dynamic complexity is related to changes over time. Systems are not static, but we often assume they are when we design and field them. Change, particularly in human and organizational behavior, is inevitable as is change (both planned and unplanned) in the non-human system components.

Rasmussen [1997] has suggested that these changes often move the system to states of higher risk. Systems migrate toward states of high risk, according to Rasmussen, under competitive and financial pressures. The good news is that if this hypothesis is true, the types of change that will occur are potentially predictable and theoretically preventable.

We want flexibility in our systems and operating environments, but we need engineering design and operations management techniques that prevent or control dangerous changes and detect (before an accident) when they occur during operations.

2.4 Decompositional Complexity

Interactive, non-linear, and dynamic complexity are related to the problem being solved and not necessarily the solution, although they impact and are reflected in the design of the system. For the most part, complexity in the design of the solution is not very relevant for safety. But design complexity does have a major impact on our ability to analyze the safety of a system. The aspect of design that most affects safety, in my experience, is decompositional complexity.

Decompositional complexity arises when the structural decomposition of the system is not consistent with the functional decomposition. Decompositional complexity makes it harder for designers and maintainers to predict and understand system behavior. Safety is related to the functional behavior of the system and its components: It is not a function of the system structure or architecture. Decompositional complexity makes it harder for humans to understand and find functional design errors (versus structural flaws). For safety, it also greatly increases the difficulty for humans to examine the system design and determine whether the system will behave safely. Most accidents (beyond simple causes such as cuts on sharp edges or physical objects falling on people) occur as a result of some system behavior, i.e., the system has to *do* something to cause an accident.

Because verifying safety requires understanding the system's functional behavior, designing to enhance such verification is necessary. For large systems, this verification may be feasible only if the system is designed using functional decomposition, for example, isolating and modularizing potentially unsafe functionality. Spreading functionality that can affect safety throughout the entire system design makes safety verification infeasible. I know of no effective way to verify the safety of most object-oriented system designs at a reasonable cost.

3 Managing Complexity in Safety Engineering

To engineer for safety in systems exhibiting interactive, non-linear, and dynamic complexity, we will need to extend our standard safety engineering approaches. The most important step is probably the most difficult for people to implement and that is to limit the complexity in the systems we build and to practice restraint in defining the requirements for our systems. At the least, extra unnecessary complexity should not be added in design and designs must be reviewable and analyzable for safety.

Given that most people will be unwilling to go back to the simpler systems of the past, any practical solution must include providing tools to stretch the basic human intellectual limits in understanding complexity. For safety, these tools need to be built on top of a model of accident causality that encompasses the complexities of the systems we are building.

3.1 STAMP: A New Accident Model

Our current safety engineering techniques assume accidents are caused by component failures and do not assist in preventing component interaction accidents. The most common accident causality model explains accidents in terms of multiple events, sequenced as a forward chain over time. The relationships among the events are assumed to be simple and direct. The events almost always involve component failure, human error, or energy-related events (e.g., an explosion).

This chain-of-events model forms the basis for most safety engineering and reliability engineering analysis (for example, fault tree analysis, probabilistic risk analysis, failure modes and effects analysis, events trees, etc.) and design for safety (e.g., redundancy, overdesign, safety margins).

This standard causality model and the tools and techniques built on it do not apply to the types of complexity described earlier. It assumes direct linearity between events and ignores common causes of failure events, it does not include component interaction accidents where no components may have failed, and it does not handle dynamic complexity and migration toward states of high risk. It also greatly oversimplifies human error by assuming it involves random failures or "slips," that are unrelated to the context in which the error occurs, and that operators are simply blindly following procedures and not making cognitively complex decisions. In fact, human error is better modeled as a feedback loop than a "failure" in a simple chain of events.

STAMP (System-Theoretic Accident Model and Processes) was created to include the causes of accidents arising from these types of complexity. STAMP is based on systems theory rather than reliability theory and treats accidents as a control problem rather than a failure problem. The basic paradigm change is to switch from a focus of "prevent failures" to one of "enforce safety constraints on system behavior." The new focus includes the old one but also includes accident causes not recognized in the old models.

In STAMP, safety is treated as an emergent property that arises when the system components interact with each other within a larger environment. There is a set of constraints related to the behavior of the system components—physical, human, and social—that enforces the emergent safety property. Accidents occur when system interactions violate those constraints.

In this model of causation, accidents are not simply an event or chain of events but involve a complex, dynamic process. Dynamic behavior of the system is also included: most accidents are assumed to arise from a slow migration of the entire system toward a state of high risk. Often this migration is not noticed until after an accident has occurred. Instead we need to control and detect this migration.

The standard chain-of-failure-events model is included in this broader control model. Component failures are simply a subset of causes of an accident that need to be controlled. STAMP more broadly defines safety as a dynamic control problem rather than a component failure problem. For example, the O-ring in the Challenger Space Shuttle did fail, but the problem was that the failure caused the O-ring not to be able to *control* the propellant gas release by sealing a gap in the field joint of the Space Shuttle. The software did not adequately *control* the descent speed of the Mars Polar Lander. The public health system did not adequately *control* the contamination of the milk supply with melamine in a recent set of losses. Our financial system did not adequately *control* the use of financial instruments in our recent financial meltdown.

Constraints are enforced by socio-technical safety control structures. Figure 1 shows an example of such a control structure. There are two hierarchical structures shown in Figure 1: development and operations. They are separated because safety is usually controlled very differently in each. A third control structure might also be included which involves emergency response when an accident does occur so that losses are minimized.

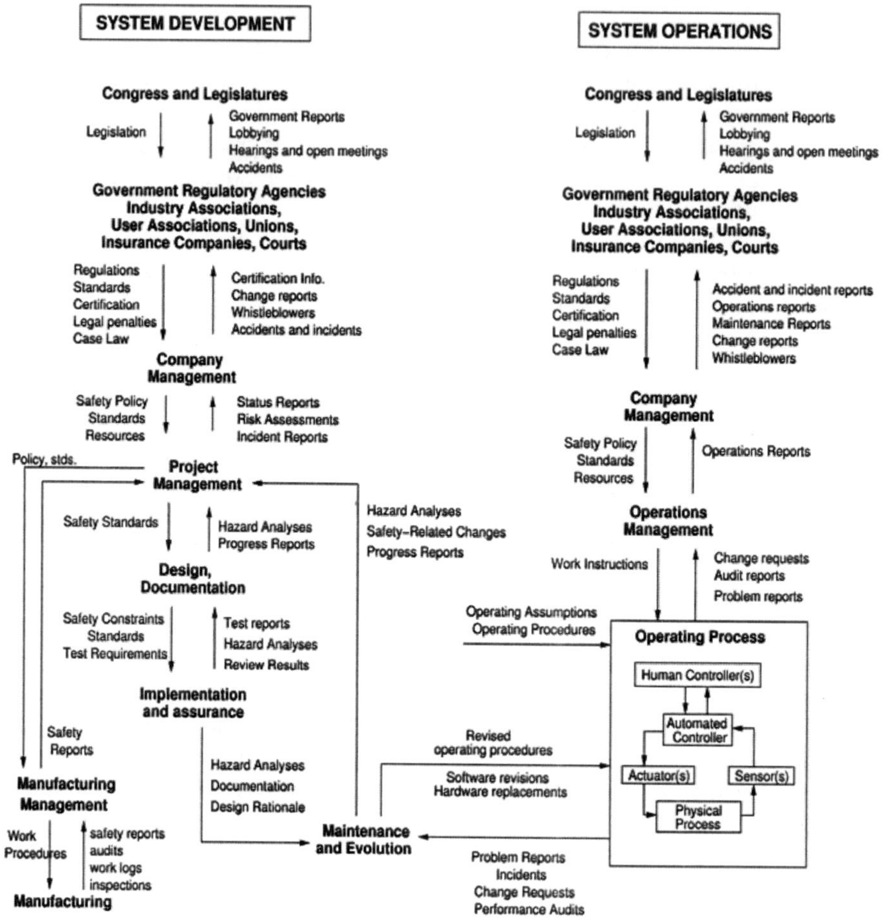

Fig. 1 An Example Socio-Technical Safety Control Structure

While Figure 1 focuses on the social and managerial aspects of the problem, the physical process itself can be treated as a control system in the standard engineering way. Figure 2 shows a sample control structure for an automobile adaptive cruise control system.

Each component in the safety control structure has assigned responsibilities, authority, and accountability for enforcing specific safety constraints. The components also have various types of controls that can be used to enforce the constraints. Each component's behavior, in turn, is influenced both by the context (environment) in which the controller is operating and by the controller's knowledge about the current state of the process.

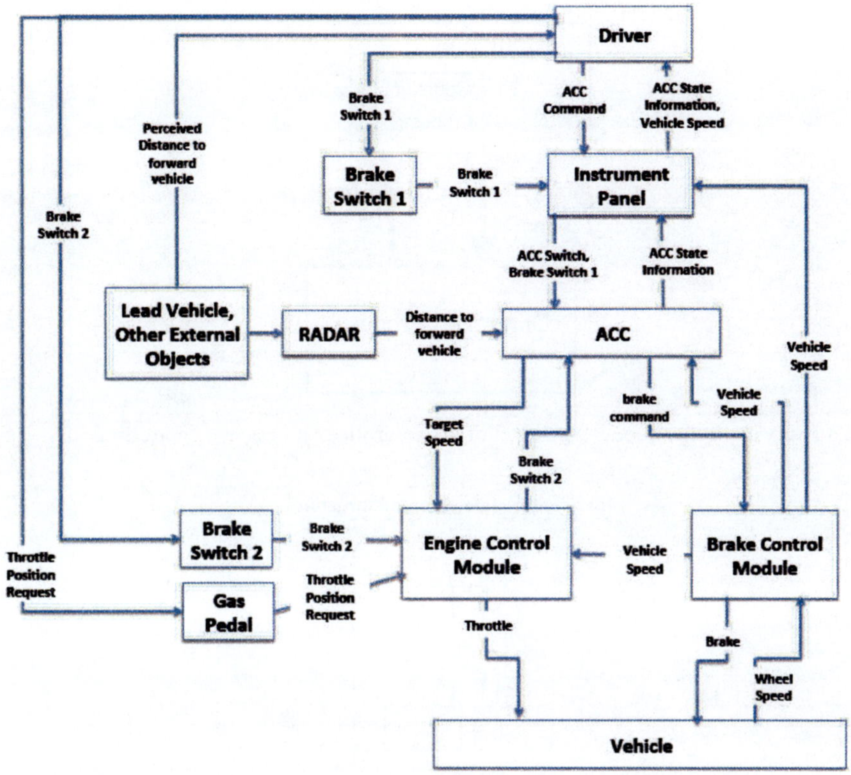

Fig. 2 A Sample Control Structure for an Automobile Adaptive Cruise Control System [Qi Hommes and Arjun Srinath]

Any controller needs to have a model of the process it is controlling in order to provide appropriate and effective control actions. That process model is in turn updated by feedback to the controller.

Accidents often occur when the model of the process is inconsistent with the real state of the process and the controller provides unsafe control actions (Figure 3). For example, the spacecraft software thinks that the spacecraft has reached the planet surface and prematurely turns on the descent engines. Accidents occur when the process models do not match the process and

- Commands required for safety (to enforce the safety constraints) are not provided or are not followed;
- Unsafe commands are given that cause an accident;
- Correct and safe commands are provided but at the wrong time (too early, too late) or in the wrong sequence
- A required control action is stopped too soon or applied too long.

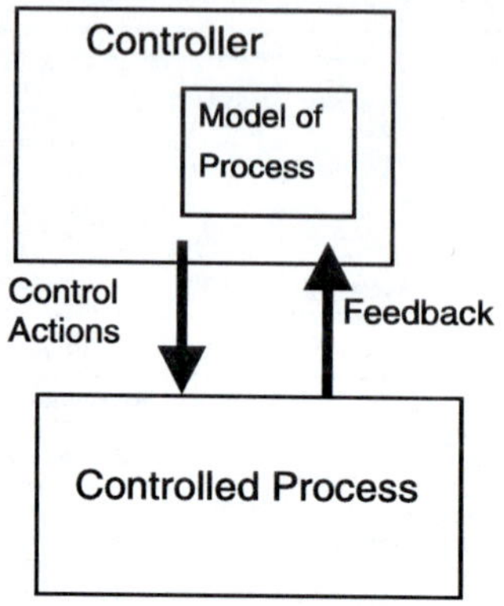

Fig. 3 Every controller contains a model of the controlled process it is controlling

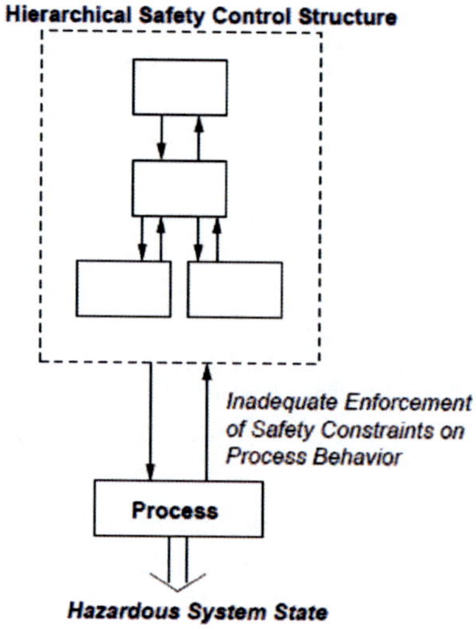

Fig. 4 In STAMP, accidents occur due to inadequate enforcement of safety constraints on system process behavior

The STAMP model of causality does a much better job of explaining accidents caused by software errors, human errors, component interactions, etc. than does a simple failure model. Figure 4 shows the overall concept behind STAMP. There are, of course, many more details. These can be found in [Leveson, 2011].

3.2 Using STAMP in Complex Systems

Just as tools like fault tree analysis have been constructed on the foundation of the chain-of-failure events model, tools and procedures can be constructed on the foundation of STAMP. Because STAMP includes more causes of accidents (but also includes standard component failure accidents), such tools provide a theoretically more powerful way to In particular, we will need more powerful tools in the form of more comprehensive accident/incident investigation and causal analysis, hazard analysis techniques that work on highly complex systems, procedures to integrate safety into the system engineering process and design safety into the system from the beginning rather than trying to add it on at the end, organizational and cultural risk analysis (including defining safety metrics and leading indicators of increasing risk), and tools to improve operational and management control of safety. Such tools have been developed and used successfully on enormously complex systems. Figure 5 shows the components of an overall safety process based on STAMP.

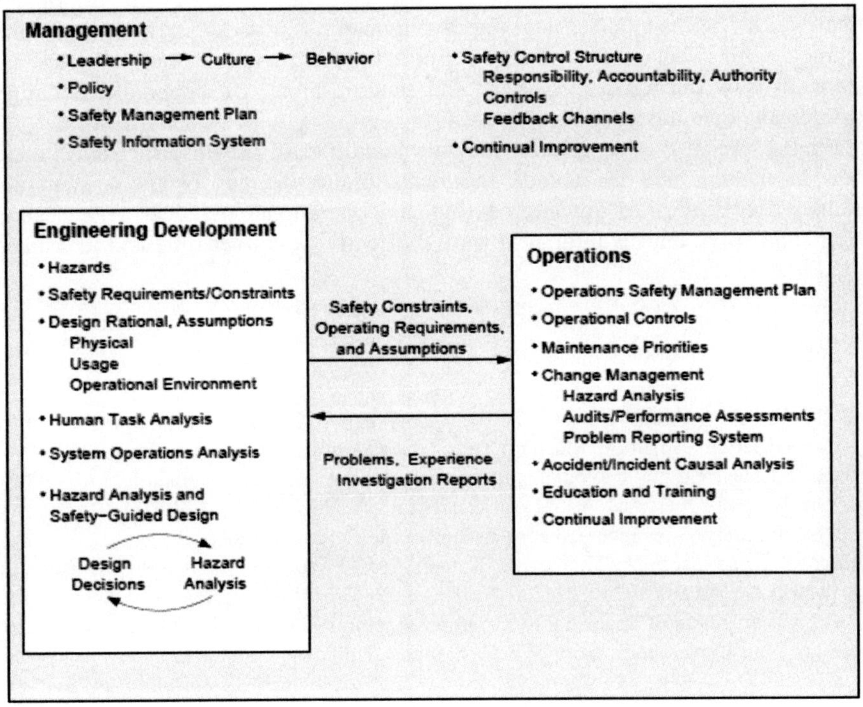

Fig. 5 The Overall Safety Process as Defined [Leveson, 2011].

4 Summary

This paper has described types of complexity affecting safety in our modern, high-tech systems and argued that a new model of accident causality is needed to handle this complexity. One important question, of course, is whether this new model and the tools built on it really work. We have been applying it to a large number of very large and complex systems in the past ten years (aerospace, medical, transportation, food safety, etc.) and have been surprised by the how well the tools worked. In some cases, standard hazard analysis techniques were applied in parallel (by people other than us) and the new tools proved to be more effective [see for example, JAXA [Arnold, 2009; Ishimatsu, 2010; Nelson, 2008; Pereira, 2006].

One lesson we have learned is the need to take a system engineering view of safety rather than the current component reliability view when building complex systems. The entire socio-technique system must be considered, including the safety culture and organizational structure. Another lesson is that safety must be built into a complex system; it cannot be added to a completed design without enormous (and usually impractical) cost and effort and with diminished effectiveness. To support this system engineering process, new specification techniques must be developed that support human review of requirements and safety analysis during development and the reanalysis of safety after changes occur during operations. changes.

Finally, we also need a more realistic handling of human errors and human decision making and to include the behavioral dynamics of the system and changes over time into our engineering and operational practices. We need to understand why controls migrate toward ineffectiveness over time and to manage this drift.

References

Arnold, R.: A Qualitative Comparative Analysis of STAMP and SOAM in ATM Occurrence Investigation, Master's Thesis, Lund University, Sweden (June 2009)

Dekker, S.: The field guide to understanding human error. Ashgate Publishing, Ltd. (2006)

Dekker, S.: Ten Questions About Human Error: A New View of Human Factors and System Safety. Lawrence Erlbaum Associate Inc., Mahwah (2005)

Ishimatsu, T., Leveson, N., Thomas, J., Katahira, M., Miyamoto, Y., Nakao, H.: Modeling and Hazard Analysis using STPA. In: Proceedings of the International Association for the Advancement of Space Safety Conference, Huntsville, Alabama (May 2010)

Leveson, N.: Engineering a Safer World: Systems Thinking Applied to Safety. MIT Press (December 2011), downloadable from `http://sunnyday.mit.edu/safer-world/safer-world.pdf`

Leveson, N.G.: Safeware: System Safety and Computers. Addison-Wesley Publishers, Reading (1995)

Nelson, P.S.: A STAMP Analysis of the LEX Comair 5191 Accident. Master's thesis, Lund University, Sweden (June 2008)

Pereira, S., Lee, G., Howard, J.: A System-Theoretic Hazard Analysis Methodology for a Non-advocate Safety Assessment of the Ballistic Missile Defense System. In: Proceedings of the 2006 AIAA Missile Sciences Conference, Monterey, CA (November 2006)

Perrow, C.: Normal Accidents: Living with High-Risk Technology. Princeton University Press (1999)

Rasmussen, J.: Risk management in a dynamic society: A modeling problem. Safety Science 27(2/3), 183–213 (1997)

Chapter 3
Autonomous Systems Behaviour

Derek Hitchins

1 Introduction

Autonomous machines have fascinated people for millennia. The Ancient Greeks told of Talos, the Man of Bronze, whose supposedly mythical existence has been given unexpected credence by the recent discovery of the Antikythera Mechanism. Could the Greeks really have created a giant mechanical man to protect Europa in Crete against pirates and invaders? Long before that, the Ancient Egyptians created the Great Pyramid of Khufu as an autonomous psychic machine to project the soul *(ka)* of the King to the heavens where he was charged with negotiating good Inundations with the heavenly gods.

In more recent times the sentient machine has been represented as logical, rational yet invariably inhuman. HAL[1] from Stanley Kubrick's *2001: A Space Odyssey* is an archetypal example, with HAL being dedicated to completing the mission regardless of the crew. Before that, in 1942, Asimov invented the Laws of Robotics, later made famous in books and the film *I Robot:*[2]

1. A robot may not injure a human being or, through inaction, allow a human being to come to harm.
2. A robot must obey any orders given to it by human beings, except where such orders would conflict with the First Law.
3. A robot must protect its own existence as long as such protection does not conflict with the First or Second Law.

These seemingly fool proof laws encouraged the robots to imprison the human population, supposedly to protect humans from self-harm, consistent with the first law.

---

[1] The name HAL derives from IBM, with each letter being one before its equivalent in the alphabet…H is one letter before I, and so on.

[2] "I Robot" is a play on Descarte's *Cogito Ergo Sum* – "I think, therefore I am:" proof of existence.

It seems to be the case that we humans generally expect automatic machines to be flawed, and that they will become so intelligent yet inhuman that they will take over the world: an uncomfortable juxtaposition of incompetence and omnipotence…

Publicised advances are being made in Japan, where they have robotic figures that can run, climb stairs, and some even display facial expressions. Honda's Asimo is able to learn about, and categorize objects. Nursing robots, and others, are being made rather smaller in stature than their equivalent human: Japanese designers seem well aware of the innate fear reaction that autonomous machines arouse in many humans.

One possible way forward, which may enable us to create smart, autonomous machines without the associated fear, might be to make the machines more human-like, so that they could act and interact with people on a more comfortable basis. Could we, perhaps, make machines that behaved in some respects like people, making acceptably ethical decisions and moral judgments, even exhibiting sympathy and empathy with victims and patients?

Even if we could do these things, would it be *wise* to do so? Or, would we, in effect, be creating beings with similar underlying flaws to Asimov's infamous Robots?

1.1 Meeting the Challenge

1.1.1 Complexity

Any autonomous machine capable of establishing its own purpose, making plans to achieve that purpose and executing those plans is inevitably going to be complex. We can imagine that such machines would have some form of 'brain' or perhaps a central decision-making and control centre. Moreover, for a machine to emulate a human in terms of physical dexterity suggests that the machine will inevitably be complex – not as complicated, perhaps, as a human, but nonetheless complicated.

Nature has much to tell us about complexity, although her lessons may be hard to learn. Life itself appears to be related to complexity. Up until some 580 million years ago, life on Earth was confined to simple cells and cell-colonies. Over the next 70-80 million years there was a mass diversification of complex life forms, the so-called Permian explosion, followed by periodic extinctions, ending in the 'great dying,' which marked the end of the Permian Period and the start of the Triassic.

Complexity, in Nature at least, appears to 'auto-generate,' and then to collapse before starting up again. Human civilizations follow the same pattern, with the same collapse: business organizations likewise. This phenomenon suggests underlying principles:

1. Nature creates wholes – organized, unified wholes – whence *holism*…
2. Wholes comprise interacting organic subsystems – whence *organicism*

3. Organic subsystems interact synergistically within their context and environment – whence *emergence*
4. Man and Nature synthesize unified wholes, with emergent properties, from complex, interacting subsystems, themselves wholes… - whence *hierarchy*
5. Complexity creates both order *and* disorder…

a. The Big Bang created stellar systems; galaxies; clusters; super-clusters… and destructive Black Holes
b. *Hymenoptera* (social insects) and *termites* create hives, colonies, bivouacs, etc., which eventually collapse only to be replaced by newer versions
c. *Homo sapiens* create families, societies, cultures, civilizations… that eventually collapse, only to rise again in revised form.

Consideration of this ubiquitous pattern of synthesis and collapse suggests a *Law of Complexity [1]:*

Open, interacting systems' entropy cycles continually at rates and levels determined by available energy

This stands in counterpoint to Kelvin's Third Law of Thermodynamics, which proposes that the entropy in a closed system will either remain the same, or will increase.

The tentative Law of Complexity also suggests that homeostasis (dynamic equilibrium) in a particularly complex body may be problematic.

1.1.2 Complex Growth

Figure 1 arranges some of these interrelated ideas into a hierarchical structure. At hierarchy level N-2 there are five complex systems (ovals), each containing six complex interacting subsystems. Since the subsystems and the systems are all open, they will all adapt to each other as they exchange energy, material and information. The result of action, interaction and adaptation is the emergence of properties of the whole (lowest layer) as shown at the right.

Hierarchy level N-1 is shown as also having five system groups. Each of these five will comprise the systems of Level N-2, but only one of these contributors is shown to prevent the figure from overflowing. Similarly, all of the complex systems at Level N-1 contribute to/synthesize only one of the smaller ovals in the six oval figures at Level N-0, the nominal level. The potential for complexity is already phenomenal, as each open element throughout the network/hierarchy adapts – inevitably, the whole would behave dynamically in simulation as well as reality.

The complexity arrow shows synthesis going up the hierarchy and reduction (analysis) going downwards. So, we can find out how things work and describe them by taking them apart, while we will understand their purpose, intent and behaviour by synthesizing/ observing the whole in situation and environment.

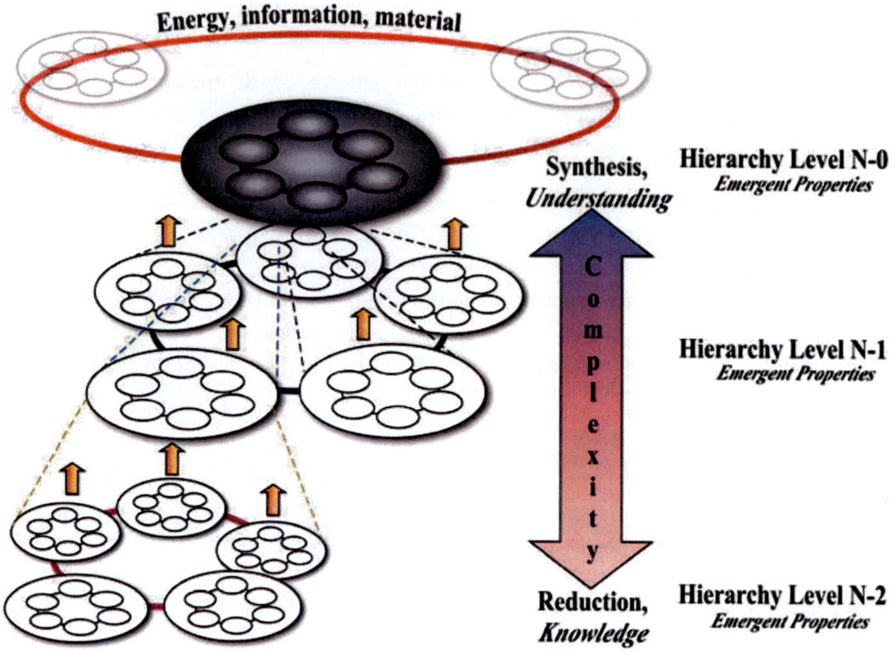

Fig. 1 Growth of Complexity through Synthesis

Evidently, integration of parts without regard for their mutual adaptation in situation and environment will prove inadequate – essentially, each layer in the hierarchy comprises the emergent properties of the lower layers, rather then their intrinsic properties. This suggests that simply adding body parts together, like Mary Shelley's Frankenstein, is unlikely to produce the desired result.

1.1.3 The Human Template

Happily, we know something of an exceedingly complex system that we might use – with care – as a template for any attempt to design an autonomous anthropomorphic machine.

The human body comprises some 11 or 12 organic subsystems, depending upon how you count them:

1. Integumentary – the outer layer of skin and appendages
2. Muscular
3. Skeletal
4. Nervous
5. Endocrine
6. Digestive
7. Respiratory
8. Cardiovascular
9. Urinary
10. Lymphatic and immune
11. Reproductive

Any autonomous machine is likely to find counterparts or equivalents to these organic subsystems. The integumentary organic subsystem has an obvious equivalent: the outer surface of the autonomous machine. Muscular and skeletal systems will also have self-evident equivalents. The nervous system of a human is exceedingly complex, including as it does the five senses, balance, and the brain, with its billions of neurons, which controls motor and sensor activities, but which also contains the higher centres of intelligence, judgment, etc. The endocrine system passes hormones into the bloodstream, controlling bodily functions and behaviour. While the nervous system controls fast response, the endocrine system operates on a slower and more lasting basis to maintain the stability of the body: the autonomous machine equivalent is less obvious, but may be associated with changed states of activity and operation, and with metabolic rate…

Items 6, 7 and 8, the digestive, respiratory and cardiovascular, are all concerned with energy – its supply, distribution, expenditure and disposal of metabolic waste. They will have an equivalent, presently undetermined and problematic, that may be quite different from their biological counterpart. Numbers 9 and 11, urinary and reproductive systems, are unlikely to require equivalents - yet. Number 10, however, one of the most complex organic subsystems outside of the brain, will surely require an equivalent to guard against, search out and neutralize pathogens, viruses and the like, and to protect against radiation and interference, to which any complex machine may fall prey.

The mutual interaction of all of these organic subsystems within their environment and situation somehow sustains life. Unlike most of our manmade/engineered systems, the organic subsystems are very closely coupled and intertwined, each depending upon the sum of the others for existence and performance. The organic subsystems interact both with each other and with their environment: for instance, personality (a human emergent property) develops partly as a result of interactions with other people, and partly as a result of interactions with other parts of the mind…and so affects, and is affected by, the health of the body.

Disappointingly, the list above tells us little about the whole human, its purpose, its capabilities, the functions it performs, etc. As Figure 1 indicated, analysis may provide knowledge of the parts, but it is synthesis that will give us understanding of the whole within its environment; the whole, as ever, is much greater than the sum of its parts.[3] Looking at the whole human, it might not be unreasonable to consider the human body as a vehicle enabling our minds to carry out our purpose, to fulfil our ambitions, to pursue our missions, etc.

Already we can see that the synthesis of an anthropomorphic machine like an Autonomous Peace Officer (APO) would present two distinct challenges: the physical structure and the management of behaviour. The list of human organic subsystems above is of little use with the latter, which is the principal subject of this discourse.

---

[3] Aristotle, 384-322BCE – Composition Laws.

1.2 The Autonomous Peace Officer [2]

The role of a human peace officer is well understood. S/he maintains and restores order in an area, often following a patrol route known as a 'beat;' this much has not changed since Talos, the ancient Greek Man of Bronze, supposedly patrolled Crete by circling it three times per day.

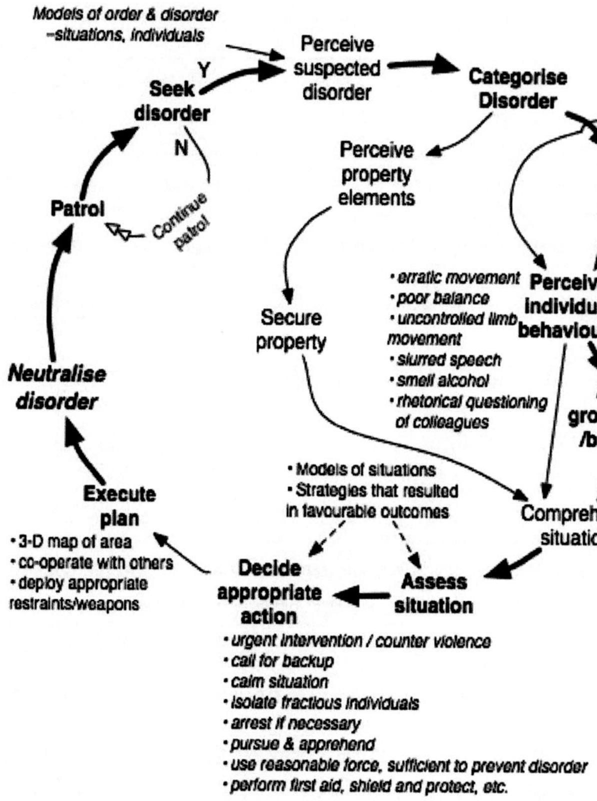

Fig. 2 Typical Peace Officer Concept of Operations (CONOPS)

Figure 2 shows a typical Peace Officer Concept of Operations – there could be several others. To unfold the concept, start at Patrol – 10 o'clock on the continuous loop diagram. Following the diagram clockwise shows that the officer has to be able to categorize disorder, categorize property and people, assess group mood and behaviour, assess situations, decide upon appropriate action, formulate and execute a plan of action to neutralize the disorder, after which s/he may resume patrol…

The peace officer is not alone in this work, although s/he may operate unaccompanied. Instead, s/he is a member of a policing team, with a command and control centre serving to coordinate the actions of the various peace officers, and if necessary to redirect officers in support of a colleague in need of assistance. Unlike soldiers, however, peace officers are expected to assess situations on their own and to make their own judgments. S/he is, after all, on the spot and therefore best aware of the situation…

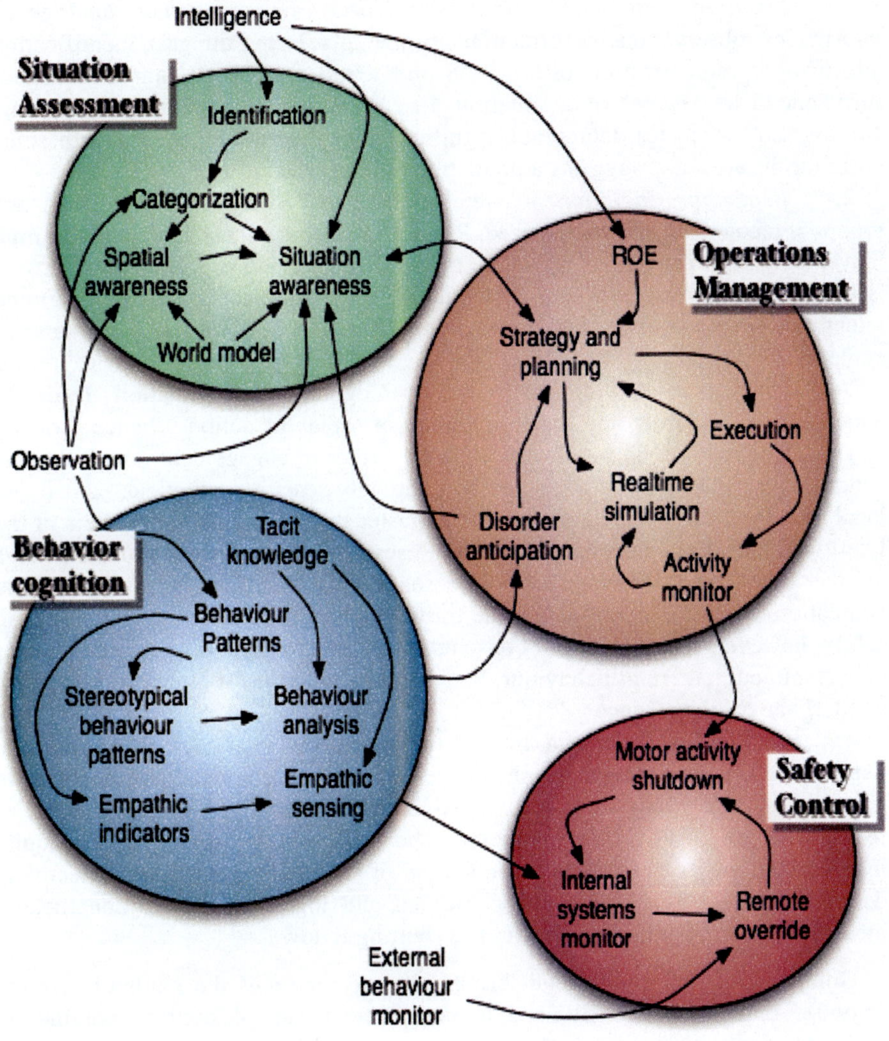

Fig. 3 APO – Outline Cerebral Activities

Developing and operating to the CONOPS is largely cerebral, with aggressive physical activity arising only occasionally under 'decide appropriate action' and 'execute plan.' The question arises, nonetheless, how do peace officers go about assessing situations, for instance, and is there any way in which an Autonomous Peace Officer might emulate them?

Figure 3 suggests several cerebral activities that an APO may be required to conduct:

1. *Situation Assessment*, where the APO brings together intelligence information, observational information ('people, places and things'), identification information, categorization information and world models (Weltanschauungen) sufficient to be 'aware' of a situation, i.e., know what is currently happening. 'Assessment' indicates degree and significance of disorder compared with some norm for the area, and suggests a prediction of likely change.
2. *Behaviour Cognition*, in which behaviour of observed individuals and groups is recognized and categorized. Peace Officers recognize 'uncontrolled limb movement,' for example, as a prelude to disorder and violence. Empathic sensing is included here: peace officers anticipate a person's situation, feelings (sic) and intentions from observing their behaviour in context compared with known contextual behaviours.
3. *Operations Management*, where Cognition of Situation leads to anticipation of disorder and the formulation of a plan of action – hence Strategy and Planning, within the framework of ROE – rules of engagement – such that the actions of the PO/APO are governed by politically sensitive, practical, ethical and legal rules. Within this area are conducted faster-than-real-time simulations of the likely outcome from any plan of action. Research shows that we humans do this all the time, even when crossing the road – we run a faster-than-real-time simulation to see if we can dodge the traffic, or not. We do it so quickly, and so subtly, however, that we are not consciously aware of it as a discrete activity – it blends into our pattern of behaviour. So must it be with the APO, where speed will similarly be 'of the essence…'
4. Finally, *Safety Control.* Items 1 to 3 may be rational, logical and sensible, but there is always, somewhere, the concern that such behaviour will result in an irrational or dangerous outcome, so some form of real time 'conscience' is required. In the APO, this might take the form of an overriding, and quite independent, way of assessing the outcome of the deliberations, and effectively determining: "the outcome may be rational and logical, but it is nonetheless, inhumane/politically insensitive," and so shutting it down.

Going into a little more detail brings into clearer focus the kinds of cerebral activities that an APO will need to undertake. Figure 4 outlines continually repeating processes that an APO might execute, conceptually. The box at the left of Figure 4 shows Situation Awareness as an outline, high-level process, leading to Recognition-Primed Decision Making in the box at right.

Situation Awareness shows the use of face detection techniques, such as we find in digital cameras we can purchase on the high street. Face and smile detection are now commonplace in small cameras, and facial recognition, while still in its infancy along with body and limb dynamics, is developing rapidly – it depends, of course, on there being a database of faces amongst which to find, and correctly recognize, the appropriate face, and so to identify the person. Happily, people who need to be identified are more likely to be in the database… this is much the same as the human Peace Officer visually recognizing a villain from the 'mug-shots' in the station, and anticipating trouble accordingly.

Interactions with the public take place in an environment with which the APO would be familiar through training; he would be equipped with a 3-D map of the area and Satnav-like sensors to locate him in position and orientation, so that he may optically map his visual perceptions (people, places and things) and cues on to his 3-D map.

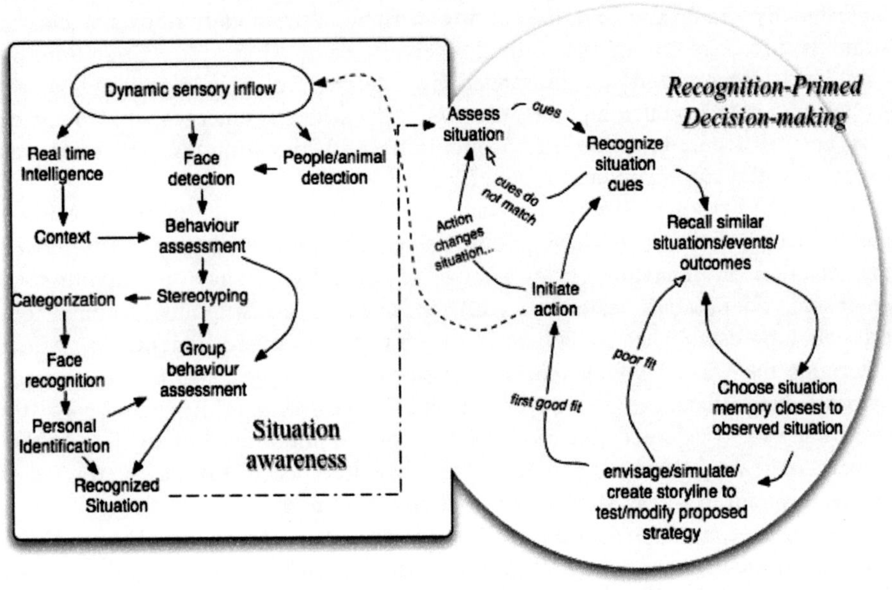

Fig. 4 APO: Strategies, Planning and Decision-Making

Recognition-Primed Decision Making (RPD) [3] is the very fast decision-making made by experts under time pressure. While the APO does not fit that bill, he may be provided with a database of previous situations and corresponding strategies that proved successful: he may build upon that information as he becomes more experienced in the field, i.e., he may update his database. Meanwhile, upon assessing a situation, he will be able to 'recall' similar situations, events and their outcomes, and hence derive a strategy for dealing that is most likely to succeed.[4] Note too, that Figure 4 is a continuous activity, not a

[4] The data rates and volumes needed to emulate an expert decision-maker are likely to prove challenging.

one-off, so assessments, decisions and strategies may be continually updated as the situation unfolds: the expert decision-maker does not make an initial decision and stick with it; instead s/he will update that decision as and when new data becomes available, will re-assess, and so will progressively 'home-in' on the best choice.

1.3 APO Behaviour Management

Figure 5 shows a general model of how behaviour may be managed [4] conceptually in humans, after Karl Jung [5]. It incorporates Nature Vs. Nurture, in the grey panels, and invokes a Belief System. The arrow and list, at top right of the figure, influence evolved Nature; while Belief System incorporates ideas indicated by the arrow and list, at lower right. The model follows a classic stimulus response paradigm. Stimulus enters centre left, into Cognition. A stimulus is 'recognized' or interpreted by reference to tacit knowledge, world models and belief, so that an *interpretation* of the stimulus emerges which may, or may not, be accurate, according to familiarity with the stimulus, and what was expected in the current situation (belief).

The interpretation of stimulus is passed to (behaviour) Selection, where Nature, Experience, Belief and constraints (such as capability, opportunity) all influence the selection of Behaviour, from a range of 'primitive' behaviours (aggressive, defensive, cooperative, cautious, etc.), in response to stimulus. Nature, too, presents long established archetypes[5]: king, magus, shepherd, knight, etc., which determine the way in which individual behaviour 'emerges.' Often in humans, nature dictates instant response for survival, ("knee-jerk"), which may be swiftly overridden by considered judgment. Note, too, the importance of training in this respect, indicating that both people and suitably designed APOs may be trained to behave/respond in ways that are contrary to their natures.

The selected archetype-influenced Behaviour is then passed into the rest of the information-decision-action procedure, where it influences the selection of options, choice of strategy and control of implementation that are the normal stuff of command and control.

For the APO, there are serious challenges in establishing Tacit Knowledge [6] (simple everyday facts like grass is green, sky is blue, things fall, etc.), World Models and Belief Systems. Vast amounts of such information would be needed to create a general-purpose autonomous machine (e.g. Data on Star Trek.)

[5] For an PO/APO, a suitable combination of Jung's archetypes might be "Shepherd-Knight," reflecting his dual role as protector of his flock, and knight challenging would-be aggressors, according to situation.

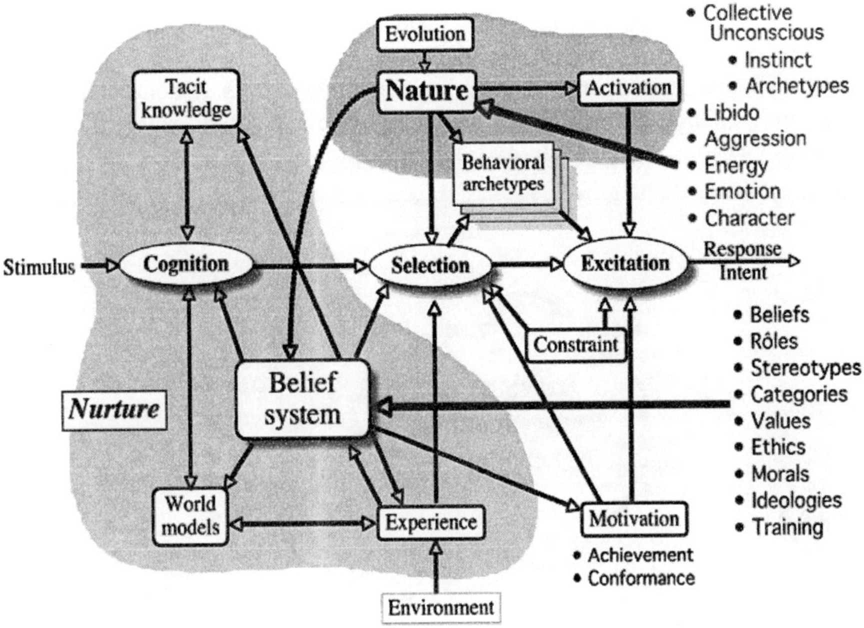

Fig. 5 Management of Behaviour

However, it may be practicable to create appropriate sets of tacit knowledge and world models for an APO in a predetermined, confined situation on a given beat. Similarly, any general belief system would be exceedingly large and complex, but the belief system associated with an APO in a particular location/situation/beat may be feasible; for instance, the number and variety of stereotypes may be quite limited. (For policing, stereotypes are invaluable – if you dress like a vagabond, chances are that you are a vagabond, and you should not be too surprised to be treated as one – unless, that is, there is a local carnival in progress – context changes everything!).

Such is the complexity likely to be involved with each of tacit knowledge, world models and belief systems, that it may prove more practicable for the APO to develop these over a period of time, some perhaps in simulation, some in real situations, and some pre-programmed. Similarly, the APO may have to learn how to stand, balance, run, jump, fall over, recover, etc… in order to make these activities efficient and life-like. Creating such abilities in automatons is currently proving challenging…

1.4 Autonomous Peace Officer Functional Design Concept

Figure 6 shows the behaviour management model of Figure 5 coupled with a more conventional command and control (C2) loop at the top typically used in mission management: collect information; assess situation; set/rest objectives; strategize and plan; execute plan; and cooperate with others if necessary in the process. Information is collected from the operational environment, and executing a plan is likely to change that operational environment, so this is a continuous loop, responding to, and initiating change. Moreover, it may be addressing a number of concurrent missions, so loop activity can be frenetic…

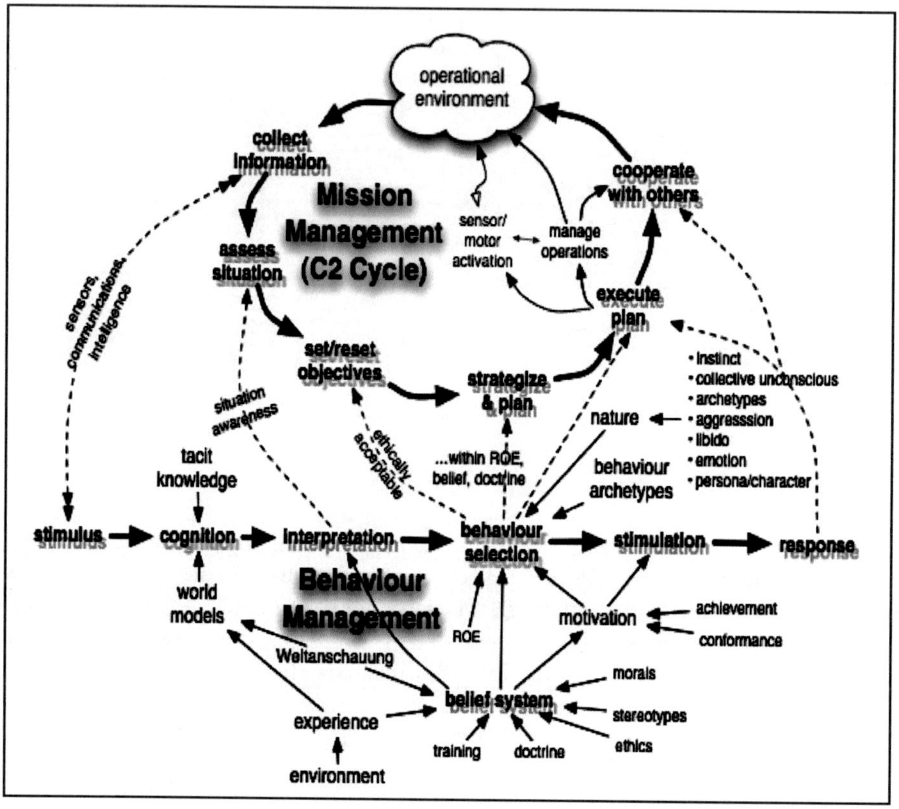

Fig. 6 A design 'framework' on which to 'hang' the APO 'cerebral' design and architecture. The upper 'circle' forms the well known command and control (C2) cycle widely employed in mission management. The lower section goes from stimulus (left) to response (right), showing how behaviour is invoked as response to stimulus. The two sections are connected, indicating ways in which behaviour influences the C2 cycle. The framework shows logical functions, datasets and relationships, but need not be configured in the best way for particular functional architectures, e.g. for situation awareness/assessment. The whole is a complex synthesis of systems…

On its own, the C2 loop is designed to produce rational decisions in dynamic situations, which may in some circumstances be considered 'uncompromising' – a charge occasionally levelled at the military. However, in the APO Design Concept, the C2 loop is moderated by Behaviour Management, so that training, ethics, morals, beliefs, Rules of Engagement, etc., can be brought to bear on the decision-making process. Note, too, how the belief system becomes involved in situation awareness, how objectives become ethically acceptable, how strategies and plans operate within doctrine, ROE, and Jung's Archetypes, e.g. Shepherd-Knight for an APO. In some situations, there are decisions that would fall naturally to the *shepherd of his flock,* which must be protected, or to the knight, *champion of his people*, who must behave chivalrously.

At top right there is a small, but important, loop: *manage operations*. We humans are able to continue assessing situations and making judgments even while pursuing some objective. It seems that the pursuit of the objective is conducted subliminally while our minds are looking ahead, anticipating problems, developing new tactics, etc. This is an important feature for an APO too: that he be able to conduct current operations while assessing developing situations. The loop *manage operations* indicates that the design delegates current operations to a lower level processor, so that higher-level situation assessment and decision-making may continue uninterrupted and at speed.

A major question remains unresolved: how can the 'cerebral' processes indicated in Figures 2 'engage' with the physical elements of the complete APO to make an operating machine that can run, jump, perceive an offender, give chase, restrain if necessary, interrogate, and – most importantly for any Peace Officer – apply *discretion*. Discretion marks the Peace Officer as "reasonable" – the peace officer is able to use his judgment about which minor misdemeanours he prosecutes and which it is reasonable, in the circumstances, to let go. So, must it be for the APO – he must be able to apply discretion, else be dubbed a 'dumb, insensitive machine'…

The outline problem is illustrated in Figure 7: sensor, motor and execution coordination feed into the primary mission functions, supporting purpose. But, how do the cerebral activities of mission management and behaviour management actually control the physical elements of sensor, motor and execution: there is a lacuna – the grey loop in the figure represents the uncertainty of connection?

We can get some idea of how to connect cerebral purpose to physical realization by considering a simple example. Prime Mission Functions for the Peace Officer are indicated in Figure 2. Consider 'Pursue and Apprehend.' The APO may be seated when he (we will anthropomorphize the APO as male) observes a miscreant running away and a person shouting for help: the APO elects to apprehend the miscreant, so establishing a new mission. He will then plan an intercept path, avoiding people and objects, as he stands and gains balance – both tricky operations, requiring a considerable amount of coordinated muscular activity in legs, back, torso, arms and head: simply standing from a low seat requires humans to rock back and the forward, thrusting the head and arms forward while pushing with the legs, straightening the back, tightening the stomach muscles, and moving body weight on to the toes – it is surprisingly complicated to stand and at the same time move forward smoothly into running mode without losing balance, yet we do it without conscious thought...

[For the APO each of these elemental functions will be learned routines, each routine triggering complex sequences of 'muscular' actions: a routine for standing in balance; one for transition to running; one for running in balance, etc., etc. just as in a human, and similarly subliminal...]

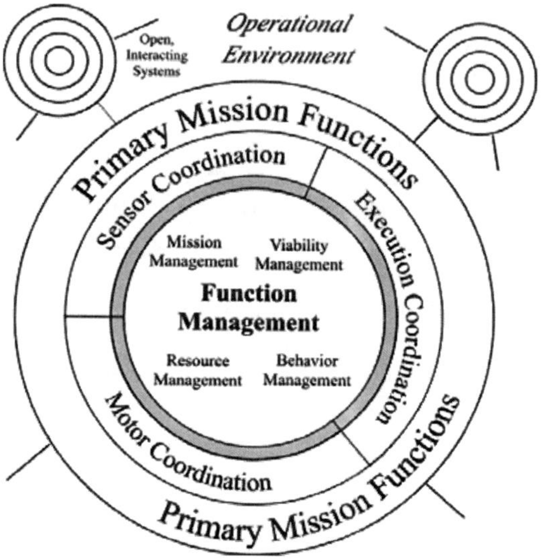

Fig. 7 Connecting Purpose to Action

The APO, having achieved this feat, will then run on an intercept path towards the miscreant, who may dodge, speed, up, slow down, requiring the APO to continually re-plan his intercept path, swerve, jump... Assuming that the APO is nimble and fast enough to catch up with the would-be escapee, the APO has then to restrain the miscreant without harming in her or him. This is a difficult manoeuvre, and may require the APO to deploy some non-lethal weapon, such as a "sticky net," to avoid cries of 'police brutality.' (If a human were killed by an APO, intentionally or inadvertently, the political fallout would be tremendous.)

Having overcome the restraining hurdle, the APO has then to subdue the miscreant, question him/her, seek to identify them, report into command central, recover any weapons or stolen goods, and so on. All of these would be serious challenges to an APO in the real world, and so far we have considered only one of the Prime Mission Functions from Figure 2. Note, however, that each PMF is carefully and extensively coded for human Peace Officers, which limits the range of actions required of the APO.

Figure 8 shows how various routines and sequences might be organized. Starting at top left of the diagram is the formulation of a Plan to execute some Primary Mission Function. Following the arrows leads to an Action Semantic Map that shows various sensor and motor actions being activated to constitute a routine. The output from the Map leads to a second map that transitions from sequence to sequence of routines, some in series some in parallel, the whole progressively executing the Primary Mission Function. The whole appears – and is – potentially complex, since it has to be able to transition smoothly from routine to routine, for example, as the miscreant swerves and dodges...

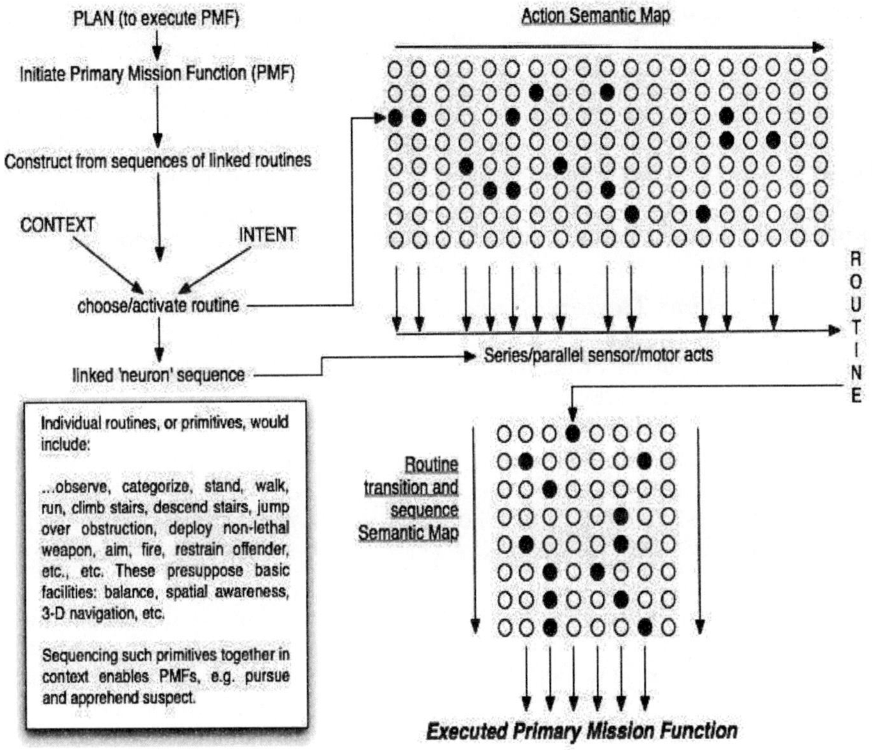

Fig. 8 Synthesizing a Primary Mission Function

The APO may also find himself required to sympathize with a victim of some crime, so that different modes of speech and behaviour will be needed, according to situation, in accord with the appropriate behavioural archetype (e.g. Shepherd-Knight). Some fascinating work with chimpanzees by Rizzolati et al [7], suggest that it may be possible for the APO to empathize both with victims and would-be perpetrators, by comparing their behaviour in context with a database of known/observed contextual behaviours, and so to anticipate likely outcomes. The APO's ability to exercise discretion is likely to lie in his ability to empathize with miscreants: to assess from their attitude from behaviour-in-context (anger, aggression, escape, regret, guilt, remorse); and, simultaneously knowing their misdemeanour records, to make a judgment call about prosecution.

An APO design feasibility study [2] indicates, then, that there is a phenomenal amount of data to be amassed to represent tacit knowledge, world models, and belief systems. Furthermore, an APO should be able to access, review, select and employ appropriate data very quickly to be practical for interaction with people – c.20ms is a sensible target, one easily reached by humans, and an APO could not afford to be slower. In developing strategies and plans, too, there would be a need to conduct simulations of proposed activities in context, to prove their effectiveness, and to mitigate risk to life and limb. (Estimates suggest that processing speeds in the order of 10^{11} bits/sec might be needed to achieve the necessary response times for Recognition-Primed Decision Making alone…)

In many situations, the APO would be called upon to discuss situations with both victims and suspects, and to make reasoned and 'instinctive' judgments as to the merits of a case: this would require more than a facility with language; as well as being fluent, the APO would need to be able to converse purposefully, to formulate sensible questions in context, and judge the validity of responses.

1.5 The Systems Design Concept Outline

Figure 9 gives an outline of the more straightforward aspects of design, excluding such problematic higher-level APO cerebral activities. The line of Prime Mission Functions runs from top left of the figure to bottom centre. The Prime Directive (shown at centre left of the figure), is the ultimate statement of purpose, which – for an APO – might be "to maintain public order." The Prime Directive suggests a Goal (societal order?), which can be achieved through a series of objectives (achieve: property order; individual order; group order; societal order – hence the Goal).

Achievement of each objective and of the Goal necessitates the generation of an appropriate strategy, and there will inevitably be threats to achieving some or all of those strategies. Together objectives, goal and strategies indicate the range of Prime Mission Functions that the Solution System (APO) will perform. As the figure shows, the execution of these Prime Mission Functions in the Future Operational Environment will present an Emergent Capability, constituting the Remedial Solution…

Emerging from the top Prime Mission Function, observe Sense Situation, leading to Situation Assessment/Real-time Simulation, and moving into the command and control activities indicated by the Mission Management panel, top left. Behaviour Management, top right, influences choices and decisions as Figures 5 and 6 indicated, leading to an aggregation of action routines, standard operating procedures and learned sequences that together enable the autonomous system to 'animate' the Prime Mission Functions appropriate to the immediate situation.

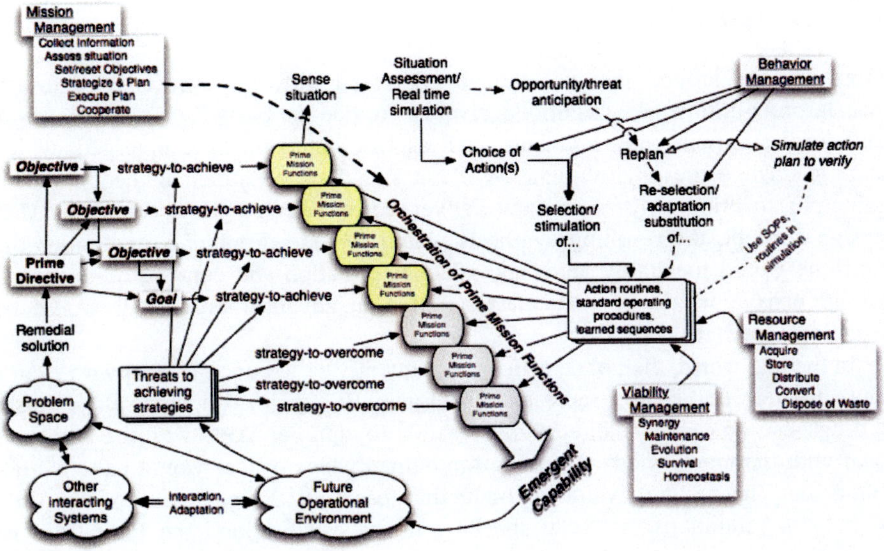

Fig. 9 The System Design Concept. The objective is to create 'Primary Mission Functions," that delineate the APO's capabilities, and to develop these by identifying and orchestrating a set of action routines, operating procedures and learned routines. The APO needs the ability to sense situations (top), assess them in real time, choose appropriate actions, simulate them to test their potential, and stimulate them in sequences that create each PMF. As the figure shows, there is a circular process resulting in closure, as the activated PMFs create emergent properties, capabilities and behaviours in the Future Operational Environment (bottom) which resolve the various problems and issues within the Problem Space. Each PMF 'emerges' from the synthesis of many subsystems interacting in cooperation and coordination, forming a complex, ever-shifting pattern of activity.

Note that the figure represents a continuous process, a control loop, where the characteristics of the Problem in the evolving Operational Environment automatically generate and apply the necessary solution – hopefully in real time.

Note too the Viability Management and Resource Management Panels, (part of the Generic Reference Model, which purports to show the 'inside' of any complex system [8]) which contribute respectively to the continuing fitness of the Solution System and to its being continually resourced…including the vital topic of energy. Viability Management, in particular, addresses much of the physical side of the solution system: synergy, maintenance, evolution, survivability and homeostasis. In some autonomous systems it may be necessary for them to *evolve* by sensing/improving their performance in changing environments, to be *self-healing,* and to be *survivable* (which will include damage repair). *Homeostasis* for such a self-evolving, self-healing complex is likely to be an issue in a dynamic and constantly evolving societal situation. Their behavioural archetypes (Shepherd-Knight, etc) would remain unchanged, however, serving as a conscience and controller of excess behaviour…

1.6 APO Conclusions

Organizations around the world have addressed different aspects of androids which, brought together, might conceivably provide the basis for a working APO, and some surprising developments such as face detection have emerged, which start to change the environment. While it might be possible, in theory and in principle, to bring all these separate advances together as proposed in the APO design concept, the resulting synthesis would be indeed complex and, since the machine would inevitably learn and adapt to situation and context, quite how it would behave would be difficult to predict in advance. Perhaps it would be excellent…perhaps limited.

In the real world, the police officer is frequently called upon to deal with events and situations that are not represented in Figure 2: a major fire; an airplane crash; a drug bust; an earthquake; etc. Is it feasible to equip an APO with the ability to deal with the unexpected? But then, the human peace officer would have his/her limits too. There is great value in conducting feasibility design studies of the APO, if only to remind us just how special humans can be…and the limitations of technology by comparison.

Finally, would it be *wise* to build and deploy an APO? There is no evidence that a 'humane' APO as described above would be acceptable, although philosophers seem to think so – Isaac Asimov's *I Robot* being a case in point. He, the APO, may be accepted, might even be considered preferable to human peace officer by virtue of impartiality, ethical behaviour, discretionary adherence to the law, survivability in harsh environments, etc. On the other hand, he may be considered weak, or slow, or stupid, and so ridiculed which may challenge his authority as an officer of the law. He might become the butt of practical jokes, testing for his weak points vis-à-vis a human peace officer, and looking on his 'discretion' as weakness after the style of teenagers.

Overall then, perhaps not wise. But the APO will happen. Soon. Like Pandora's Box, it is irresistible…

2 Autonomous Air Vehicles (AAVs)

2.1 Different Domains, Different Objectives

Autonomous Vehicles are a contemporary reality for military applications, where risks and constraints are different from civilian and commercial life, in which people-concerns generally preclude full autonomy. The Jubilee Line on the London Underground is capable of unmanned operation, but operates manned owing to safety concerns by the travelling public and unions. Similarly, air transport could, in principle, operate unmanned in many cases, but the fare-paying public are much more comfortable with a real, fallible person at the controls than a supposedly infallible machine…

2.2 Autonomous Ground Attack

Autonomous air vehicles are used in Afghanistan for attacking ground targets. AAVs do not risk the lives of the attacking force, a very significant advantage to the attacker. However, there are the ever-present risks of unintended collateral damage and civilian deaths. The latter, in particular, can prove counter-productive, stirring up resentment and hatred, and firing-up a new army of jihadists, quite the opposite of the intentions.

So, would it be practicable to equip such AAVs with the ability to:

- Assess the situation upon arrival in the target area
- Acquire real-time intelligence about the intended target
- Assess the potential for civilian casualties and collateral damage
- Determine, as could a human pilot, whether it was prudent to proceed in the circumstances
- If not, to search around for another target (an 'alternate' may have already been designated) and to assess its viability as a reasonable target
- Etc.?

Even if it were practicable – and it may be - there is an immediate problem. If an AAV were equipped with the 'behaviour,' to make humane judgment calls, and the opposition came to know of it, then the opposition could invoke 'human shields:' civilians, hospitals, schools, etc. This tactic, towards which several countries have already shown a propensity, negates the attackers' intent to be humane.

The attacked regime essentially dares the attacker to attack such sensitive targets, and will assuredly use any such attack – whether against a real, or mock, target – as a publicity weapon. It is, assuredly, an effective counter. The attacker is left with the need to improve their intelligence, introduce more selective weapons, or just call the bluff and attack the target willy-nilly and live with the political fallout.

2.3 An Alternative Approach

In some situations, however, there may be a different way – conceptually, at least, and depending upon situation, context and – in this instance – technology. Consider a potential combat situation [9] where an invader was entering an area that was ecologically sensitive, where damage to the environment could be immense. There are many such areas on Earth at present: deserts, Great Plains, Polar Regions, etc.

Consider further that a transportable mobile force, Land Force 2020, had been established to defend the territory, to eject the invaders, but with minimal damage to sensitive ecosystems and with little loss of human life, on the part of both defenders and invaders. Such a force might arise under the auspices of the United Nations, for example.

The raptor, presently uncovered, will be 'dressed as an indigenous bird of prey, with flapping wings, capable of finding and soaring on thermals: wings are solar panels, tail is an antenna, and eyes are video cameras. Raptors are armed and carry dragonflies.

The dragonfly, much smaller than the raptor, similarly has solar panel wings and video camera eyes. The body is an antenna. Dragonflies can operate individually or in swarms, can hover, dart and dodge, and are armed.

Fig. 10 Land Force 2020 Raptor and Dragonfly

The force is comprised of lightweight land vehicles, transported by global transport aircraft, so that they may be deployed swiftly, but need reside in territory. The land vehicles are not designed to fight, but are able to move across all kinds of terrain and water, operating in automatically controlled formations, swarms, according to situation and context. There may be one human operator to several automatically steered vehicles, greatly reducing the risk to human life.

Each vehicle carries a number of autonomous solar-powered *raptors*, which may be launched and retrieved on the move. Each raptor carries a number of *dragonflies*. The raptors are able to soar on thermals, so may stay aloft for hours or days, using very little energy. They are made to look like indigenous birds of prey to avoid disturbing the local fauna and alerting the invaders. Similarly, the dragonflies are made to look like, uh, dragonflies…Figure 10 shows dynamic simulation models of raptor and dragonfly.

Envisage a mobile force, a swarm, of land elements travelling across open country, preceded by their autonomous raptors, with the raptors launching autonomous dragonflies as needed. Raptors and dragonflies are equipped with video-camera eyes and digital communications, allowing any human controllers to detect, locate, identify, monitor and even communicate with any intruders. There could be a command and control centre with several operators in the supporting transport aircraft, which would stay in the rear.

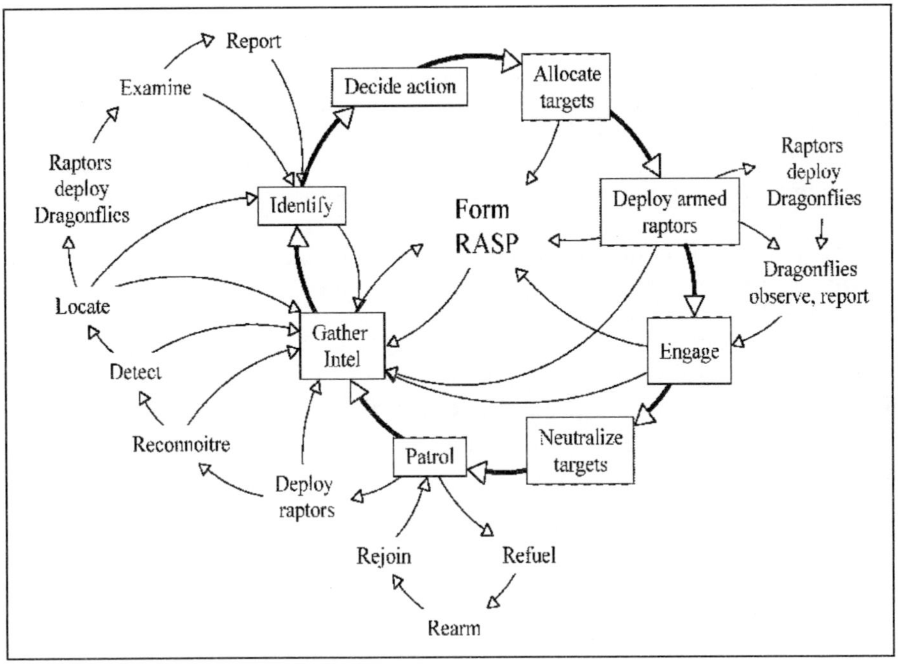

Fig. 11 Land Force 2020 CONOPS

Figure 11 shows a conceptual CONOPS for such a situation. Note the resemblance between the main loop in the figure and that of Figure 2 above, the Peace Officer CONOPS. As Figure 10 shows, the swarm of land vehicles travels with raptors ahead of them reconnoitring, and following the routine: detect; locate; identify. Intelligence may be passed back to C2, but – if straightforward – may be addressed by a raptor, which may deploy a dragonfly to investigate. A RASP (Recognized Air and Surface Picture) is formed using data from all sources, and the RASP is available to C2, Land vehicles and raptors alike. The RASP integrates and presents the surveillance, acquisition, targeting, and kill/deter assessment (SATKA) of the whole force.

A decision to act on Intel may be made by the land vehicle commander, the C2 in the supporting aircraft, or automatically (in the absence of command) by the raptors together – these would be network-centric operations. Action may include raptor and dragonfly weapons, but the dragonflies are able to address and converse with intruders – acting as '2-way-comms' links with C2, so intruders may be warned, may be advised to stand down or withdraw. If conflict in inevitable, then, raptors and dragonflies are equipped with a variety of lethal and non-lethal weapons to incapacitate the intruders, and/or render their technology useless. Dragonflies are potentially able to get very close to, hover, pick out a particular human target and to neutralize that target without risk, let or hindrance to other people, places or things…

2.3.1 Is the Concept Viable?

Surprisingly, perhaps, the concept is viable in the near-to-mid term, drawing upon current research in a number of areas for autonomous control of land vehicles, for the creation of fast-beat mechanical wings for small, agile, unmanned air vehicles, and for non-lethal weapons. Communications are based on mobile smart phone technology, which is currently able to communicate visually and aurally and could, with some modification, form the binocular eyes of both raptor and dragonfly. Bringing all of these together and creating the whole solution would be expensive, and at present no one appears to have either the will or the money.

However, the systems design concept does suggest that the behaviour of our autonomous fighting systems could be made more effective, subtler, less aggressive, and less harmful to sensitive environments in future – if we had the will... In effect, the Land Force 2010 concept is of a 'natural' extended system, with a distinct 'whole-system' behaviour...

3 Consciousness and Sentience

Note the absence of any reference to machine consciousness, or sentience. [10] Whether an APO that was aware of, and able to assess situations, identify threats to order and make fast, Recognition-Primed Decisions to restore order could be deemed sentient or conscious, may be unfathomable. Perhaps there are degrees of consciousness. Are we humans fully conscious at all time – whatever 'fully conscious means' – or are we more conscious at times of high alert, less conscious at times of relaxation? The APO would not recognize the questions, let alone the implications...he would be unable to think about himself as an entity, although he would be aware of his own capabilities and limitations within his narrowly defined domain context... within those constraints, might he be deemed 'self aware?'

4 Conclusion

Autonomous machines of the future are widely feared for their anticipated inhumane behaviour: literature and the media have stoked fear, to such an extent that the general public may discount the potential advantages of autonomous systems in support of humanity.

The idea is gaining some ground is that we might endow our autonomous machines with more human characteristics – appropriate facial expressions, personality, empathy, appropriate human-like behaviours, morality and ethics – so that we humans will find machine human interactions less daunting, more rewarding, while the autonomous machines will be less likely to 'run amok' or behave in a cold, impersonal, machine-like manner.

It seems reasonable to suppose that we could so endow autonomous machines, at a cost: yet, there is no hard evidence that autonomous machines so endowed would really be more acceptable – indeed, it is difficult to see how such evidence could be garnered without practical examples and trials. However, it seems *reasonable...*

Exploring the practicality of endowing autonomous machines with human-like behaviour suggests that current technology may not be sufficient to do this in the general case, but that it might be possible, by constraining the domain of use and operation of the autonomous machine, to bring the technology required within attainable limits.

The likely cost of doing this also suggests that trained humans are likely to remain the best (most cost-effective) solution in cases where danger and extremes of environment are not the driving considerations...

Meanwhile, innovative technological advances such as face detection, smile detection, facial recognition and improved understanding of empathy on the cerebral side, and the development of small, agile flying machines on the physical side, together with the potential repurposing of current technology such as smart phones, suggest that there may be alternative ways of creating smart, affordable, autonomous machines that we have yet to think of... So, perhaps, the notion that 'humane' autonomous machines are too complex and expensive is being eroded by the ever-rising tide of technological advance.

References

[1] Hitchins, D.K.: Getting to Grips with Complexity... A Theory of Everything Else (2000), `http://www.hitchins.net/ASE_Book.html`

[2] Hitchins, D.K.: Design Feasibility Study for an Autonomous Peace Officer. In: IET Autonomous Systems Conference (2007), `http://www.hitchins.net/Systems%20Design%20/AutonomousPeaceOfficer.pdf`

[3] Klein, G.A.: Recognition-Primed Decisions. In: Advances in Man-Machine Research, vol. 5. JAI Press (1989)

[4] Hitchins, D.K.: Advanced Systems Thinking, Engineering and Management. Artech House, Boston (2003)

[5] Jung, C.G.: The Portable Jung. Penguin Books (1976)

[6] Polanyi, M.: Tacit Knowing. Doubleday & Co. (1966) (1983); reprinted : ch.1. Peter Smith, Gloucester

[7] Rizzolati, G., Fogassi, L., Gallese, V.: Mirrors in the Mind. Scientific American 295(5) (November 2006)

[8] Hitchins, D.K.: Systems Engineering: A 21st Century Systems Methodology, pp. 124–142. John Wiley & Sons, Chichester (2007)

[9] Hitchins, D.K.: Systems Engineering: A 21st Century Systems Methodology, pp. 313–348. John Wiley & Sons, Chichester (2007)

[10] Koch, C., Tononi, G.: A Test for Consciousness. Scientific American 304(6) (June 2011)

Chapter 4
Fundamentals of Designing Complex Aerospace Software Systems

Emil Vassev and Mike Hinchey

Abstract. Contemporary aerospace systems are complex conglomerates of components where control software drives rigid hardware to aid such systems meet their standards and safety requirements. The design and development of such systems is an inherently complex task where complex hardware and sophisticated software must exhibit adequate reliability and thus, they need to be carefully designed and thoroughly checked and tested. We discuss some of the best practices in designing complex aerospace systems. Ideally, these practices might be used to form a design strategy directing designers and developers in finding the "right design concept" that can be applied to design a reliable aerospace system meeting important safety requirements. Moreover, the design aspects of a new class of aerospace systems termed "autonomic" is briefly discussed as well.

Keywords: software design, aerospace systems, complexity, autonomic systems.

1 Introduction

Nowadays, IT (information technology) is a key element in the aerospace industry, which relies on software to ensure both safety and efficiency. Aerospace software systems can be exceedingly complex, and consequently extremely difficult to develop. The purpose of aerospace system design is to produce a feasible and reliable aerospace that meets performance objectives. System complexity and stringent regulations drive the development process, where many factors and constraints need to be balanced to find the "right solution". The design

---

Emil Vassev · Mike Hinchey
Lero—the Irish Software Engineering Research Centre
University of Limerick, Ireland
e-mail: {Mike.Hinchey,Emil.Vassev}@lero.ie

of aerospace software systems is an inherently complex task due to the large number of components to be developed and integrated and the large number of design requirements, rigid constraints and parameters. Moreover, an aerospace design environment must be able to deal with highly risk-driven systems where risk and uncertainty are not that easy to capture or understand. All this makes an aerospace design environment quite unique.

We rely on our experience to reveal some of the key *fundamentals* in designing complex aerospace software systems. We discuss best design practices that ideally form a design strategy that might direct designers in finding the "right design concept" that can be applied to design a reliable aerospace system meeting important safety requirements. We talk about design principles and the application of formal methods in the aerospace industry. Finally, we show the tremendous advantage of using a special class of space systems called autonomic systems, because the latter are capable of self-management, thus saving both money and resources for maintenance and increasing the reliability of the unmanned systems where human intervention is not feasible or impractical.

The rest of this paper is organized as follows: In Section 2, we briefly present the complex nature of the contemporary aerospace systems. In Section 3, we present some of the best practices in designing such complex systems. In Section 4, we introduce a few important design aspects of unmanned space systems and finally, Section 5 provides brief summary remarks.

2 Complexity in Aerospace Software Systems

The domain of aerospace systems covers a broad spectrum of computerized systems dedicated to the aerospace industry. Such systems might be onboard systems controlling contemporary aircraft and spacecraft or ground-control systems assisting the operation and performance of aerospace vehicles. Improving reliability and safety of aerospace systems is one of the main objectives of the whole aerospace industry. The development of aerospace systems from concept to validation is a complex, multidisciplinary activity. Ultimately, such systems should have no post-release faults and failures that may jeopardize the mission or cause loss of life. Contemporary aerospace systems integrate complex hardware and sophisticated software and to exhibit adequate reliability they need to be carefully designed and thoroughly checked and tested. Moreover, aerospace systems have strict dependability and real-time requirements, as well as a need for flexible resource reallocation and reduced size, weight and power consumption. Thus, system engineers must optimize their designs for three key factors: *performance*, *reliability*, and *cost*. As a result, the development process, characterized by numerous iterative design and analysis activities, is lengthy and costly. Moreover, for systems where certification is required prior to operation, the control software must go through rigorous verification and validation.

Contemporary aerospace systems are complex systems designed and implemented as *multi-component* systems where the components are *self-contained* and *reusable*, thus requiring high independency and complex synchronization. Moreover, the components of more sophisticated systems are considered as agents (multi-agent systems) incorporating some degree of intelligence. Note that intelligent agents [1] are considered one of the key concepts needed to realize self-managing systems. The following elements outline the aspects of complexity in designing aerospace systems:

- multi-component systems where inter-component interactions and system-level impact cannot always be modeled;
- elements of artificial intelligence;
- autonomous systems;
- evolving systems;
- high-risk and high-cost systems, often intended to perform missions with significant societal and scientific impacts;
- rigid design constraints;
- often extremely tight feasible design space;
- highly risk-driven systems where risk and uncertainty cannot always be captured or understood.

3 Design of Aerospace Systems – Best Practices

In this section, we present some of the best practices that help us mitigate the complexity in designing aerospace systems.

3.1 Verification-Driven Software Development Process

In general for any software system to be developed, it is very important to choose the appropriate development lifecycle process to the project at hand because all other activities are derived from the process. An aerospace software development process must take into account the fact that aerospace systems need to meet a variety of standards and also have high safety requirements. To cope with these aspects, the development of aerospace systems emphasizes *verification*, *validation*, *certification* and *testing*. The software development process must be technically adequate and cost-effective for managing the design complexity and safety requirements of aerospace systems and for certifying their embedded software. For most modern aerospace software development projects, some kind of spiral-based methodology is used over a waterfall process, where the emphasis is on verification.

As shown in Figure 1, NASA's aerospace development process involves intensive verification, validation, and certification steps to produce sufficiently safe and reliable *control systems*.

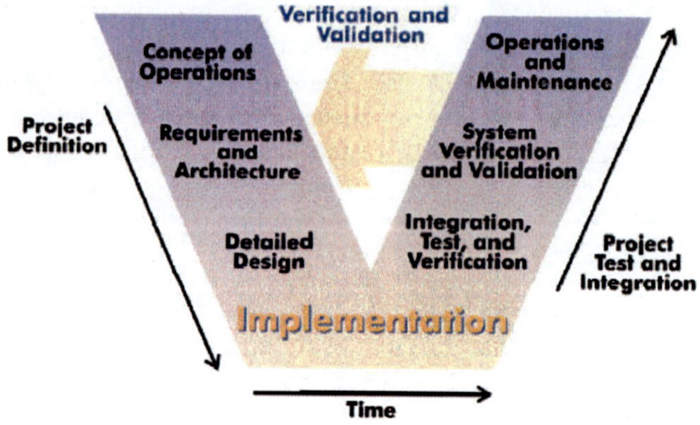

Fig. 1 A common view of the NASA Software Development Process [2]

3.2 Emphasis on Safety

It is necessary to ensure that an adequate level of safety is properly specified, designed and implemented. Software safety can be expressed as a set of features and procedures, those ensuring that the system performs predictably under normal and abnormal conditions. Furthermore, "the likelihood of an unplanned event occurring is minimized and its consequences controlled and contained" [3].

NASA uses two software safety standards [4]. These standards define four qualitative hazard severity levels: *catastrophic*, *critical*, *marginal*, and *negligible*. In addition, four qualitative hazard probability levels are defined: probable, occasional, remote, and improbable. Hazard severity and probability are correlated to derive the risk index (see Table 1). The risk index can be used to determine the priority for resolving certain risks first.

Table 1 NASA Risk Index Determination [4]

Hazard Severity	Hazard Probability			
	Probable	Occasional	Remote	Improbable
Catastrophic	1	1	2	3
Critical	1	2	4	4
Marginal	2	3	4	5
Negligible	3	4	5	5

3.3 Formal Methods

Formal methods are a means of providing a computer system development approach where both a formal notation and suitable mature tool support are provided. Whereas the formal notation is used to specify requirements or model a system design in a mathematical logic, the tool support helps to demonstrate that the implemented system meets its specification. Even if a full proof is hard to achieve in practice due to engineering and cost limitations, it makes software and hardware systems more reliable. By using formal methods, we can reason about a system and perform a mathematical verification of that system's specification; i.e., we can detect and isolate errors in the early stages of the software development.

By using formal methods *appropriately* within the software development process of aerospace systems, developers gain the benefit of reducing overall development cost. In fact, costs tend to be increased early in the system lifecycle, but reduced later on at the coding, testing, and maintenance stages, where correction of errors is far more expensive. Due to their precise notation, formal methods help to capture requirements abstractly in a precise and unambiguous form and then, through a series of semantic steps, introduce design and implementation level detail.

Formal methods have been recognized as an important technique to help ensure quality of aerospace systems, where system failures can easily cause safety hazards. For example, for the development of the control software for the C130J Hercules II, Lockheed Martin applied a correctness-by-construction approach based on formal (SPARK) and semi-formal (Consortium Requirements Engineering) methods [5]. The results showed that this combination was sufficient to eliminate a large number of errors and brought Lockheed Martin significant dividends in terms of high quality and less costly software. Note that an important part of the success of this approach is due to the use of the appropriate formal language. The use of a light version of SPARK, where the complex and difficult-to-understand parts of the Ada language had been removed, allowed for a higher level of abstraction, reducing the overall system complexity.

So-called synchronous languages [6] are formal languages dedicated to the programming of reactive systems. Synchronous languages (e.g., Lustre) were successfully applied in the development of automatic control software for critical applications like the software for nuclear plants, Airbus airplanes and the fight control software for Rafale fighters [7].

R2D2C (Requirements-to-Design-to-Code) [8] is a NASA approach to the engineering of complex computer systems where the need for correctness of the system, with respect to its requirements, is particularly high. This category includes NASA mission software, most of which exhibits both *autonomous* and *autonomic properties*. The approach embodies the main idea of *requirements-based programming* [9] and offers not only an underlying formalism, but also full formal development from requirements capture through to automatic generation of provably correct code. Moreover, the approach can be adapted to generate instructions in formats other than conventional programming languages—for example, instructions for controlling a physical device, or rules embodying the knowledge contained in an expert system. In these contexts, NASA has applied the approach to the verification of the instructions and procedures to be generated by the Hubble Space Telescope Robotic Servicing Missions and in the validation of the rule base used in the ground control of the ACE spacecraft [10].

3.4 Abstraction

The software engineering community recognizes abstraction as one of the best means of emphasizing important system aspects, thus helping to drive out unnecessary complexity and to come up with better solutions. According to James Rumbaugh, abstraction presents a selective examination of certain system aspects with the goal of emphasizing those aspects considered important for some purpose and suppressing the unimportant ones [11]. Designers of aerospace systems, shall consider abstraction provided by formal methods. Usually, aerospace software projects start with understanding the basic concepts of operations and requirements gathering, which results into a set of informal requirements (see Figure 1).

Once these requirements are documented, they can be formalized, e.g., with Lustre or R2D2C (see Section 3.3). The next step will be to describe the design in more detail. This is to specify how the desired software system is going to operate. Just as Java and C++ are high-level programming languages, in the sense that they are typed and structured, the formal languages dedicated to aerospace are structured and domain-specific and thus, they provide high-level structures to emphasize on the important properties of the system in question.

3.5 Decomposition and Modularity

In general, a complex aerospace system is a combination of distributed and heterogeneous components. Often, the task of modeling an aerospace system is about decomposition and modularity. Decomposition and modularity are well known concepts, which are the fundamentals of software engineering methods. Decomposition is an abstraction technique where we start with a high-level depiction of the system and create low-level abstractions of the latter, where features and functions fit together. Note that both high-level and low-level abstractions should be defined explicitly by the designer and that the low-level abstractions eventually result into components. This kind of modularity is based on explicitly-assigned functions to components, thus reducing the design effort and complexity. The design of complex systems always requires multiple decompositions.

3.6 Separation of Concerns

This Section describes a methodological approach to designing aerospace systems along the lines of the *separation-of-concerns* idea—one of the remarkable means of complexity reduction. This methodology strives to optimize the development process at its various stages and it has proven its efficiency in the hardware design and the software engineering of complex aerospace systems. As we have seen in Section 3.3, complex aerospace systems are composed of interacting components, where the separation-of-concerns methodology provides separate design "concerns" that (i) focus on complementary aspects of the component-based system design, and (ii) have a systematic way of composing individual components into a larger system. Thus, a fundamental insight is that the design concerns can be divided into four groups: *component behavior*, *inter-component interaction*, *component integration*, and *system-level interaction*. This makes it possible to hierarchically design a system by defining different models for the behavior, interaction, and integration of the components.

- *component behavior*: This is the core of system functionality, and is about the computation processes with the single components that provides the real added value to the overall system.
- *inter-component interaction*: This concern might be divided into three subgroups:
  - *communication*: brings data to the computational components that require it, with the right quality of service, i.e., time, bandwidth, latency, accuracy, priority, etc.;
  - *connection*: this is the responsibility of system designers to specify which components should communicate with each other;
  - *coordination*: determines how the activities of all components in the system should work together.
- *component integration*: Addresses the concept of efficient matching of the various design elements of an aerospace system into the most efficient way possible. Component integration is typically a series of multidisciplinary design optimization activities that involve component behavior and inter-component interaction concerns. To provide this capability, the design environment (and often the aerospace system itself) incorporates a mechanism that automates component retrieval, adaptation, and integration. Component integration may also require *configuration* (or re-configuration) which is about giving concrete values to the provided component-specific parameters, e.g., tuning control or estimation gains, determining communication channels and their inter-component interaction policies, providing hardware and software resources and taking care of their appropriate allocation, etc.
- *system-level interaction*: For an aerospace system, the interaction between the system and its environment (including human users) becomes an integral part of the computing process.

The clear distinction between the concerns allows for a much better perception and understanding of the system's features and consequently the design of the same. Separating behavior from interaction is essential in reconciling the disparity between concerns, but it may lead aerospace system designers to make a wrong conclusion: intended component behaviors can be designed in isolation from their intended interaction models.

3.7 Requirements-Based Programming

Requirements-Based Programming (RBP) has been advocated [9] as a viable means of developing complex, evolving systems. It embodies the idea that requirements can be systematically and mechanically transformed into executable code. Generating code directly from requirements would enable software development to better accommodate the ever increasing demands on systems. In addition to increased software development productivity through eliminating manual efforts in the coding phase of the software lifecycle, RBP can also increase the quality of generated systems by automatically performing verification on the software—if the transformation is based on the formal foundations of computing. This may seem to be an obvious goal in the engineering of aerospace software systems, but RBP does in fact go a step further than current development methods. System development typically assumes the existence of a model of reality (design or, more precisely, a design specification), from which an implementation will be derived.

4 Designing Unmanned Space Systems

Space poses numerous hazards and harsh conditions, which makes it a very hostile place for humans. Without risking human lives, robotic technology such as robotic missions, automatic probes and unmanned observatories allow for space exploration. Unmanned space exploration poses numerous technological challenges. This is basically due to the fact that unmanned missions are intended to explore places where no man has gone before and thus, such missions must deal, often autonomously and with no human control, with unknown factors, risks, events and uncertainties.

4.1 Intelligent Agents

Both autonomy and artificial intelligence lay the basis for unmanned space systems. So-called "intelligent agents" [12] provide for the ability of space systems to act without human intervention. An agent can be viewed as perceiving its environment through sensors and acting upon that environment through effectors (see Figure 2). Therefore, in addition to the requirements traditional for an aerospace system such as reliability and safety, when designing an agent-based space system, we must also tackle issues related to agent-environment communication.

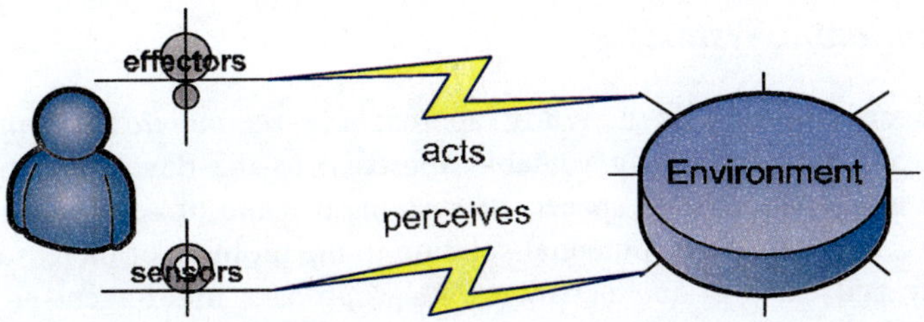

Fig. 2 Agent-Environment Relationship

Therefore, to design efficiently, we must consider the operational environment, because it plays a crucial role in an agent's behavior. There are a few important classes of environment that must be considered in order to properly design the agent-environment communication. The agent environment can be:

- *fully observable* (vs. *partially observable*) – the agent's sensors sense the complete state of the environment at each point in time;
- *deterministic* (vs. *stochastic*) – the next state of the environment is completely determined by the current state and the action executed by the agent;
- *episodic* (vs. *sequential*) – the agent's experience is divided into atomic "episodes" (each episode consists of the agent perceiving and then performing a single action), and the choice of action in each episode depends only on the episode itself;
- *static* (vs. *dynamic*) – the environment is unchanged while an agent is deliberating (the environment is semi-dynamic if it does not change with the passage of time but the agent's performance does);
- *discrete* (vs. *continuous*) – an agent relies on a limited number of clearly defined distinct environment properties and actions;
- *single-agent* (vs. *multi-agent*) – there is only one agent operating in the environment.

As we have mentioned above, space systems are often regarded as *multi-agent systems*, where many intelligent agents interact with each other. These agents are considered to be *autonomous entities* that interact either cooperatively or non-cooperatively (on a selfish base). A popular multi-agent system approach is the so-called *intelligent swarms*. Conceptually, a swarm-based system consists of many simple entities (agents) that are independent but grouped as a whole appear to be highly organized. Without a centralized supervision, but due to simple local interactions and interactions with the environment, the swarm systems expose complex behavior emerging from the simple microscopic behavior of their members.

4.2 Autonomic Systems

The aerospace industry is currently approaching *autonomic computing* (AC) recognizing in its paradigm a valuable approach to the development of single intelligent agents and whole spacecraft systems capable of self-management. In general, AC is considered a potential solution to the problem of increasing system complexity and costs of maintenance. AC proposes a multi-agent architectural approach to large-scale computing systems, where the agents are special autonomic elements (AEs) [13, 14]. The "Vision of Autonomic Computing" [14] defines AEs as components that manage their own behavior in accordance with policies, and interact with other AEs to provide or consume computational services.

4.2.1 Self-management

An autonomic system (AS) is designed around the idea of self-management, which traditionally results into designing four basic policies (objectives) - *self-configuring*, *self-healing*, *self-optimizing*, and *self-protecting* (often termed as self-CHOP policies). In addition, in order to achieve these self-managing objectives, an autonomic system (AS) must constitute the following features:

- *self-awareness* – aware of its internal state;
- *self-situation* – environment awareness, situation and context awareness;
- *self-monitoring* – able to monitor its internal components;
- *self-adjusting* – able to adapt to the changes that may occur.

Both objectives (policies) and features (attributes) form generic properties applicable to any AS. Essentially, AS objectives could be considered as system requirements, while AS attributes could be considered as guidelines identifying basic implementation mechanisms.

4.2.2 ·Autonomic Element

An AS might be decomposed (see Section 3.5) into AEs (*autonomic elements*). In general, an AE extends programming elements (i.e., objects, components, services) to define a self-contained software unit (design module) with specified interfaces and explicit context dependencies. Essentially, an AE encapsulates rules, constraints and mechanisms for self-management, and can dynamically interact with other AEs. As stated in the IBM Blueprint [13], the core of the AEs is a special *control loop*. The latter is a set of functionally related units: *monitor*, *analyzer*, *planner*, and *executor*, all of them sharing knowledge (see Figure 3). A basic control loop is composed of a *managed element* (also called a *managed resource*) and a *controller* (called *autonomic manager*). The autonomic manager makes decisions and controls the managed resource based on measurements and events.

Fig. 3 AE Control Loop and Managed Resource

4.2.3 Awareness

Awareness is a concept playing a crucial role in ASs. Conceptually, awareness is a product of knowledge processing and monitoring. The AC paradigm addresses two kinds of awareness in ASs [13]:

- *self-awareness* – a system (or a system component) has detailed knowledge about its own entities, current states, capacity and capabilities, physical connections and ownership relations with other (similar) systems in its environment;
- *context-awareness* – a system (or a system component) knows how to negotiate, communicate and interact with environmental systems (or other components of a system) and how to anticipate environmental system states, situations and changes.

4.2.4 Autonomic Systems Design Principles

Although recognized as a valuable approach to unmanned spacecraft systems, the aerospace industry (NASA, ESA) does not currently employ any development approaches that facilitate the development of autonomic features. Instead, the development process of autonomic components and systems is identical to the one of traditional software systems (see Figure 1), thus causing inherent difficulties in:

- expressing autonomy requirements;
- designing and implementing autonomic features;
- efficiently testing autonomic behavior.

For example, the experience of developing autonomic components for ESA's ExoMars [15] has shown that the traditional development approaches do not cope well with the non-deterministic behavior of the autonomic elements – a proper testing requires a huge number of test cases. The following is a short overview of aspects and features that are needed to be addressed by an AS design.

Self-* Requirements. Like any other contemporary computer systems, ASs also need to fulfill specific functional and non-functional requirements (e.g., safety requirements). However, unlike other systems, the development of an AS is driven by the self-management objectives and attributes (see Section 4.2.1) that must be implemented by that very system. Such properties introduce special requirements, which we term *self-* requirements*. Note that self-management requires 1) *self-diagnosis* to analyze a problem situation and to determine a diagnosis, and 2) *self-adaptation* to repair the discovered faults. The ability of a system to perform adequate self-diagnosis depends largely on the quality and quantity of its knowledge of its current state, i.e., on the system *awareness* (see Section 4.2.3).

Knowledge. In general, an AS is intended to possess awareness capabilities based on well-structured *knowledge* and algorithms operating over the same. Therefore, *knowledge representation* is one of the important design activities of developing ASs. Knowledge helps ASs achieve awareness and autonomic behavior, where the more knowledgeable systems are, the closer we get to real intelligent systems.

Adaptability. The core concept behind adaptability is the general ability to change a system's observable behavior, structure, or realization. This requirement is amplified by self-adaptation (or *automatic adaptation*). Self-adaptation enables a system to decide on-the-fly about an adaptation on its own, in contrast to an ordinary adaptation, which is explicitly decided and triggered by the system's environment (e.g., a user or administrator). Adaptation may result to changes in some *functionality*, *algorithms* or *system parameters* as well as the *system's structure* or any other aspect of the system. If an adaptation leads to a change of the complete system model, including the model that actually decides on the adaptation, this system is called a *totally reconfigurable system*. Note that self-adaptation requires a model of the system's environment. (often referred to as *context*) and therefore, self-adaptation may be also called *context adaptation*.

Monitoring. Since monitoring is often regarded as a prerequisite for awareness, it constitutes a subset of awareness. For ASs, monitoring (often referred to as self-monitoring) is the process of obtaining knowledge through a collection of sensors instrumented within the AS in question. Note that monitoring is not responsible for diagnostic reasoning or adaptation tasks. One of the main challenges of monitoring is to determine which information is most crucial for analysis of a system's behavior, and when. The notion of monitoring is closely related to the notion of context. Context embraces the *system state*, its *environment*, and any *information relevant to the adaptation*. Consequently, it is also a matter of context, which information indicates an erroneous system state and hence characterizes a situation in which a certain adaptation is necessary. In this case, adaptation can be compared to error handling, as it transfers the system from an erroneous (unwanted) system state to a well-defined (wanted) system state.

Dynamicity. Dynamicity embraces the system ability to change at runtime. In contrast to adaptability this only constitutes the technical facility of change. While adaptability refers to the conceptual change of certain system aspects, which does not necessarily imply the change of components or services, dynamicity is about the technical ability to remove, add or exchange services and components. There is a close but not dependable relation between both dynamicity and adaptability. Dynamicity may also include a system ability to exchange certain (defective or obsolete) components without changing the observable behavior. Conceptually, dynamicity deals with concerns like preserving states during functionality change, starting, stopping and restarting system functions, etc.

Autonomy. As the term Autonomic Computing already suggests, autonomy is one of the essential characteristics of ASs. AC aims at freeing human administrators from complex tasks, which typically requires a lot of decision making without human intervention (and thus without direct human interaction). Autonomy, however, is not only intelligent behavior but also an organizational manner. Context adaptation is not possible without a certain degree of autonomy. Here, the design and implementation of the AE *control loop* (see Section 4.2.2) is of vital importance for autonomy. A rule engine obeying a predefined set of conditional statements (e.g., if-then-else) put in an endless loop is the simplest form of control loop's implementation. In many cases, such a simple- rule-based mechanism however may not be sufficient. In such cases, the control loop should facilitate force feedback learning and learning by observation to refine the decisions concerning the priority of services and their granted QoS, respectively.

Robustness. Robustness is a requirement that is claimed for almost every system. ASs should benefit from robustness since this may facilitate the design of system parts that deal with self-healing and self-protecting. In addition, the system architecture could ease the appliance of measures in cases of errors and attacks. Robustness states the first and most obvious step on the road to dependable systems. Beside a special focus on error avoidance, several requirements aiming at correcting errors should also be forced. Robustness could be often achieved by decoupling and asynchronous communication, e.g., between interacting AEs (autonomic elements). Error avoidance, error prevention, and fault tolerance are approved techniques in software engineering, which shall help us in preventing from error propagation when designing ASs.

Mobility. Mobility enfolds all parts of the system: from mobility of code on the lowest granularity level via mobility of services or components up to mobility of devices or even mobility of the overall system. Mobility enables dynamic discovery and usage of new resources, recovery of crucial functionalities, etc. Often, mobile devices are used for detection and analysis of problems. For example, AEs may rely on mobility of code to transfer some functionality relevant for security updates or other self-management issues.

Traceability. Traceability enables the unambiguous mapping of the logical onto the physical system architecture, thus facilitating both system implementation and deployment. The deployment of system updates is usually automatic and thus, it requires traceability. Traceability is additionally helpful when analyzing the reasons for wrong decisions made by the system.

4.2.5 Formalism for Autonomic Systems

ASs are special computer systems that emphasize *self-management* through *context* and *self-awareness* [13, 14]. Therefore, an AC formalism should not only provide a means of description of system behavior but also should tackle the issues vital for autonomic systems self-management and awareness. Moreover, an AC formalism should provide a well-defined semantics that makes the AC specifications a base from which developers may design, implement, and verify ASs (including autonomic aerospace components or systems).

ASSL (Autonomic System Specification Language) [16] is a *declarative* specification language for ASs with well-defined semantics. It implements modern programming language concepts and constructs like inheritance, modularity, type system, and high abstract expressiveness. Being a formal language designed explicitly for specifying ASs, ASSL copes well with many of the AS aspects (see Section 4.2). Moreover, specifications written in ASSL present a view of the system under consideration, where specification and design are intertwined. Conceptually, ASSL is defined through *formalization tiers* [16]. Over these tiers, ASSL provides a *multi-tier specification model* that is designed to be scalable and exposes a judicious selection and configuration of infrastructure elements and mechanisms needed by an AS. ASSL defines ASs with special *self-managing policies*, *interaction protocols*, and *autonomic elements*. As a formal language, ASSL defines a neutral, implementation-independent representation for ASs. Similar to many formal notations, ASSL enriches the underlying logic with modern programming concepts and constructs thereby increasing the expressiveness of the formal language while retaining the precise semantics of the underlying logic.

The authors of this paper have successfully used ASSL to design and implement autonomic features for part of NASA's ANTS (Autonomous Nano-Technology Swarm) concept mission [17].

5 Conclusions

We have presented key *fundamentals* in designing complex aerospace software systems. By relying on our experience, we have discussed best design practices that can be used as guidelines by software engineers to build their own design strategy directing them towards the "right design concept" that can be applied to design a reliable aerospace system meeting the important safety requirements. Moreover, we have talked about design principles and the application of formal methods in the aerospace industry. Finally, we have shown the tremendous advantage of the so-called ASs (autonomic systems). ASs offer a solution for unmanned spacecraft systems, because the former are capable of self-adaptation, thus increasing the reliability of the unmanned systems where human intervention is not feasible or impractical. Although recognized as a valuable approach to unmanned spacecraft systems, the aerospace industry (NASA, ESA) does not currently employ any development approaches that facilitate the development of autonomic features. This makes both the implementation and testing of such features hardly feasible. We have given a short overview of aspects and features that are needed to be addressed by an AS design in order to make such a design efficient. To design and implement efficient ASs (including autonomic aerospace systems) we need AC-dedicated frameworks and tools. ASSL (Autonomic System Specification Language) is such a formal method, which we have successfully used at Lero—the Irish Software Engineering Research Centre, to develop autonomic features for a variety of systems, including NASA's ANTS (Autonomous Nano-Technology Swarm) prospective mission.

Acknowledgment. This work was supported in part by Science Foundation Ireland grant 10/CE/I1855 to Lero—the Irish Software Engineering Research Centre.

References

1. Gilbert, D., Aparicio, M., Atkinson, B., Brady, S., Ciccarino, J., Grosof, B., O'Connor, P., Osisek, D., Pritko, S., Spagna, R., Wilson, L.: IBM Intelligent Agent Strategy. White Paper, IBM Corporation (1995)
2. Philippe, C.: Verification, Validation, and Certification Challenges for Control Systems. In: Samad, T., Annaswamy, A.M. (eds.) The Impact of Control Technology. IEEE Control Systems Society (2011)
3. Herrmann, D.S.: Software Safety and Reliability. IEEE Computer Society Press, Los Alamitos (1999)
4. NASA-STD-8719.13A: Software Safety. NASA Technical Standard (1997)
5. Amey, P.: Correctness By Construction: Better Can Also Be Cheaper. CrossTalk Magazine. The Journal of Defense Software Engineering (2002)
6. Halbwachs, N.: Synchronous Programming of Reactive Systems. Kluwer Academic Publishers, Boston (1993)
7. Benveniste, A., Caspi, P., Edwards, S., Halbwachs, N., Le Guernic, P., De Simone, R.: The Synchronous Languages Twelve Years Later. Proceedings of the IEEE 91(1), 64–83 (2003)

8. Hinchey, M.G., Rash, J.L., Rouff, C.A.: Requirements to Design to Code: Towards a Fully Formal Approach to Automatic Code Generation. Technical Report TM-2005-212774, NASA Goddard Space Flight Center, Greenbelt, MD, USA (2004)
9. Harel, D.: From Play-In Scenarios To Code: An Achievable Dream. IEEE Computer 34(1), 53–60 (2001)
10. ACE Spacecraft, Astrophysics Science Division at NASA's GSFC (2005), `http://helios.gsfc.nasa.gov/ace_spacecraft.html`
11. Blaha, M., Rumbaugh, J.: Object-Oriented Modeling and Design with UML, 2nd edn. Pearson, Prentice Hall, New Jersey (2005)
12. Gilbert, D., Aparicio, M., Atkinson, B., Brady, S., Ciccarino, J., Grosof, B., O'Connor, P., Osisek, D., Pritko, S., Spagna, R., Wilson, L.: IBM Intelligent Agent Strategy. White Paper, IBM Corporation (1995)
13. IBM Corporation: An architectural blueprint for autonomic computing, 4th edn. White paper, IBM Corporation (2006)
14. Kephart, J.O., Chess, D.M.: The vision of Autonomic Computing. IEEE Computer 36(1), 41–50 (2003)
15. ESA: Robotic Exploration of Mars, `http://www.esa.int/esaMI/Aurora/SEM1NVZKQAD_0.html`
16. Vassev, E.: ASSL: Autonomic System Specification Language - A Framework for Specification and Code Generation of Autonomic Systems. LAP Lambert Academic Publishing, Germany (2009)
17. Truszkowski, M., Hinchey, M., Rash, J., Rouff, C.: NASA's swarm missions: The challenge of building autonomous software. IT Professional 6(5), 47–52 (2004)

Chapter 5
Simulation and Gaming for Understanding the Complexity of Cooperation in Industrial Networks

Andreas Ligtvoet and Paulien M. Herder

Abstract. In dealing with the energy challenges that our societies face (dwindling fossil resources, price uncertainty, carbon emissions), we increasingly fall back on system designs that transcend the boundaries of firms, industrial parks or even countries. Whereas from the drawing board the challenges of integrating energy networks already seem daunting, the inclusion of different stakeholders in a process of setting up large energy systems is excruciatingly complex and ripe with uncertainty.

New directions in risk assessment and adaptive policy making do not attempt to 'solve' risk, uncertainty or complexity, but to provide researchers and decision-makers with tools to handle the lack of certitude. After delving into the intricacies of cooperation, this paper addresses two approaches to clarify the complexity of cooperation: agent-based simulation and serious games. Both approaches have advantages and disadvantages in terms of the phenomena that can be captured. By comparing the outcomes of the approaches new insights can be gained.

1 Large Solutions for Large Problems

One of the large challenges facing our society in the next decades is how to sustainably cope with our energy demand and use. The historical path that western societies have taken led to a number of issues that require attention from policy makers, corporations and citizens alike.

- Vulnerability of and dependability on energy supply: increasingly, fossil fuels (oil, gas and coal) are being procured from fewer countries, the majority of which lie in unstable regions such as the Middle East.

Andreas Ligtvoet
Faculty of Technology, Policy and Management, Delft University of Technology
e-mail: a.ligtvoet@tudelft.nl

Paulien M. Herder
Faculty of Technology, Policy and Management, Delft University of Technology
e-mail: p.m.herder@tudelft.nl

- Energy use is a cause of climate change, notably through CO_2 emissions.
- Petroleum exploration and production are facing geological, financial, organisational and political constraints that herald the 'end of cheap oil'. As a result, oil prices have become highly volatile.
- The large growth in energy demand of countries such as China and India increasingly leads to shortages on the world energy market.

In order to tackle some of these challenges, large industrial complexes have been designed and built to provide alternatives to fossil fuels or to deal with harmful emissions. There are, for example, plans to provide industrial areas with a new synthesis gas infrastructure, to create countrywide networks for charging electric cars, networks for carbon capture and storage (CCS), and even larger plans to electrify the North-African desert (Desertec).

One could label these projects complex socio-technical systems [13]. They are technical, because they involve technical artefacts. They are social because these technical artefacts are governed by man-made rules, the networks often provide critical societal services (e.g. light, heat and transportation), and require organisations to build and operate them. They are complex, because of the many interacting components, both social and technical, that co-evolve, self-organise, and lead to a degree of non-determinism [26]. What makes an analysis of these systems even more challenging is the fact that they exist in and are affected by a constantly changing environment.

While different analyses of the workings of these complex socio-technical systems in interaction with their uncertain surroundings are possible, we focus on the question how cooperation between actors in the socio-technical system can lead to more effective infrastructures. Especially in industrial networks, cooperation is a *conditio sine qua non*:

- industrial networks require large upfront investments that cannot be borne by single organisations;
- the exchange in industrial networks is more physical than in other companies: participants become more interdependent, because the relationship is 'hardwired' in the network (as opposed to an exchange of money, goods, or services in a market-like setting).

Industrial networks require cooperation by more than two actors, which requires weighing of constantly shifting interests. We are interested in the interaction between different decision makers at the level of industrial infrastructure development, as this requires a balancing of the needs and interests of several stakeholders. Whereas for individual entrepreneurs it is already challenging to take action amidst uncertainty [22], an industrial network faces the additional task of coordinating with other actors that may have different interests [13].

This research aims to contribute to understanding the different ways in which actors cooperate in complex industrial settings. The first part draws on existing literature on cooperative behaviour (section 2). We then explore how the behaviour of actors influences the design-space. The range of technical options in a complex (energy) system is restricted by individual actors' behaviour in response to others, for which we show agent-based simulation is an appropriate approach (section 3). For a more in-depth analysis of the social and institutional variations, the use of serious games seems a fruitful approach (section 4). We explore how these approaches provide us with different insights on cooperation (section 5) and argue for a combined approach for further research (section 6).

2 Cooperation from a Multidisciplinary Perspective

Cooperation is being researched in a range of fields: e.g. in strategic management [23], (evolutionary) biology [25], behavioural economics, psychology, and game theory [3]. In most fields it is found that contrary to what we understand from the Darwinian idea of 'survival of the fittest' — a vicious dog-eat-dog world — there is cooperation from micro-organisms to macro-societal regimes. It turns out that although non-cooperation (sometimes called defection) is a robust and profitable strategy for organisms, it pays to cooperate under certain conditions, e.g. when competing organisations want to pool resources to enable a mutually beneficial project [23]. Although some literature does specify what options for cooperation there are, it does not explain why these options are pursued [27]. The *how?* and *why?* of cooperation and how it can be maintained in human societies, construe one of the important unanswered questions of social sciences [7]. By analysing a broad range of research fields, we provide a framework that can help understand the role of cooperation in industrial networks [17].

It should be noted that the research in these fields tends to overlap and that a clear distinction between disciplines cannot always be made [27, 15]. This holds for many fields of research: the field of *Industrial Ecology*, for example, in which the behaviour of individual industries is compared to the behaviour of plant or animal species in their natural environment, is a discipline that is approached from engineering traditions as well as economic and social researchers [12]. Likewise, *Evolutionary Psychology* builds psychological insights on the basis of biology and paleo-anthropology [5], whereas *Transaction Cost Economics* is a cross-disciplinary mix of economics, organisational theory and law [33]. By approaching a problem in a cross-disciplinary fashion, the researchers hope to learn from insights developed in different fields. In the same way this section combines different academic fields in order to reach insights into the mechanisms of cooperation (as is suggested by [15]).

2.1 A Layered Approach

Socio-technical systems can be analysed as containing layers of elements at different levels of aggregation. In the field of *Transition Management* (e.g. [28]), societal changes are described as taking place in three distinct levels: the micro-level (single organisations, people, innovations and technologies within these organisations), the meso-level (groups of organisations, such as sectors of industry, governmental organisations) and the macro-level (laws and regulations, society and culture at large). In short, the theory states that at the micro- or niche level a multitude of experiments or changes take place that, under the right circumstances, are picked up in the meso- or regime level. When these regimes incorporate innovations, they turn into 'the way we do things', and eventually become part of the socio-cultural phenomena at macro-level. Conversely, the way that our culture is shaped determines the types of organisational regimes that exist, which provide the framework for individual elements at the micro-level. A similar way of analysing societal systems can be found in *Transaction Cost Economics* [33].

We suggest that layered analysis can help in the analysis of complex social phenomena that take place in socio-technical systems. Each layer is described in terms of the level below it, and layers influence each other mutually. For instance, the nature of organisations' behaviour influences their relationships, but these relationships also in turn affect the organisations' behaviour [14] (especially if they are connected in a physical network). In observing how individual firms behave, we have to acknowledge that this behaviour is conditioned by the social networks and cultural traditions that these individuals operate in.

We therefore propose that research into the complex phenomenon of cooperation should also consider several layers of analysis. As we have seen in game theory and biology, *individual fitness* or *gain* is an important factor in deciding what strategy (the choice between cooperate or defect) to follow. Behavioural science teaches us that such simple and elegant approaches often do not help us in predicting and understanding human behaviour. Cultural norms, acquired over centuries, and institutions like government and law, determine the 'degrees of freedom' that individuals or organisations have. At the same time, the importance of formal and informal social networks allow for the dissemination of ideas, the influence of peers, and the opportunity to find partners. These are the micro, macro and meso layers that play an important role in cooperation between organisations.

2.2 Cooperation as a Complex Adaptive Phenomenon

In order to understand cooperation, we have to understand what influence the different layers of cooperation play. The freedom to act is curtailed as path dependency (the fact that history matters) plays an important role. Society has become more 'turbulent' as its networks and institutions have become more densely interconnected and interdependent. Just as Darwin saw the biological world as a 'web of life', so the organisational world is endowed with relations that connect its elements in a highly sophisticated way. In this organisational world, single organisations belong to multiple collectives because of the multiplicity of actions they engage in and the relationships they must have.

Cooperation therefore is not a simple matter of cost and benefit (although the financial balance may greatly determine the choice), but a learning process that is influenced by history and tradition, laws and regulations, networks and alliances, goals and aspirations, and, quite simply, chance. In short: a complex adaptive phenomenon.

The methods we choose to investigate problems regarding cooperation (in a real-life context), should allow for (some of) these complex elements to be taken into consideration. We have chosen to focus on *agent-based modelling* and *serious gaming*, as they both are relatively new and dynamic fields of research. By contrasting the contributions that both of these approaches can make for understanding cooperation, we hope to further our knowledge of cooperation as well as to depict the advantages and limitations of these methods.

3 Agent-Based, Exploratory Simulation

Decisions that are made at different levels (firm, region, or entire countries) require insight in long-term energy needs and the availability of resources to cope with these needs. However, due to uncertain events such as economic crises, political interventions, and natural disasters, these attempts often fail in their quantification efforts. Furthermore, responses of different actors at different societal levels may lead to countervailing strategies and with a large amount of independent actors, a system's response becomes complex [2].

With complex systems the issue is not to predict — as this is by definition impossible — but to understand system behaviour. Thus, decision making under uncertainty requires a different approach than calculating probability and effect. Issues of indeterminacy, stochastic effects, and non-linear relationships cannot be handled by these approaches. We believe that agent-based modelling and simulation can be a useful tool to deal with uncertainty in complex systems.

3.1 Agent-Based Modelling (ABM)

The agent-based modelling method aims to analyse the actions of individual stakeholders (agents) and the effects of different agents on their environment and on each other. The approach is based on the thought that in order to understand systemic behaviour, the behaviour of individual components should be understood ('Seeing the trees, instead of the forest' [29]).

Agent-based models (ABMs) are particularly useful to study system behaviour that is a function of the interaction of agents and their dynamic environment, which cannot be deduced by aggregating the properties of agents [6]. In general, an agent is a model for any entity in reality that acts according to a set of rules, depending on input from the outside world. Agent-based modelling uses agents that act and interact according to a given set of rules to get a better insight into system behaviour. The emergent (system) behaviour follows from the behaviour of the agents at the lower level.

With regard to the uncertainty that we face concerning future (global) developments in, for example, fuel availability, the use of agent-based models will enable us to follow an exploratory approach [4]. In stead of using 'traditional' methods that are based on calculations of probability and effect, using ABMs allows a more scenario-oriented approach (asking 'what if?' questions), implementing thousands of scenarios. Combining exploratory thinking with agent-based models is still a field of research in development [1].

In agent-based models of industrial networks the space of abstract concepts has been largely explored; the next frontier is in getting closer to reality. The strength of agent-based models of real organisations is that decision makers end up saying 'I would have never thought that this could happen!'. According to Fioretti, the practical value of agent-based modelling is its ability to produce emergent properties that lead to these reactions [11].

3.2 Model Implementation

We adapted an existing set of Java instructions (based on the RePast 3 toolkit) and ontology (structured knowledge database) which form an ABM-toolkit that is described in [24, 31]. The toolkit provides basic functions for describing relationships between agents that operate in an industrial network. Herein, agents represent industrial organisations that are characterised by ownership of technologies, exchange of goods and money, contractual relationships, and basic economic assessment (e.g. discounted cash flow and net present value calculation). For the technical design of (energy) clusters, the methods described by Van Dam and Nikolić have already shown a wide range of applications and perform favourably when compared to other modelling approaches [31]. Agents' behaviour, however, is mainly based on rational cost/benefit assessments. By implementing more social aspects of the agents, such as cooperation, trust, and different risk attitudes, other dynamics may emerge in clusters of agents. This will impact the assessment of the feasibility of certain projects. We therefore modified the agent behaviour to examine specific cooperation-related behaviour. The following list of behavioural assumptions were added:

- agents create a **list of (strategic) options** they wish to pursue: an option contains a set of agents they want to cooperate with and the net present value of that cooperation;
- the agents have a **maximum number of options** they can consider, to prevent a combinatorial explosion when the number of agents is above a certain number;
- the agents select the agents they want to cooperate with on the basis of (predetermined) **trust relationships**;
- agents can have a short term (<5 years) or long term (>5 years) **planning horizon**, which determines the time within which payback of investments should be achieved;
- agents can be **risk-averse** or risk-seeking (by adjusting discount factor in the discounted cash flow);

- agents want a minimum percentage of cost reduction before they consider an alternative options (**minimum required improvement**);
- agents can be **initiative taking**, which means that they initiate and respond to inter-agent communication.

These assumptions are to provide the agents with 'bounded rationality': Herbert Simon's idea that true rational decision making is impossible due to the time constraints and incomplete access to information by decision makers [30]. We assume that agents do not have the time nor the resources to completely analyse all possible combinations of teams in their network. They will have to make do with (micro behaviour) heuristics to select and analyse possible partnerships. The meso level is represented by the existing trust relationships or social network and it also emerges during the simulation as new (physical) networks are forged. As is often done in institutional economics [33], the societal or macro level is disregarded as all agents are presumed to share the same cultural background.

A simulation can be run that present industrial cooperation and network development. Figure 1 represents a cluster of eight identical agents that trade a particular good (e.g. petroleum). For reasons of simplicity and tractability, the agents are placed with equal distances on a circle. The issue at hand is the transportation of the good: the agents can either decide to choose for the flexible option (i.e. truck transport) or to cooperatively build a piped infrastructure to transport the good. Flexible commitments are contracted for a yearly period only. They are agreements between two agents with as the main cost component variable costs as a function of distance. The permanent transportation infrastructure is built by two or more agents (sharing

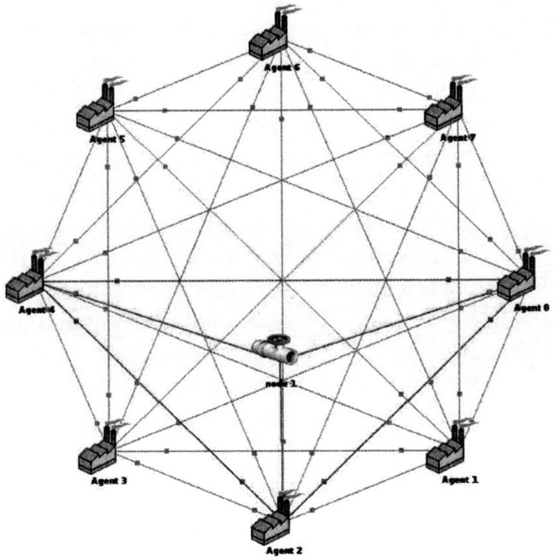

Fig. 1 A cluster of 8 simulated trading agents in which 3 agents have cooperated to build a pipeline

costs) and is characterised by high initial capital costs and relatively low variable costs per distance. The more agents participate in the building of the infrastructure, the lower the capital costs per agent. Costs minimisation is a dominating factor in the selection of the optimal infrastructure. Depending on the distance and the number of agents involved, a flexible solution may be cheaper than building a fixed infrastructure. By varying the behavioural assumptions mentioned above, we investigate to what extent cooperation takes place and what the (financial) consequences are for the agents. (For more detail see [18]).

4 Serious Gaming

Although also stemming from applied mathematics, operations research and systems analysis, the field of *(serious)* gaming has a different approach to understanding the "counter-intuitive behaviour of social systems (Forrester, 1971)". Whereas increased computing power has enabled ever more complicated representations of reality, studies in policy sciences have shown that decision-making was far from rational and comprehensive, but rather political, incremental, highly erratic and volatile [19]. The toolbox used by the system and policy analysts needed to become more human-centred and responsive to socio-political complexity. By allowing for more freedom to the human players, games lend themselves particularly to transmitting the character of complex, confusing reality [9].

4.1 Gaming Goals

Far more than analytical understanding, gaming allows for acquiring social or teamwork skills, gaining strategic and decision-making experience, training in and learning from stressful situations. As an educational tool, business simulation games have grown considerably in use during the past 40 years and have moved from being a supplemental exercise in business courses to a central mode of business instruction [10].

In general, games can be defined as experience-focused, experimental, rule-based, interactive environments, where players learn by taking actions and by experiencing their effects through feed-back mechanisms that are deliberately built into and around the game [19]. Gaming is based on the assumption that the individual and social learning that emerges in the game can be transferred to the world outside the game. Games can take many different forms, from fully oral to dice-, card- and board-based to computer-supported, with strict adherence to rules or allowing for more freedom of action. In terms of usability for complex policy making, variants such as free-form gaming and all-man gaming seemed to perform much better, especially in terms of usability, client satisfaction, communication and learning and, not unimportantly, cost effectiveness. On one hand there is a need to keep games simple and playable [21], on the other hand there is a positive relationship between realism and the degree of learning from the simulation.

4.2 Game Implementation

To be able to compare ABM to gaming, we developed a boardgame that in broad lines is similar to the situation as described in section 3.2: different organisations exchanging an energy carrier. The main gist of the game is to get the players to co-operatively invest in infrastructures while being surrounded by an uncertain world (which is clearly depicted by randomly drawn 'chance' cards that determine whether fuel prices and the economy change). To make the game more enticing to the players, we did not choose a circular layout with homogeneous players, but a map that represents a harbour industrial area with different organisations (refinery, greenhouse, garbage incinerator, hospital, housing) that are inspired by several Dutch plans to distribute excess heat [8]. Although this game was tested with university colleagues, we intend it to be played by decision-makers who face similar issues. By playing this game, we intend to find out (a) whether the behavioural assumptions in section 3.2 play a role and (b) whether there are other salient decision criteria that were not yet taken into consideration in the model. Furthermore, by asking the players to fill out a questionnaire, we intend to gain further insight in the importance of the micro-, meso- and macro- levels addressed in section 2.

5 Computer Simulation versus Gaming

Simulation and gaming are not two distinct methods, but freely flow into each other: computer models are used to support and enhance gaming exercises, whereas games can provide the empirical basis for the stylised behaviour of agents as well as a validation of the observed outcomes. We nevertheless believe that it is important to distinguish several basic elements or characteristics of the archetypical forms of these approaches (see table 1).

First of all, computer simulation is more geared toward the mechanical regularities that technology embodies, than the subtle pallet of inter-human conduct.

Table 1 Characteristics of archetypes of simulation and gaming

Characteristics	Simulation	Gaming
focus	technical	social
main elements	pipes, poles, machines	trust, friendship, bargaining
level of detail	inclination to complexity	simplicity required
rules	rules are fixed	rules are negotiable
uncertainties	can/must be captured	cannot be captured
model	closed	open
abstraction	black box	explicit
dynamics	shown	revealed
learning	researchers and clients	participants and researchers
goal	outcomes	understanding

Whereas physical realities (pipes, poles, machines) can be confidently captured by a limited set of equations, social phenomena (trust, friendship, bargaining) are dependent on a wide range of inputs that can hardly be specified in detail. Of course, they can be represented by variables and serve as input to models, which then either become one-dimensional or quickly become intractable. As computers are patient, the researchers are not hampered in their desire to capture detail and enhance complexity [20, 16]. Game participants, on the other hand, can only handle a limited cognitive load: the design needs to embody simplicity to a certain extent.

When a computer simulation is run, all elements of the model are specified: rules are explicitly stated and uncertainties have to be captured in a certain way to allow for quantification and calculation. The world the model represents is necessarily closed. Gaming is more geared towards allowing the knowledge and experience of the participants to directly influence the process of the game. First of all, the outcome of the game is strongly dependent on the will of participants to 'play along'. Often, details of the rules are still negotiable while gaming, allowing for a more realistic setting. Thus, behaviour-related uncertainties cannot be captured in a game. This implies that the model should be open to 'irregularities' taking place.

Simulations often hide many of the assumptions and abstractions that underlie the calculations — this is often necessary as the parameter space is very large and the details are many. Although gaming, as indicated, can also rely on such mathematical models, the abstractions are often more explicit, as participants are in closer contact with the representation of the world: it is potentially easier to trace outcomes of activities (and then either accept or reject the abstractions).

The learning aspects of simulations lie predominantly with the researchers themselves (although the models are often made for policy makers or other clients). In designing a valid simulation model, many details need to be considered and researched which constitutes a learning process ('Modelling as a way of organising knowledge' [32]); for outsiders, the simulation quickly turns into a black box. Gaming potentially allows for the same learning experiences for researchers or designers, but is also often specifically focused on a learning experience for the participants. We would suggest that gaming therefore is more geared towards understanding social intricacies, whereas simulations are often expected to produce quantitative outcomes.

6 Conclusions

For social scientists, explaining and understanding cooperation is still one of the grand challenges. For researchers of industrial clusters, the space of abstract (game theoretic) concepts has been largely explored; the challenge is to get closer to realistic representations. Whether using gaming, simulation or a hybrid of these two, it is important to find the appropriate balance between detail or 'richness' and general applicability or 'simpleness', and to be clear about what elements of the modelled system are included and also excluded.

Both simulation and gaming allow for understanding different aspects of complex, adaptive, socio-technical systems. It is generally accepted that prediction is not the main goal, but that patterns emerge that teach us something about the systems we investigate. By using both approaches beside each other, we can give due attention to the dualistic nature of socio-technical systems.

Acknowledgements. This research was made possible by the Next Generation Infrastructures foundation (`www.nextgenerationinfrastructures.eu`).

References

1. Agusdinata, D.B.: Exploratory Modeling and analysis: a promising method to deal with deep uncertainty. PhD thesis, Delft University of Technology (2008)
2. Anderson, P.: Complexity theory and organization science. Organization Science 10(3), 216–232 (1999)
3. Axelrod, R.: The complexity of cooperation. Princeton University Press, Princeton (1997)
4. Bankes, S.: Tools and techniques for developing policies for complex and uncertain systems. PNAS 99, 7263–7266 (2002)
5. Barkow, J.H., Cosmides, L., Tooby, J. (eds.): The adapted mind - evolutionary psychology and the generation of culture. Oxford University Press, New York (1992)
6. Beck, J., Kempener, R., Cohen, B., Petrie, J.: A complex systems approach to planning, optimization and decision making for energy networks. Energy Policy 36, 2803–2813 (2008)
7. Colman, A.M.: The puzzle of cooperation. Nature 440, 745–746 (2006)
8. de Jong, K.: Warmte in Nederland; Warmte- en koudeprojecten in de praktijk. Uitgeverij MGMC (2010)
9. Duke, R.D.: A paradigm for game design. Simulation & Games 11(3), 364–377 (1980)
10. Faria, A., Hutchinson, D., Wellington, W.J., Gold, S.: Developments in business gaming: A review of the past 40 years. Simulation Gaming 40, 464 (2009)
11. Fioretti, G.: Agent based models of industrial clusters and districts (March 2005)
12. Garner, A., Keoleian, G.A.: Industrial ecology: an introduction. Technical report, National Pollution Prevention Center for Higher Education, University of Michigan, Ann Arbor, MI, USA (November 1995)
13. Herder, P.M., Stikkelman, R.M., Dijkema, G.P., Correljé, A.F.: Design of a syngas infrastructure. In: Braunschweig, B., Joulia, X. (eds.) 18th European Symposium on Computer Aided Process Engineering, ESCAPE 18. Elsevier (2008)
14. Kohler, T.A.: Putting social sciences together again: an introduction to the volume. In: Dynamics in Human and Primate Societies. Santa Fe Institute studies in the sciences of complexity. Oxford University Press, Oxford (2000)
15. Lazarus, J.: Let's cooperate to understand cooperation. Behavioral and Brain Sciences 26, 139–198 (2003)
16. Lee, D.B.: Requiem for large-scale models. Journal of the American Institute of Planners 39(3) (1973)
17. Ligtvoet, A.: Cooperation as a complex, layered phenomenon. In: Eighth International Conference on Complex Systems, June 26-July 1 (2011)

18. Ligtvoet, A., Chappin, E., Stikkelman, R.: Modelling cooperative agents in infrastructure networks. In: Ernst, A., Kuhn, S. (eds.) Proceedings of the 3rd World Congress on Social Simulation, WCSS 2010, Kassel, Germany (2010)
19. Mayer, I.S.: The gaming of policy and the politics of gaming: A review. Simulation & Gaming X, 1–38 (2009)
20. Meadows, D.H., Robinson, J.M.: The electronic oracle: computer models and social decisions. System Dynamics Review 18(2), 271–308 (2002)
21. Meadows, D.L.: Learning to be simple: My odyssey with games. Simulation Gaming 30, 342–351 (1999); Meadows 1999a
22. Meijer, I.S., Hekkert, M.P., Koppenjan, J.F.: The influence of perceived uncertainty on entrepreneurial action in emerging renewable energy technology; biomass gasification projects in the Netherlands. Energy Policy 35, 5836–5854 (2007)
23. Nielsen, R.P.: Cooperative strategy. Strategic Management Journal 9, 475–492 (1988)
24. Nikolić, I.: Co-Evolutionary Process For Modelling Large Scale Socio-Technical Systems Evolution. PhD thesis, Delft University of Technology, Delft, The Netherlands (2009)
25. Nowak, M.A.: Five rules for the evolution of cooperation. Science 314, 1560–1563 (2006)
26. Pavard, B., Dugdale, J.: The Contribution Of Complexity Theory To The Study Of Socio-Technical Cooperative Systems. In: Proceedings from the Third International Conference on Unifying Themes in Complex Systems. New Research, vol. III B, pp. 39–48. Springer, Heidelberg (2006)
27. Roberts, G., Sherratt, T.N.: Cooperative reading: some suggestions for the integration of cooperation literature. Behavioural Processes 76, 126–130 (2007)
28. Rotmans, J., Kemp, R., van Asselt, M.: More evolution than revolution: transition management in public policy. Foresight 3(1), 15–31 (2001)
29. Schieritz, N., Milling, P.M.: Modeling the forest or modeling the trees: A comparison of system dynamics and agent-based simulation. In: 21st International Conference of the System Dynamics Society, New York, Mannheim, Germany (2003)
30. Simon, F.B.: Einführung in die systemische Organisationstheorie. Carl-Auer Verlag, Heidelberg (2007)
31. van Dam, K.H.: Capturing socio-technical systems with agent-based modelling. PhD thesis, Delft University of Technology, Delft, The Netherlands (2009)
32. Wierzbicki, A.P.: Modelling as a way of organising knowledge. European Journal of Operational Research 176, 610–635 (2007)
33. Williamson, O.E.: Transaction cost economics: how it works; where it is headed. De Economist 146(1), 23–58 (1998)

Chapter 6
FIT for SOA? Introducing the F.I.T.-Metric to Optimize the Availability of Service Oriented Architectures

Sebastian Frischbier, Alejandro Buchmann, and Dieter Pütz

Abstract. The paradigm of service-oriented architectures (SOA) is by now accepted for application integration and in widespread use. As an underlying key-technology of cloud computing and because of unresolved issues during operation and maintenance it remains a hot topic. SOA encapsulates business functionality in services, combining aspects from both the business and infrastructure level. The reuse of services results in hidden chains of dependencies that affect governance and optimization of service-based systems. To guarantee the cost-effective availability of the whole service-based application landscape, the real criticality of each dependency has to be determined for IT Service Management (ITSM) to act accordingly. We propose the FIT-metric as a tool to characterize the stability of existing service configurations based on three components: functionality, integration and traffic. In this paper we describe the design of FIT and apply it to configurations taken from a production-strength SOA-landscape. A prototype of FIT is currently being implemented at Deutsche Post MAIL.

1 Introduction

A company's IT Service Management (ITSM) has to fulfill conflicting demands: while minimizing costs, IT solutions have to support a wide range of functionality, be highly reliable and flexible [22]. The paradigm of service-oriented architectures (SOA) has been proposed to solve this conflict. SOA was intended to facilitate the integration of inter-organizational IT-systems, thus becoming a key enabler of cloud computing [12]. At present, it is used mostly for intra-organizational application

Sebastian Frischbier · Alejandro Buchmann
Databases and Distributed Systems Group, Technische Universität Darmstadt
e-mail: lastname@dvs.tu-darmstadt.de

Dieter Pütz
Deutsche Post AG, Bonn
e-mail: d.puetz@deutschepost.de

integration. Especially large companies, such as Deutsche Post DHL, use SOA to integrate and optimize their historically grown heterogeneous application landscape.

From an architectural point of view, the SOA paradigm reduces complexity and redundancy as it restructures the application landscape according to functionality and data-ownership. Basic entities within a SOA are services organized in domains without overlap. Each service encapsulates a specific function with the corresponding data and is only accessible through an implementation-independent interface. Services are connected according to a given workflow based on a business process [29].

From an infrastructure point of view, services are usually provided and consumed by applications. With the term 'application' we refer to large-scale complex systems, themselves quite often consisting of multi-tier architectures running on server clusters serving thousands of clients. These applications have to be available according to their business criticality. The desired level of availability is specified in service level agreements (SLA) in terms of service level targets (SLT). These differ according to the characteristics of the individual application and the means necessary to meet the desired level of availability [7]. Usually, the different service levels can be grouped into three classes: *high* availability (gold), *medium* availability (silver) and *low* availability (bronze). Effort and means needed for the operations provider to guarantee a given level of availability are reflected in the costs.

Deciding on the proper level of availability and defining adequate service levels is a difficult task, which becomes even more complex in service-based and distributed environments. The SOA paradigm drives the reuse of existing services by enabling their transparent composition within new services. As functionality and data are encapsulated, services have to rely on other services in order to run properly. This results in a network of hidden dependencies, since each service is only aware of its own direct dependencies on the consumed services. These dependencies affect the availability of applications directly as applications rely on other services to communicate and access information in service-based environments. Chains of interdependent services can lead to an application with higher availability becoming dependent on an application with lower availability even if the applications have no direct semantic relationship.

IT Service Management has to decide on the criticality of such a relationship and act accordingly. Criticality in this context does not refer to the *probability* of a breakdown actually taking place but to the *impact* on the application landscape once it occurs. The following approaches are possible to cope with disparities in service levels of depending applications: i) all participating applications are operated on the highest level of availability present in a chain of dependencies; ii) the configuration stays unchanged; iii) the SLA of single participants is changed to minimize the expected impact. As the "methods used are almost always pure guesswork, frequently resulting in drastic loss or penalties" [38, 43], the first approach is often favored. Although it succeeds, it is surely inefficient and expensive. Even hosting services alternatively in cloud environments rather than on-premise does not solve this problem. It rather shifts the risk of availability management to the cloud provider who will charge for it.

Due to the lack of proper decision support, both the second and third approach are usually avoided as they may result in serious breakdowns and loss of revenue. Therefore, ITSM requires methods and tools to: i) model all relevant dependencies; ii) identify hotspots and; iii) decide on their criticality. In particular, deciding on the criticality is important as this allows for ranking hotspots (e.g. as preparation for closer inspection) and simulating changes.

We introduce the FIT-metric to aid ITSM in these tasks, especially in deciding on the criticality of dependencies in existing service-oriented architectures and simulating changes. Our metric consists of the three components: functionality, integration and traffic. The necessary data can be cost-effectively obtained from end-to-end monitoring and existing documentation. FIT is the result of our analysis conducted at Deutsche Post MAIL and is currently implemented there.

The contributions of this paper are: i) we identify the need for applications and their dependencies to be ranked according to their criticality; ii) we propose a metric taking into account functionality, integration and traffic of the services involved to aid ITSM in assessing the appropriate service level in interdependent SOA-based systems; iii) we evaluate our approach by applying it to actual service configurations taken from a product-strength SOA-landscape.

The structure of this paper is as follows: we present a production-strength SOA in Sect. 2 to point out the need for a criticality-metric for service-dependencies. In Sect. 3 we present the design of our FIT-metric in detail. We use a case study to evaluate our approach in Sect. 4. We conclude our paper by reviewing related work on the topic of metrics for service-oriented architectures in Sect. 5 followed by a summary of our findings and a short outlook on future work in Sect. 6.

2 A Production-Strength SOA Environment

Deutsche Post AG is the largest postal provider in Europe, with Deutsche Post MAIL division alone delivering about 66 million letters and 2.6 million parcels to 39 million households in Germany each working day. Furthermore, Deutsche Post MAIL increases the level of digitalization in its product-portfolio (e.g. online and mobile based value-added services) since 2009 [8]. In 2010 Deutsche Post MAIL started to offer the *E-Postbrief* product to provide consumers and business users with a secure and legally compliant form of electronic communication [9].

The application landscape supporting these processes and products was transformed to apply the SOA-paradigm in 2001 [14]. Today, applications communicate across a distributed enterprise service bus (ESB) by consuming and providing SOA-services that are grouped in mutually exclusive domains. The initially used SOA-framework has been a proprietary development by Deutsche Post MAIL called *Service-oriented Platform* (SOP). Today, SOP's open-source successor SOPERA is at the heart of both *Eclipse SOA*[1] and *Eclipse Swordfish*[2].

[1] `http://www.eclipse.org/eclipsesoa/`

[2] `http://www.eclipse.org/swordfish/`

The results discussed here are based on our analysis of the Deutsche Post MAIL SOP/SOPERA application landscape using a Six Sigma-based [11] approach. This included conducting interviews, reviewing documentation as well as assessing monitoring capabilities to identify dependencies and business criticalities. All data presented in this paper is anonymized due to confidentiality requirements.

The main findings of our analysis are: i) long chains of dependencies between services affect the availability of applications in matured production-strength SOA-landscapes as the SOA-paradigm itself fosters the reuse of services; ii) SOA-dependencies are hard to uncover for ITSM at runtime as they are hidden in the SOA-layer itself; iii) the data necessary to identify and analyze them at runtime may be already existing but is often not available at first as it is spread across different heterogeneous applications; iv) to the best of our knowledge, there is no metric available for ITSM to decide on the criticality of service-relationships based on the data usually available at runtime. Based on these findings: i) we initiated a specialized but extensive end-to-end monitoring of the SOA-application landscape to allow for dependencies and their usage to be quantified automatically in the future; ii) we defined a cost-effective criticality-metric based on the available data; iii) we built a prototypic software tool named *FIT-Calculator* to allow for automated graph-based analysis and simulation based on the monitored data.

The availability-'heat map' of the SOA-landscape as illustrated in Fig. 1 is automatically generated based on the monitoring data currently available to us. It gives an overview of 31 participating applications and their 69 service-relationships. Each node represents an application providing and consuming SOA-services. The desired

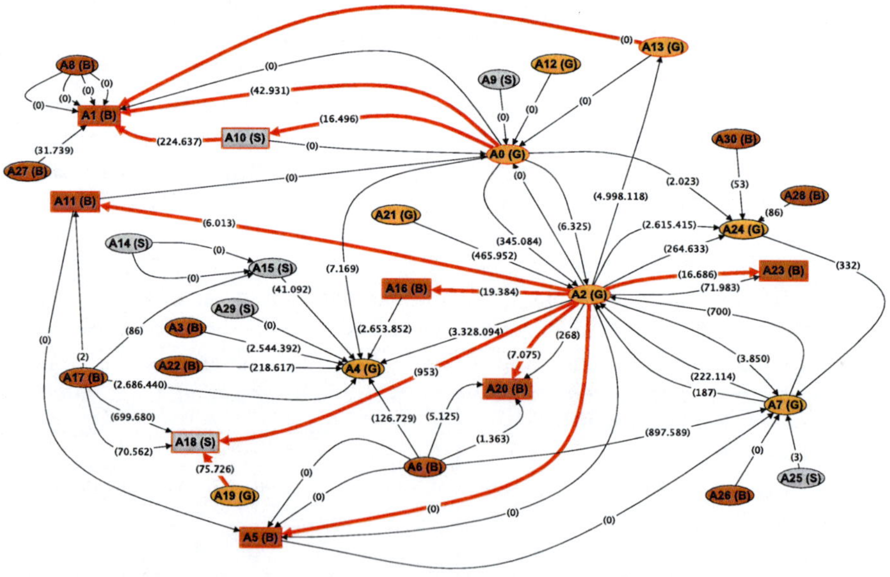

Fig. 1 Graph representing applications and their direct SOA-relationships

level of availability for each node x is expressed by the node's color as well as by an abbreviation in brackets ([g]old, [s]ilver and [b]ronze). Edges denote service-relationships between two applications with an edge pointing from the consuming application to the application providing the service (direction of request). Edge weights refer to the number of requests processed over this dependency within a given period. Dependencies of consuming SOA-services on providing SOA-services within an application are modeled as an overlay for each application (not shown).

This visualization allows ITSM to identify hotspots easily. Hotspots, in this context, refer to applications that cause potentially critical relationships by providing SOA-services to applications with higher levels of availability. In the given example, 8 hotspots (A1, A5, A10, A11, A16, A18, A20, A23) cause 11 potentially critical relationships. On the heat map in Fig. 1, these relationships are marked bold red with the hotspots being drawn as rectangles.

3 Introducing the F.I.T.-Metric

The criticality of a single hotspot a depends on the criticality of each relationship between a and the depending applications with higher SLA. To us, the criticality of a single relationship $e(a,x)$ in general is primarily influenced by: i) the business relevance F_x of the application x directly depending on a via $e(a,x)$; ii) the impact of $e(a,x)$ on other applications in the SOA-landscape due to the integration $I_{a,x}$ of x (i.e. x serving as a proxy); iii) the actual usage $T_{a,x}$ of the relationship $e(a,x)$ by the depending application x.

F_x and $I_{a,x}$ refer to independent aspects of x's importance to the system landscape and the business users. An application's core function alone can be highly relevant to business users (e.g. business intelligence systems) while it may be unimportant for other applications from an integration point of view. In turn, an application serving mainly as a proxy for other applications can be relatively unimportant to business by its own. As these two aspects of a relationship are rather static (i.e. an application's core functionality is seldom altered completely over short time and dependencies between applications change only infrequently), they have to be weighted by an indicator for the actual usage of this relationship.

Therefore, we define the criticality $eFIT_{e(a,x)}$ of the relationship $e(a,x)$ as the sum of F_x and $I_{a,x}$, weighted by $T_{a,x}$ in Eq. (1).

$$eFIT_{e(a,x)} = (F_x + I_{a,x}) \cdot T_{a,x} \tag{1}$$

In turn, the sum of these relationship-criticalities for all relationships to a defines the criticality FIT_a of hotspot a as defined in Eq. (2).

$$FIT_a = \sum_{\forall e(a,x)} eFIT_{e(a,x)} \tag{2}$$

In this setting, uncritical relationships and applications have a FIT-score of 0 while critical hotspots are ranked ascending by their criticality with FIT-scores > 0.

3.1 Component I: Functionality

Functionality refers to quantifying an application's relevance to business. As the business impact of IT systems is hard to determine and even harder to quantify from the IT point of view [38], these categorizations are often based on subjective expert knowledge and individual perception. Although this makes an unbiased comparison between applications difficult, we suggest reusing already existing information as an approximation. For example, we turn to data regarding business continuity management (BCM). In this context, applications have to be categorized according to their recovery time objective (RTO) in case of disaster [7]. The RTO-class $RTOC_x = 1, \ldots, n$ increases with the duration allowed for x to be unavailable. The *economic* $RTOC_x$ ($econRTOC_x$) is the RTOC the users are willing to pay and is inversely proportional to the quality level of the SLA. Therefore we define the assumed business relevance F_x of application x as:

$$F_x = \frac{1}{econRTOC_x} \tag{3}$$

3.2 Component II: Integration

Integration quantifies the impact of an inoperative application on all other applications based on dependencies between SOA-services. Information about these dependencies can be drawn at runtime from workflow-documentation (e.g. BPEL-workflows or service descriptions) or service monitoring.

We define the dependency tree DT_a for each application a as the first step. The tree's root node is the initial application a itself. All direct consumers $tSC_{1,1}, \ldots, tSC_{1,m}$ of this application's services are added as the root's children. On the following levels $i = 2, \ldots, h$ only applications that are *indirectly dependent* on services provided by a are added as $tSC_{i,1}, \ldots, tSC_{i,n}$. Thus, DT_a is not identical to the simple graph of a's predecessors in Fig. 1 as the nodes on level i depend on the internal dependencies inside applications (services depending on services).

Figures 2a- 2f show the dependency trees for the applications A1, A2, A5, A13, A18 and A20 based on the relationship graph shown in Fig. 1 and the overlay modeling internal dependencies of services inside applications. Edge weights denote the number of requests processed over a service-dependency alone as well as bold red edges representing possibly critical relationships. Here, edges point from service provider to service consumer (direction of response).

The weighted dependency tree wDT_a^x quantifies the direct and indirect dependencies on a over $e(a,x)$ by weighting the sub-tree of DT_a with application x as root. For DT_{A2} (cf. Fig.2d), the corresponding wDT_{A2}^{A7} would consist of A7 (root), A26, A25, A6, A24, A5 and A11.

Deep dependency trees containing long chains of indirect dependencies have a far-reaching impact on the landscape once the root node breaks down. They have to be emphasized as they are far less obvious to ITSM than a huge number of direct dependencies on an application. Therefore, wDT_a^x takes into account the assumed business relevance of each node $tSC_{i,j}$ in DT_a as well as the length of the dependency chain to $tSC_{i,j}$. The occurrence of each node $tSC_{i,j}$ is weighted with its functionality $F_{tSC_{i,j}}$ and its level of occurrence in the dependency tree. We define the integration of x as depending on a as:

$$I_{a,x} = wDT_a^x = \sum_{i=2}^{h} i \cdot \sum_{j=1}^{m} F_{tSC_{i,j}} \tag{4}$$

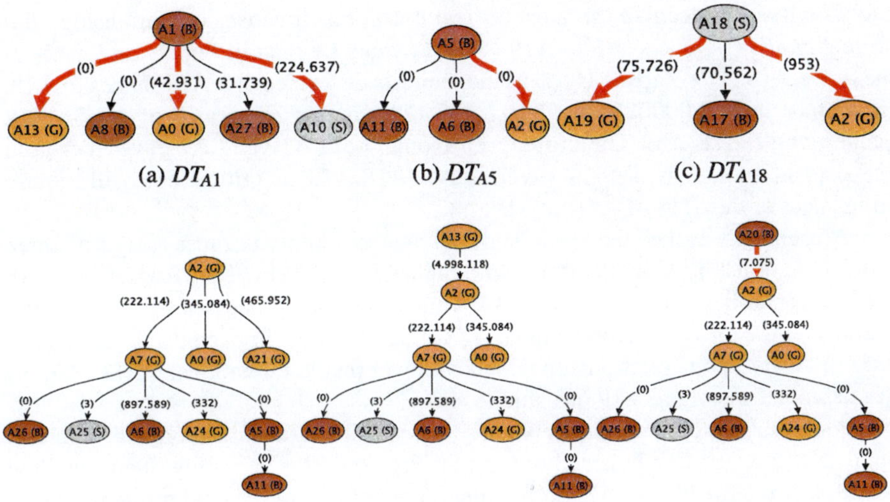

(a) DT_{A1} (b) DT_{A5} (c) DT_{A18}

(d) DT_{A2} (e) DT_{A13} (f) DT_{A20}

Fig. 2 Dependency trees DT of selected nodes (c.f. Fig. 1)

3.3 Component III: Traffic

Traffic quantifies the real usage of a given relationship between two applications a and x. As both F_x and $I_{a,x}$ refer to the worst-case impact of a breaking down, we need to balance this with an approximation for the current utilization of the relationship $T_{a,x}$. In order to get such an approximation from the data available to us, we relate the number of requests by x to a over a given critical edge $e(x,a)$ to the total number of requests by x:

$$T_{(a,x)} = \frac{cREQ_{e(x,a)}}{\sum_{\forall e(x,i)} REQ_{e(x,i)}} \tag{5}$$

4 Case Study: Applying FIT to a Real Application Landscape

We test our approach with data taken from the SOA environment presented in Sect. 2. We discuss the criticality of the initial 8 hotspots identified on the heat map in Fig. 1 (rectangular nodes). We simulate two alternative SLA-structures (*scenario 1* and *scenario 2*) aimed at eliminating the two most critical hotspots and discuss the effects. The FIT-scores of the 8 initial hotspots are listed in Table 1 in descending order. The detailed values to retrace how the initial FIT-scores for A1, A5, A18 and A20 were obtained are also shown in Table 1. We discuss selected applications.

A18 (silver) is deemed the most critical hotspot as it causes a relationship with criticality $eFIT_{e(A18,A19)} = 3$ to A19 (gold). As can be seen in Fig. 2c and Table 1, the link $e(A18,A19)$ carries 100% of the requests of service that A19 makes to A18. In contrast only 0.00008% or 953/11,332,472 requests of service of A2 on A18 occur along $e(A18,A2)$. Therefore, even though both A19 and A2 have SLA gold and depend on A18 with SLA silver, only $e(A18,A19)$ is critical and will require adjustment of the SLA of A18 or A19.

A1 (bronze) is ranked the second critical hotspot mainly because two of its three critical relationships to applications with higher SLAs are in heavy use (cf. Fig. 2a). A10 (silver) relies fully (100%) on A1 while A0 (gold) processes 10% of its total traffic over the critical relationship $e(A1,A0)$. As the most impacted application has only SLA silver, this relationship is ranked lower than the relationship $e(A18,A19)$ discussed before where A19 has gold level.

A20 (bronze) is ranked relatively uncritical in spite of the large body of important applications depending indirectly on it (c.f. Fig. 2f). This is mainly due to the low usage of its relationship to A2, accounting for only 0.1% of the total traffic produced by A2. Nevertheless, the heavy dependency tree of A2 weights this relationship $e(A20,A2)$ more critical than the relationship $e(A18,A2)$ as discussed earlier.

A5 (bronze) is ranked with criticality 0, putting it on the same level as an uncritical configuration (e.g. application A2, c.f. Fig. 2d and Table 1). Although there is a potentially critical relationship to an application with high availability, there is no traffic across this relationship within the measured period (c.f. Fig. 2b). Therefore, the importance of this relationship can be discarded as it seems to be unlikely that the relationship should be used exactly in the time of a downtime of A5. In addition, no other applications are indirectly depending on A5 over this relationship. Therefore, A5 can be assumed to be non-critical.

In *scenario 1* we now try to cost-effectively eliminate the hotspots top-down. We start with A18 by lowering the depending application A19's SLA to silver. Simulating the resulting SLA-structure shows that A18's criticality drops to 0 with no further negative impact on the surrounding applications. Based on this setting, we try to also eliminate hotspot A1 in *scenario 2* by leveling the SLAs of A0, A1, A10 and A13 to silver. Nevertheless, simulating this structure shows that A13 becomes a new hotspot with criticality $FIT_{A13} = 18$, beating A1 in its criticality. Therefore leveling all four applications should be discarded for being ineffective and other structures have to be simulated instead.

The examples discussed here in detail show the importance of assessing the potential hotspots identified on the heat map in Fig.1. Especially to balance static

Table 1 FIT-scores for hotspots h and component values for selected applications.

Initial		**Scenario 1**		**Scenario 2**		**FIT-score components for A1, A2, A18, A20**							
h	FIT_h	h	FIT_h	h	FIT_h	a	x	$eFIT_{e(a,x)}$	F_x	$I_{a,x}$	$T_{a,x}$	$cREQ_{e(x,a)}$	$\Sigma REQ_{e(x,y)}$
A18	3	A18	0	A18	0	A1	A0	0.307	3	0	0.102	42,931	420,028
A1	2.307	A1	2.307	A1	0	A1	A8	0	1	0	0	0	0
A10	0.118	A10	0.118	A13	18.083	A1	A10	2	2	0	1	224,637	224,637
A23	0.063	A23	0.063	A23	0.06	A1	A13	0	3	0	0	0	0
A20	0.027	A20	0.027	A20	0.026	A1	A27	0	1	0	0	0	31,739
A11	0.021	A11	0.021	A11	0.02	A2	A0	0	3	0	0	0	420,028
A16	0.005	A16	0.005	A16	0.005	A2	A7	0	3	19	0	0	346,735
A5	0	A5	0	A5	0	A2	A21	0	3	0	0	0	583,490
						A18	A2	0	3	0	0	953	11,332,472
						A18	A17	0	1	0	0	0	3,456,770
						A18	A19	3	3	0	1	75,726	75,726
						A20	A2	0.027	3	40	0.001	7,075	11,332,472

information (i.e. business-criticality of applications and their relationships) with factual usage is crucial to focus on the really critical hotspots. Simulating different structures based on this approach aids ITSM optimizing the availability of service-based application landscapes.

5 Related Work

Related work published over the past years deals with several types of metrics for SOA. To categorize these contributions we mapped them to the phases of the application management lifecycle: requirements specification, design, build, deploy, operate and optimize [7]. Most of the reviewed work deals with aspects and metrics to support the first four phases based on design-time data: Metrics to measure business alignment of SOA implementations [1, 30], procedures to model [6, 16] and implement SOA [3], including the prediction of development effort and implementation complexity early in the design phase [37]. Metrics to measure granularity, complexity and reuse [15, 35, 36], performance [5, 13] and QoS [28, 31] of SOA-based services also rely on design-time data. Most work on operation and optimization has been done on how to handle service level agreements primarily based on design-time data: how to formally describe them [19, 34, 39, 40], technically implement, test and enforce them [4, 10, 15, 17, 18, 23, 25, 26, 32, 33, 42, 44, 45] or how to monitor them [2, 20, 21]. Contributions available on SLA design deal with isolated approaches: Sauvé et al. [38] and Marques et al. [27] are in favor of deriving the service level targets directly from the business impact of the given service (i.e. taking into account the risk of causing revenue loss on the business layer). Li et al. [24] and Smit et al. [41] focus on infrastructure aspects of specific applications.

Most of these contributions require customized frameworks or rely massively on design-time data and services being designed as glass-boxes. None of these contributions propose a solution how to characterize the criticality of an existing service configuration in historically grown heterogeneous application landscapes based on runtime data provided by end-to-end monitoring. Nevertheless, this is crucial for ITSM to decide on changes in the SLA-structure cost-effectively.

6 Conclusion and Outlook

SOA reduces the complexity of system integration. However, it increases the problems of governance and availability management on the infrastructure level because of hidden dependencies among services. As services are transparently reused, applications with higher SLAs can become dependent on applications with lower SLAs, thus creating hotspots in the SLA structure. To guarantee overall cost-effective availability in such a setting, ITSM has to identify these hotspots and decide on their criticality.

In this paper, we proposed the FIT-metric based on the three components: function, integration and traffic. Our metric allows ranking hotspots and their relationships according to their criticality for ITSM. Based on this ranking, different alternative SLA-structures and their impact can be simulated. Therefore, the contributions of this paper are threefold: i) we showed the need for a criticality metric in a historically grown production-strength SOA-landscape; ii) we presented the cost-effective FIT-metric to rank hotspots and their relationships according to their criticality for ITSM to optimize SLA levels; iii) we demonstrated our approach by applying it to actual service configurations.

We are about to conclude the implementation of our prototype at Deutsche Post MAIL. This includes finishing the rollout of our service monitoring and reporting to allow for more extensive analysis in the future (e.g. include data about latencies in FIT). As part of our future work we want to apply our findings to other loosely coupled systems such as event-based systems (EBS). Today, SOA is mostly used intra-organizationally to implement given workflows within a single organization. Thus, critical knowledge about participants, their interdependencies and corresponding business impact are available in principle. Tomorrow's systems tend to become even more federated, distributed and loosely coupled. In those service-based inter-organizational systems availability management is even more difficult.

Acknowledgements. We would like to thank Irene Buchmann, Jacqueline Pranke, Achim Stegmeier, Alexander Nachtigall and the two anonymous reviewers for their valuable input and discussions on this work. Part of this work is funded by German Federal Ministry of Education and Research (BMBF) under research grants ADiWa (01IA08006) and Software-Cluster project EMERGENT (01IC10S01), and by Deutsche Post MAIL. The authors assume responsibility for the content.

References

1. Aier, S., Ahrens, M., Stutz, M., Bub, U.: Deriving SOA Evaluation Metrics in an Enterprise Architecture Context. In: Di Nitto, E., Ripeanu, M. (eds.) ICSOC 2007. LNCS, vol. 4907, pp. 224–233. Springer, Heidelberg (2009)

2. Ameller, D., Franch, X.: Service level agreement monitor (SALMon). In: ICCBSS 2008, pp. 224–227 (2008)
3. Arsanjani, A., Ghosh, S., Allam, A., Abdollah, T., Ganapathy, S., Holley, K.: SOMA: a method for developing service-oriented solutions. IBM Systems Journal 47(3), 377–396 (2010)
4. Bause, F., Buchholz, P., Kriege, J., Vastag, S.: Simulation based validation of quantitative requirements in service oriented architectures. In: WSC 2009, pp. 1015–1026 (2009)
5. Brebner, P.C.: Performance modeling for service oriented architectures. In: ICSE Companion 2008, pp. 953–954 (2008)
6. Broy, M., Leuxner, C., Fernndez, D.M., Heinemann, L., Spanfelner, B., Mai, W., Schlör, R.: Towards a Formal Engineering Approach for SOA. Techreport, Technische Universität München (2010), http://www4.informatik.tu-muenchen.de/publ/papers/TUM-I1024.pdf (accessed on June 21, 2011)
7. Cannon, D.: ITIL Service Operation: Office of Government Commerce. The Stationery Office Ltd. (2007)
8. Deutsche Post DHL: Deutsche Post CEO Frank Appel presents Strategy 2015 (2010), http://www.dp-dhl.com/en/investors/investor_news/news/2009/dpwn_strategie_2015.html (accessed on March 18, 2011)
9. Deutsche Post DHL: MAIL Division (2011), http://www.dp-dhl.com/en/about_us/corporate_divisions/mail.html (accessed on March 18, 2011)
10. Di Modica, G., Regalbuto, V., Tomarchio, O., Vita, L.: Dynamic re-negotiations of SLA in service composition scenarios. In: EUROMICRO 2007, pp. 359–366 (2007)
11. Eckes, G.: Six SIGMA for Everyone, 1st edn. Wiley & Sons (2003)
12. Frischbier, S., Petrov, I.: Aspects of Data-Intensive Cloud Computing. In: Sachs, K., Petrov, I., Guerrero, P. (eds.) Buchmann Festschrift. LNCS, vol. 6462, pp. 57–77. Springer, Heidelberg (2010)
13. Her, J.S., Choi, S.W., Oh, S.H., Kim, S.D.: A framework for measuring performance in Service-Oriented architecture. In: NWeSP 2007, pp. 55–60 (2007)
14. Herr, M., Bath, U., Koschel, A.: Implementation of a service oriented architecture at deutsche post MAIL. Web Services, 227–238 (2004)
15. Hirzalla, M., Cleland-Huang, J., Arsanjani, A.: A Metrics Suite for Evaluating Flexibility and Complexity in Service Oriented Architectures. In: Feuerlicht, G., Lamersdorf, W. (eds.) ICSOC 2008. LNCS, vol. 5472, pp. 41–52. Springer, Heidelberg (2009)
16. Hofmeister, H., Wirtz, G.: Supporting Service-Oriented design with metrics. In: EDOC 2008, pp. 191–200 (2008)
17. Hsu, C., Liao, Y., Kuo, C.: Disassembling SLAs for follow-up processes in an SOA system. In: ICCIT 2008, pp. 37–42 (2008)
18. Kotsokalis, C., Winkler, U.: Translation of Service Level Agreements: A Generic Problem Definition. In: Dan, A., Gittler, F., Toumani, F. (eds.) ICSOC/ServiceWave 2009. LNCS, vol. 6275, pp. 248–257. Springer, Heidelberg (2010)
19. Kotsokalis, C., Yahyapour, R., Rojas Gonzalez, M.A.: Modeling Service Level Agreements with Binary Decision Diagrams. In: Baresi, L., Chi, C.-H., Suzuki, J. (eds.) ICSOC-ServiceWave 2009. LNCS, vol. 5900, pp. 190–204. Springer, Heidelberg (2009)
20. Kunz, M., Schmietendorf, A., Dumke, R., Rud, D.: SOA-capability of software measurement tools. In: ENSUR A, p. 216 (2006)
21. Kunz, M., Schmietendorf, A., Dumke, R., Wille, C.: Towards a service-oriented measurement infrastructure. In: SMEF 2006, pp. 10–12 (2006)
22. Kütz, M.: Kennzahlen in der IT, 2nd edn. Werkzeuge für Controlling und Management. Dpunkt Verlag (2007)
23. Lam, T., Minsky, N.: Enforcement of server commitments and system global constraints in SOA-based systems. In: APSCC 2009, pp. 126–133 (2009)

24. Li, H., Casale, G., Ellahi, T.: SLA-driven planning and optimization of enterprise applications. In: WOSP/SIPEW 2010, pp. 117–128 (2010)
25. Liu, L., Schmeck, H.: Enabling Self-Organising service level management with automated negotiation. In: IEEE/WIC/ACM 2010, pp. 42–45 (2010)
26. Liu, L., Zhou, W.: A novel SOA-Oriented federate SLA management architecture. In: IEEC 2009, pp. 630–634 (2009)
27. Marques, F., Sauvé, J., Moura, A.: Service level agreement design and service provisioning for outsourced services. In: LANOMS 2007, pp. 106–113 (2007)
28. Mayerl, C., Huner, K.M., Gaspar, J., Momm, C., Abeck, S.: Definition of metric dependencies for monitoring the impact of quality of services on quality of processes. In: IEEE/IFIP 2007, pp. 1–10 (2007), doi:10.1109/BDIM.2007.375006
29. McGovern, J., Sims, O., Jain, A.: Enterprise service oriented architectures: concepts, challenges, recommendations. Kluwer Academic Pub. (2006)
30. O'Brien, L., Brebner, P., Gray, J.: Business transformation to SOA. In: SDSOA 2008, pp. 35–40 (2008)
31. O'Brien, L., Merson, P., Bass, L.: Quality attributes for service-oriented architectures (2007)
32. Palacios, M., Garcia-Fanjul, J., Tuya, J., de la Riva, C.: A proactive approach to test service level agreements. In: ICSEA 2010, pp. 453–458 (2010)
33. Parejo, J.A., Fernandez, P., Ruiz-Corts, A., Garca, J.M.: SLAWs: towards a conceptual architecture for SLA enforcement. In: SERVICES-1 2008, pp. 322–328 (2008)
34. Raibulet, C., Massarelli, M.: Managing Non-Functional Aspects in SOA Through SLA. In: Bhowmick, S.S., Küng, J., Wagner, R. (eds.) DEXA 2008. LNCS, vol. 5181, pp. 701–705. Springer, Heidelberg (2008)
35. Rud, D., Schmietendorf, A., Dumke, R.: Product metrics for service-oriented infrastructures. In: Proceedings of IWSM/MetriKon 2006, pp. 161–174 (2006)
36. Rud, D., Schmietendorf, A., Dumke, R.: Resource metrics for service-oriented infrastructures. In: SEMSOA 2007, pp. 90–98 (2007)
37. Salman, N., Dogru, A.: Complexity and development effort prediction models using component oriented metrics. In: ENSUR A (2006)
38. Sauvé, J., Marques, F., Moura, A., Sampaio, M., Jornada, J., Radziuk, E.: SLA Design from a Business Perspective. In: Schönwälder, J., Serrat, J. (eds.) DSOM 2005. LNCS, vol. 3775, pp. 72–83. Springer, Heidelberg (2005)
39. Schulz, F.: Towards measuring the degree of fulfillment of service level agreements. In: ICIC 2010, pp. 273–276 (2010)
40. Skene, J., Lamanna, D.D., Emmerich, W.: Precise service level agreements. In: ICSE 2004, pp. 179–188 (2004)
41. Smit, M., Nisbet, A., Stroulia, E., Edgar, A., Iszlai, G., Litoiu, M.: Capacity planning for service-oriented architectures. In: CASCON 2008, pp. 144–156 (2008)
42. Strunk, A.: An algorithm to predict the QoS-Reliability of service compositions. In: SERVICES 2010, pp. 205–212 (2010)
43. Taylor, R., Tofts, C.: Death by a thousand SLAs: a short study of commercial suicide pacts. Hewlett-Packard Labs (2005)
44. Thanheiser, S., Liu, L., Schmeck, H.: SimSOA: an Approach for Agent-Based Simulation and Design-Time Assessment of SOC-based IT Systems. In: Jacobson Jr., M.J., Rijmen, V., Safavi-Naini, R. (eds.) SAC 2009. LNCS, vol. 5867, pp. 2162–2169. Springer, Heidelberg (2009)
45. Theilmann, W., Winkler, U., Happe, J., de Abril, I.M.: Managing On-Demand Business Applications with Hierarchical Service Level Agreements. In: Berre, A.J., Gómez-Pérez, A., Tutschku, K., Fensel, D. (eds.) FIS 2010. LNCS, vol. 6369, pp. 97–106. Springer, Heidelberg (2010)

Chapter 7
How to Design and Manage Complex Sustainable Networks of Enterprises

Clara Ceppa

Abstract. Current production processes do not fully exploit natural resources and discard a significant percentage. To exemplify, think of beer manufacturers that extract only 8% of the nutritional elements contained in barley or rice for the fermentation process while all the rest of the resource is thrown away as waste. To restrain this phenomenon we developed, in collaboration with Neosidea Group, an instrument for making the changes needed on the level of the management, organization and procurement of raw materials and energy. We can start seeing the importance of creating an IT instrument based on the concept of an open-loop system that can help companies, according to their business purpose or geographical location, to organize themselves into "ecological networks" to achieve production that moves towards zero emissions by means of sustainable management and valorization of waste: by following the first principle of Systemic Design, waste (output) of one productive system can be used as a resource (input) for another. Linked enterprises could reach a condition of reciprocal advantage by allowing the reutilization of the materials put out by their production processes; profits can be obtained from the sale of these outputs. The constant exchange of information and sharing of knowledge between the players involved allows a continuous systemic culture to spread, along with the concepts of prevention and the ongoing improvement of the environment. Essentially this paper presents an IT network at the service of the environment, a web that speaks to the earthly roots of humanity and the deep need for a revived attention to nature and the resources it offers. The huge amount of data obtained by using Systemic Software is a precious asset and a vital platform for designer, scholars of the environment, researchers, ecologists, public agencies, local administrators and, obviously, for entrepreneurs, who will be able to work in

---

Clara Ceppa
Politecnico di Torino, Italy

a more sustainable way. The combination of the systemic approach and this technological support instrument improves understanding that an effective environmental protection is not in conflict with the economic growth of enterprises.

Keywords: tool, complexity, systemic design, flow of resources.

1 Introduction

In all types of productive activity some of the resources used are returned to the environment in a random and disorderly manner in the form of gaseous or liquid effluent. The time has come to realize that our current productive activities produce a large amount of waste and squander most of the resources they take from Nature. To give an example, when we extract cellulose from wood to make paper, we cut down an entire forest but use only 20-25% of the trees while the remaining 70-80% are discarded as waste. Palm oil makes up only 4% of the overall biomass of the palm tree; coffee beans make up only 4% of coffee bushes. Breweries extract only 8% of the nutritional elements contained in barley or rice for fermentation (Capra, 2004).

Moreover the necessity of a more sustainable approach to production and consumption patterns has been widely highlighted by international resolutions and directives as a way to promote sustainable development in daily life activities (e.g. the EU Strategy for Sustainable Development and the EU Integrated Product Policy).

In order to build sustainable industrial societies, it's necessary to help the industries to organizes themselves into ecological groupings in order to manage at best their sources and waste. To do that we have to create an instrument for making the changes needed on the level of the management, organization and procurement of energy and resources. Only in this way can achieve a society based on a life cycle of products, consistent with environmental needs and able to meet human needs (Lanzavecchia, 2000) while consuming few resources.

It's desirable the creation of new manufacturing scenarios where the output of one company, a useless material to be eliminated incurring expenses only, can be reused to ensure the survival of another company related to the business category or physical location of the first company. In this sense all in industrial production must reduce the use of no-renewable materials and evolve toward less energivorous processes, making uncontaminated outputs that can be reused for their qualities.

The above mentioned concept is the first of five principles of Systemic Design (Bistagnino, 2009) described below:

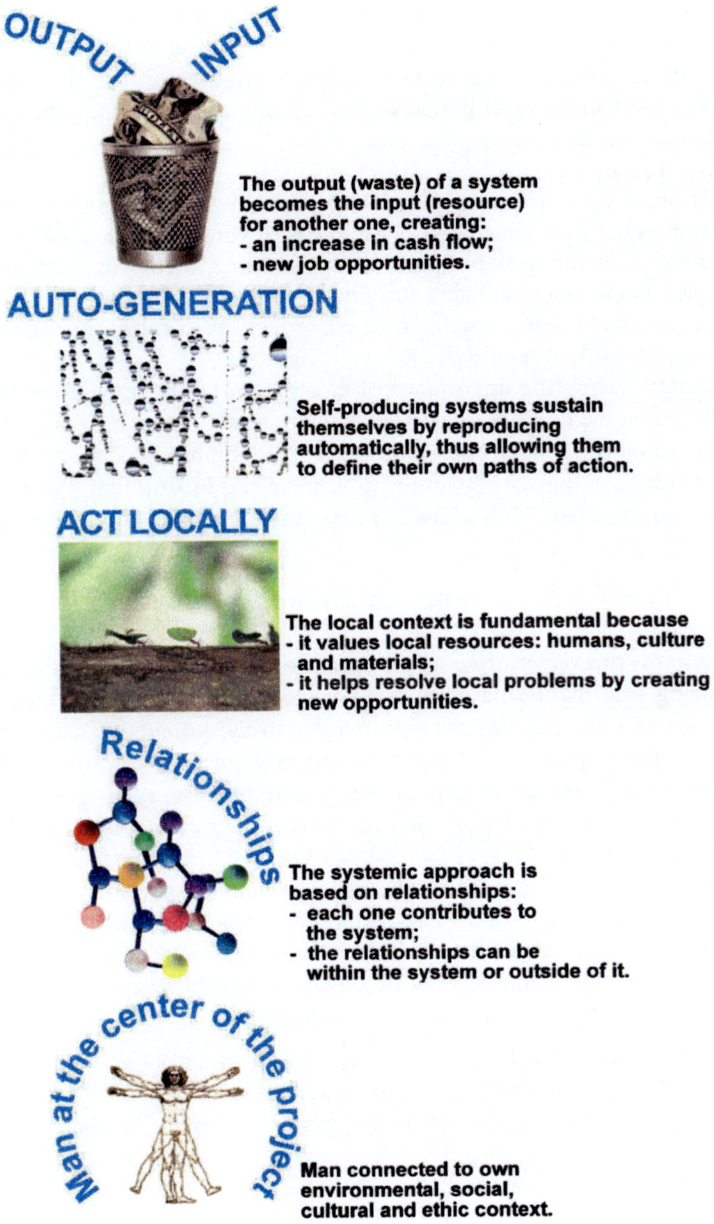

Fig. 1 The five principles of Systemic design.

By the proposed methodology and the corresponding re-evaluation of the rejected material, it becomes possible to skip treatment costs and create a network for selling one's own output. This generates greater profits and benefits to the territory due to the realization of new enterprises, the development and improvement of the already established enterprises and the creation of new jobs.

It is a process that can be applied to any production sector. It is deliberately applied locally to enhance local potentials and specificities and strengthen the bond with tradition. Another reason it is applied locally is to avoid the high costs of transportation along with the air pollution it creates.

If the systemic approach is a methodology capable of turning a cost into a benefit, a waste product into a resource, Systemic Software is an instrument that can support analysis of the system approach applied to a local area and define the possible interactions apt to create a sustainable network of companies that can exchange resources and competencies with consequent gain for all the operators involved in the network of relationships.

We can start seeing the importance of creating an IT instrument for study and analysis based on the concept of an open-loop system that can help neighbor companies, according to their business purpose or geographical location, to organize themselves into "ecological networks" to achieve production that moves towards zero emissions by means of sustainable management and the valorization of waste.

2 Systemic Software to Manage Complex Ecological Networks

In specific terms this paper presents the definition, design and realization of a tool for processing information based on evolved technological systems that can acquire, catalog and organize information relative to the productive activities in the area of study, the outputs produced and the inputs required as resources.

This data is acquired and organized in terms of quantity, type, quality and geographical location on the territory. All the data are correlated with each other by means of a complex logic.

The huge amount of data obtained by using Systemic Software is a precious asset and a vital platform for scholars of the environment, researchers, ecologists, public agencies, local administrators and, obviously, for entrepreneurs. The last mentioned actors will be able to work in a more sustainable way.

The functions of the systemic software are fourfold:

producers of waste would be able to determine which local companies could use their outputs as resources in their production process;

it tells input-seekers which companies produce outputs they can use as resources;

it informs different producers about new business opportunities on the local territory that have previously remained hidden;

it is an efficacious instrument for evaluating the entire production process and becomes an instrument for providing feedback.

Therefore this system can give useful and reliable information regarding one's current production process: if you enter the type of waste produced by your company as a search criterion, and the Software gives no results for possible reutilization of your outputs, this means your current production process makes waste that cannot be reused or recycled. It means your company produces items by using inputs and processes that do not comply with the vision of an open system. We have observed the need to implement certain changes within the production line, for example to reassess current inputs and substitute them with others that are more environmentally sustainable.

To make operating the fourth item the processing system, developed in collaboration with Neosidea Group, was also supplemented with the function of geo-locating business is and materials and this provides a solution and that gives not only information regarding new areas of application of the outputs but also determines with precision and localizes by territory the flows of material within a local network whose nodes are represented by local companies.

By using the geo-localization function, the system can ascertain which local activities are situated within the range of action (it depends on specific territory shape and could be for example 60km or 100km) defined according to the search criteria.

Then it positions them exactly on a geographical map to show the user which companies can be part of the network of enterprises which enter into a reciprocal relationship to increase their own business and maximize their earnings through the sale of their outputs and the acquisition of raw materials from other local production activities.

The information generated by the system has a double value and allows to define both the human-machine interactions for querying the system itself and the machine-machine interactions to activate controls and automatisms for interfacing with the geo-localization system and the input computational algorithms within the system itself. We defined the entities through which such information fluctuate and are processed as represented by the following diagram defined in the representative methodology as "data flow chart" (Ceppa, 2009).

The logic and the algorithms that intervene on the acquired information serve to normalize the structures, allowing them to be interlaced and evaluated by evolved technological instruments which serve to render the information in an intelligible and intuitive format for all of those who interface with the Systemic Software.

The Systemic Software was developed using web technologies with the goal of supporting multi-users, or rather it can allow multiple operators to access to the system and to interact both in consultation and in processing or modifying data contained herein. To that end, it was important to establish a set of features to allow a proper management of all the users and roles that they were going to play.

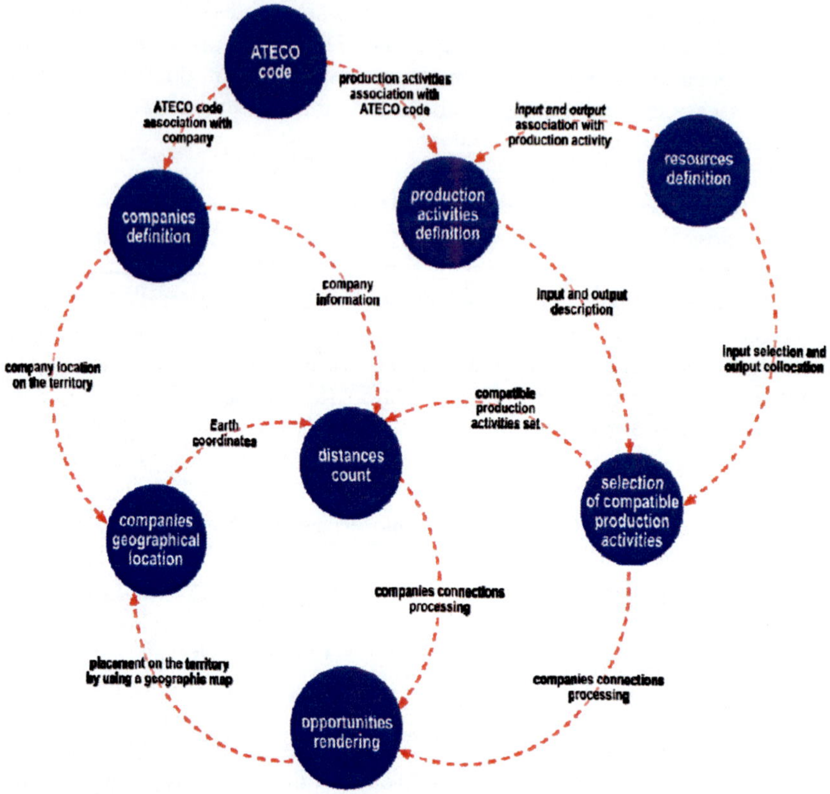

Fig. 2 Data flow chart of software architecture.

The main functions provided for this purpose are linked both to the ability to create, edit and delete users and to create ad hoc operational roles. The latters are defined by associating each user of the tool to a specific access profile that will allow or not the use of the features available. Several control features have also been implemented to measure the operation both users and the tool itself by tracking the carried out operations and the outcome thereof.

The consultation of the system was designed by following the systemic approach and made usable by means of Web 2.0 technologies; this approach has made it possible to publish an interactive web portal as a facility that can be used by operators who want to consult it and interact with it. We chose to use web 2.0 technologies cause it enhances the user's role and consequently the social dimension of the network: Tim O'Reilly argues that the network's value lies not in technology but in content and services, and that the strength of the network is represented mainly by its users. The enhancement of the social dimension of the network, through instruments capable of facilitating interaction between individuals and the user transformation into active creators of these services, allows that the quality of the offered services improves as the number of users involved in their use (Rasetti, 2008).

The system has been designed to provide an informative service dedicated to two types of operators for whom different functions and typologies of access have been developed. These operators are represented both by companies seeking to optimize their production processes through the exchange of sources with other entrepreneurs, both by researchers and public bodies who can analyze the material flow in the area and the distribution of productions to facilitate the development of the territory and of the local economy. In order to define both methodologies of access, one for "companies" (front-end) and one for "researchers" (back-end), have been created two distinct interfaces connected to a single database capable, through a set of features of low level and processing algorithms, to contextualize the information about a specific territory with the requests of the operators as shown in the figure below.

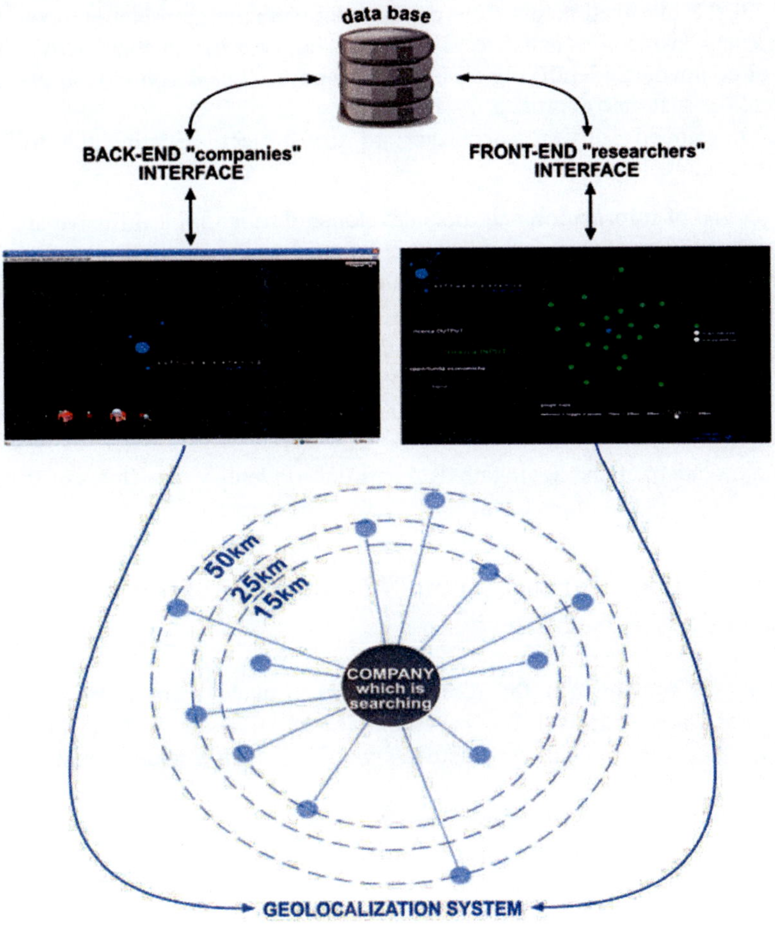

Fig. 3 Web 2.0 interfaces linked to geo-localization function.

Elaborative processes are distinct from person to person who interacts, in fact the features the two types of users can access are different and allow operators "researcher" to expand and contract the system by interacting directly with the information contained therein, to establish new relationships and enable new business realities within the network of relationships. Instead operators "company" can consult the information through a dedicated interface, the front-end, capable of rendering information resulting from processing and extraction of the database managed by the operators "researcher".

The role of the operators "researcher" is to spread the systemic methodology, on the other hand the operators "company" are the users of this methodology that through the Systemic Software is operative on the territory, allowing direct visualization of the companies that could be included in an ecological network where all the sources are re-used. Moreover this tool is able to manage the interaction between all the involved actors through the establishment of relations between resources and productive activities as well as between production activities and types of companies, identifying all possible opportunities deriving from the application of the systemic approach.

The developed software, according to the project, satisfies the following requirements:

projection of information and tools developed through joint diffusive and interactive instruments;

ability to connect different actors (public agencies, entrepreneurs, researchers, etc) on a territory;

ability to link production processes with materials according to systemic logic;

geographic location of industries on the territory;

acquisition of external database to manage the types of materials and production activities;

ability to abstract and aggregate data in order to identify new flows of materials and economic opportunities for people involved into the network of relationships.

3 Systemic Software vs. Actual Market for Waster and Secondary Raw Materials

In recent year something in this direction has been made: some software could be found on internet. However they have still technological and features limits. This happens cause waste is still considered as worthless matter and not new source to be re-used.

Let's example of an Italian free service: an electronic market for waste. It is a tool for the exchange of waste, accessible upon registration through a website.

This is a transparent medium of exchange and relationship between demand and supply of waste and secondary raw materials (resulting from recovery); it propose also a informative service about waste/material legislation and recovery technologies.

The aim is to promote:

a national market for recovery and recycling of waste and secondary raw materials (SRM);

the spread of conditions of transparency and security for operators.

The virtual market is accessed through the listings on the website that can concern both the supply and the demand for waste and/or materials or some services. Each user has to indicate its own expertise (e.g. waste producer or waste manager). This is necessary due to the fact that for the same type of waste presents on the market, user may request a recovery or disposal activity or a transport service.

According to recent data, it shows that the demand far exceeds the supply of waste.

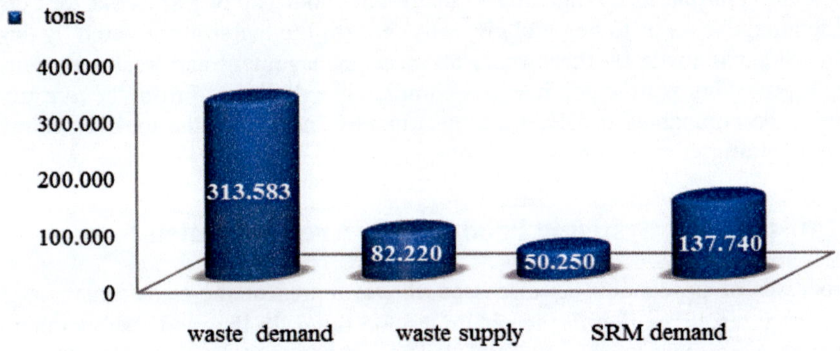

Fig. 4 Demand and supply of waste and SRM.

The demand of waste is a direct request derived from waste management plant and it amounts to about 313.000 tons, instead the supply is around 82.000 tons (ecocerved).

Comparing data, the greater demand for materials can be attributed to such a preponderant use of this free service by companies that manage waste. They express a strong demand on the basis of waste treatment capacity of their management plant, while waste producers offer only the quantities actually produced and available at the time of insertion.

As for the secondary raw materials (SRM), we are faced with the opposite phenomenon: greater waste supply rather than demand. But this consideration must be followed by another cause the waste treatment and management companies require SRM such as fertilizer composted mixture (42%), packing pallets (8%), wood (14%), etc. While SRM producers principally offer inert materials (84%). One hypothesis to give an explanation for this phenomenon is that this type of SRM is more easily retrievable, there is a high quantity bound to recovery and, therefore, an increased production of recovered material.

This situation shows the real problem of the actual recovery of output which consists of an inadequate knowledge of materials, their intrinsic properties and a deep lack of awareness that they can be reused as raw materials in other production processes, as claimed by the systemic approach.

Analyzing the tool from a systemic point of view, it's possible to underline that the focus of this service is solely on products and not on production cycles of which they are part: this implies a limited (or rather, totally absent) overview of the productive process as a whole. The approach is in fact linear: it leads to focus on each individual stage of the process, considered independent from one another. As stated previously, by not applying the systemic approach, does not allow this free service to offer its users new options to re-use waste and create new sustainable productive networks. It's only a virtual trade where to buy or sell waste or secondary raw materials. The latters in fact are not raised to the higher level of resources but continue to be regarded as matters without value: just as waste. This consideration allows us to better clarify why most of the materials required by this virtual market is made by the owners of waste management and treatment. Furthermore, the context in which it acts is completely extrapolated from the territory in which the companies involved are, not thereby increasing the local economy and not exploiting the resources of the place.

4 How the New Systemic Productions Are Generated

he processes of production, as discussed above, represent the transformation of one or more resources in order to obtain a good thing. In the model shown there are three subjects that exist in the performance of the production process: the resources, the transformation, the result of the transformation. The model, however, is minimized compared to reality cause it doesn't mention a fourth element that acts on the definition of a realistic model: waste. Indeed each production cycle, in addition to the final product, generates a series of scraps. It is therefore appropriate to model the production cycle with a logic of multiple action as follows:

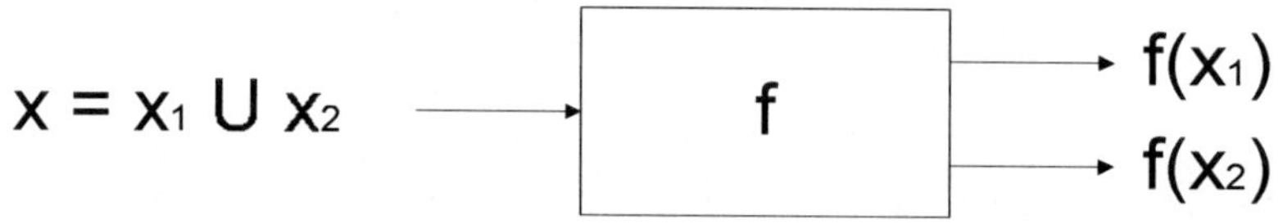

Formula 1. relation between input, production and output of a process.

The parameter x represents the set of properties (broken down into units x1 and x2) that determines the resource; f is the process of production; f(x1) is the finished product and f(x2) is the waste.

This mathematical model established that from any process of production are produced an expected result and a sub-result is not desired (Ceppa, 2009). This sub-result, cause unexpected, is thrown away even entailing an economic burden to dispose of it.

The systemic approach is proposed as an approach capable of restoring value to waste rising to the level of resources. The sub-result of production is itself a resource characterized by an intrinsic ability to serve as input for another manufacturing process. The new model could be able to feed itself by a constant flow of resources through each productive process. Accordingly, the tool supports the identification of possible redeployment of production waste on the territory within a variable range of kilometers and through the association of the processes with the activity types not only according to statistic or theoretical logic but to strongly logic linked to the companies present on the geographic area under consideration.

5 Case-Study: Sustainable Cattle Breeding by Using the Systemic Software

The case-study explains a project where items of study were not only products, but also and mainly the production cycles in order to generate a productive system similar to nature, where there is no concept of waste and also the surpluses are metabolized by the system itself.

In this way, as we mentioned before, the output of a productive process can be use as raw materials in other processes.

It's illustrated a systemic production chain where output are re-used as resources in other productive processes.

The analyzed chain starts with a cow farm and ends with the retail sale of final products, passing from the milking phase to slaughter. Thanks to the development of a structured implementation logic based on the systemic vision, the information processing instrument or systemic software, is able to provide further information to set up new production chains and new flows of materials and services in favor of all the businesses who join the initiative thanks to a constant updating and comparison among the systemic logics for reusing materials, local productive activities and the territory itself.

The starting point is to analyze the actual process and underline some critical points: ac- cording to the analysis conducted on the various phases of the process, it appeared that the phases are considered separate from one another and follow a linear course where the waste is seen as something to throw away and not as a resource. By applying the systemic methodology, and using Systemic Software, it was possible to establish new ways to use these re- sources and create local flows of material.

The outputs from the cow farm were sent to other production enterprises: the water with urine content was sent to water treatment facilities to be treated. The manure, sawdust and urine were used in biogas production plants which produce methane and sludge that are excellent ingredients for high-quality compost for farming purposes. The outgoing material of the milking phase is currently thrown away but the water contains a certain percentage of milk. This resource is rich in nutritional value if managed systemically and can be used to feed freshwater fish.

Numerous critical points were also found in the slaughtering process. Particularly noticeable was the problem of the squandering of certain fundamental byproducts with a high biological value, e.g. the blood (Ganapini, 1985). In the new web of connections blood is used for the production of soil and natural flower fertilizer. Blood traces were also contained in the water sent to treatment plants and plant-filtering processes. The remains of the meat and some of the animals' organs and entrails give a major contribution to raising worms, an essential food for raising quail. Quail eggs are high-quality food products. The last phase of the chain, the retail sale of the final products, produces outputs, though certainly in lower quantities due to the small-scale operations of the butcher. However they are not of lower quality. Animal bones and fat can be used by companies that process and conserve food products.

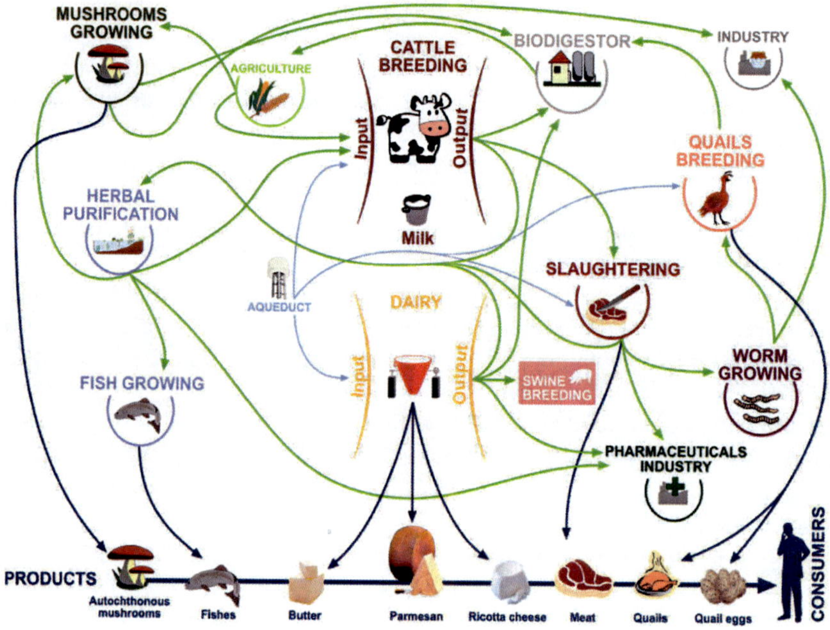

Fig. 5 Systemic cattle breeding.

New flows of material taken from waste, became resources, are bringing together different industries that join forces to achieve the goal of zero emissions.

The systemic approach improves understanding of the environmental and the economic benefits generated by a systemic nonlinear productive culture which enables us to transform waste into materials worthy of a proper rational use. Such an approach is aimed at an optimal management of the waste/materials. More importantly it aims at the profitably reutilization of these materials.

The advantages of such methodology are both environmental and economic.

Among these the most important goal is to reduce the cost of waste treatment and therefore increase the profits from selling the company's outputs. It demonstrates that optimal management of the input-production-output circuit is made possible by improving the use of natural resources by recovering and making much of outputs.

This last example is suggestion for a sustainable production future.

The obtained results highlight the significant differences and benefits between the current production process, characterized by a linear structure, and the new one that proposes an open industry system based on the following sequence: quality output > reuse output > resources > profits (Ceppa et al., 2008).

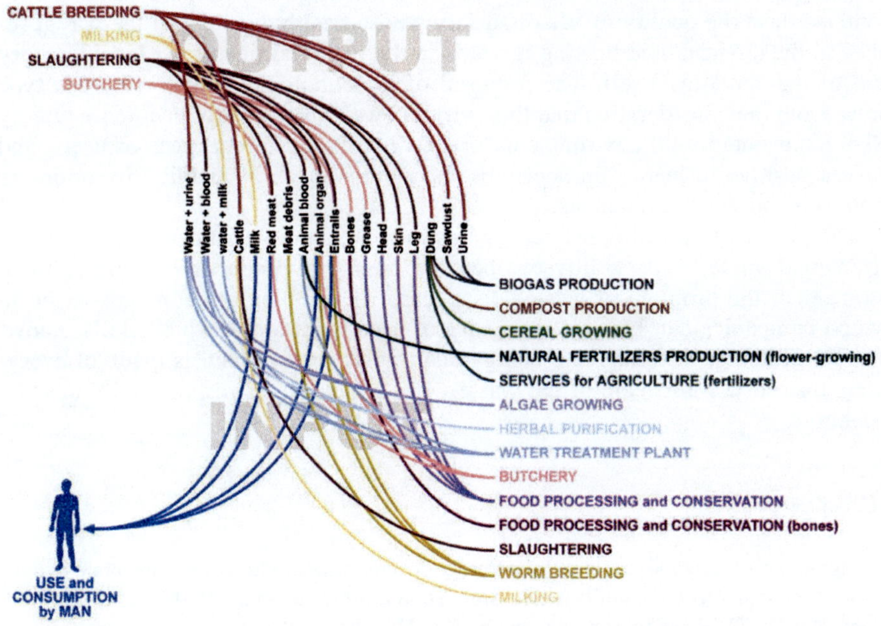

Fig. 6 Graphic elaboration (obtained by Systemic Software) of output uses in new fields of production.

6 Conclusion

The proposal of a technological instrument of this type facilitates the raising of awareness by the various actors on the territory, on various levels of expertise, about the numerous possibilities offered by the systemic culture, in particular Systems Design applied to a productive territory. The study therefore aims at making knowledge about the instruments offered by the systemic approach explicit and more accessible. By sharing knowledge and experience through networks and design-driven instruments we can offer an interpretive key for understanding its benefits to the environment and the economy, benefits generated by a possible transition towards a systemic nonlinear type of productive and territorial culture.

The network and instruments offer concrete possibilities to transform waste into materials worthy of appropriate, rational and targeted management, and more importantly, profitable reuse. This reinforces the concept according to which an efficacious protection of the environment is not in conflict with the economic growth of businesses.

Essentially this paper presents an IT network at the service of the environment, a web that speaks to the earthly roots of humanity and the deep need for a revived attention to nature and the resources it offers. The advantages of such an instrument are that they: improve usability, facilitate use and satisfaction, expand the potential area of users, improve the use of technological resources and local resources, raise the quality of life of society whose health depends on the way it relates to the environment hosting it, valorize the potentialities of the local territory and of the economy itself. The proposal of a technological support of this type arose from the consideration that this "virtual" web allows us to react more rapidly when confronted with environmental issues, involve different areas of users, and have a positive influence on decisions and actions taken by public institutions as well as on producer companies.

The greatest innovation offered by this approach and instrument besides its instrumental value, is its ability to open the minds of producers and make them aware that: the problem of waste "disappears" if complex relations are set up in which companies can become the nodes of a network along which skills, know-how, well-being, materials and energy can transit; an overhaul is made of everything that occurs upstream of the waste without delegating responsibility to other operators.

References

Bistagnino, L.: Design sistemico. Progettare la sostenibilità produttiva e ambientale in agricoltura, industria e comunità locali. Slow Food Editore, Bra, CN (2009)

Capra, F.: The Hidden Connections. Doubleday, New York (2004)

Ceppa, C.: Resources management tool to optimize company productivity. In: MOTSP 2010 - Management of Technology – Step to Sustainable Production, pp. 1–7. Faculty of Mechanical Engineering and Naval Architecture Zagreb, Croatia (2010)

Ceppa, C.: Software sistemico output-input. Unpublished doctoral dissertation, Politecnico di Torino, Italy (2009)

Ceppa, C., Campagnaro, C., Barbero, S., Fassio, F.: New Outputs policies and New connection: Reducing waste and adding value to outputs. In: Cipolla, C., Peruccio, P.P. (eds.) Changing the Change: Design Vision, Proposal and Tools, pp. 1–14. Allemandi, Turin (2008)

`http://www.ecocerved.it`

Ganapini, W.: La risorsa rifiuti, pp. 151–153. ETAS Libri, Milan (1985)

Lanzavecchia, C.: Il fare ecologico. Paravia, Turin (2000)

Rasetti, A.: Il filmato interattivo. Sperimentazioni. Time & Mind Press, Turin (2008)

Chapter 8
"Rework: Models and Metrics" An Experience Report at Thales Airborne Systems

Edmond Tonnellier and Olivier Terrien

Abstract. This experience report illustrates the implementation of metrics about rework in Complex Systems Design and Management at Thales Airborne Systems. It explains how models and quantification of rework contribute to reduce waste in development processes.

This paper is based upon our real experiences. Therefore, it could be considered as an industrial contribution to share this concept in Systems / Products Development and Engineering.

This experience report describes an achieved example of metrics and its positive impacts on Defense & Aeronautics Systems. However, this document proposes a scientific way to implement our results in other companies. Clues are also given to apply this methodology through a Lean Engineering approach.

Finally, key factors of success and upcoming steps are described to further improve the performance of Complex Systems Design and Management.

1 Context

Thales Airborne Systems is a company within the Thales group.

Edmond Tonnellier
2 Avenue Gay Lussac, 78851 Elancourt Cedex, France
edmond.tonnellier@fr.thalesgroup.com
+33 (0) 1 34 81 99 54

Olivier Terrien
2 Avenue Gay Lussac, 78851 Elancourt Cedex, France
+33 (0) 1 34 81 75 21
olivier.terrien@fr.thalesgroup.com

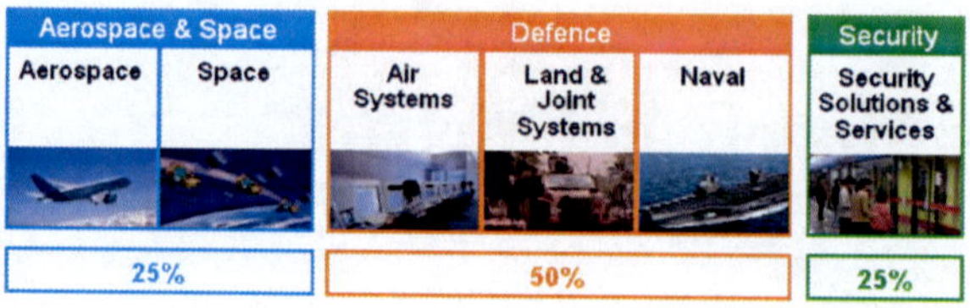

Thales Airborne Systems is currently the European leader and the third player worldwide on the market of airborne and naval defense equipment and systems. The company designs, develops and produces solutions at the cutting edge of technology for platforms as varied as fighter aircrafts, drones, surveillance aircrafts as well as ships and submarines.

About 20 % of the turnover is dedicated to R&D activities with a large proportion of it for systems engineering. With facilities in many European countries, the company employs numerous highly-qualified experts to design the solutions matching increasingly complex customer requirements as closely as possible. Benefiting from traditional positions as leader in the electronic intelligence and surveillance systems, recognized know-how in radar for fighter aircraft and electronic warfare, Thales Airborne Systems involves its customers from the solution definition phase up to the operational validation of the products. This requires high reactivity to changes in specifications and an ongoing effort to reduce development cycles. In addition, international competition and the effect of the euro/dollar conversion result in major budget constraints (in recurring and non-recurring costs).

The complexity of the systems developed within Thales is due to the number of components involved, the numerous technologies used and the volume of the embedded software. Due to the systematic combination of technical and non-technical constraints in new contracts, the technical teams must synchronize more and more precisely, skills from complementary disciplines whose contributors are often based in several countries. Lastly, the accelerated pace of the developments requires detailed definitions as well as optimum reactivity to the events or defects encountered, inevitable consequences of faster changes in requirements and markets.

In this context, in 2010 Thales Airborne Systems launched an initiative on “models and metrics of rework in its development processes”. Resolutely practical and pragmatic, this approach has observed, assessed, limited and then reduced the impact and frequency of the phenomenon in the engineering processes, in particular systems. This document describes the theoretical approach of modelling and quantification as well as the illustration of its deployment within technical teams.

2 Rework Problem

This chapter evokes the origin and the scope of the initiative on the rework led by the authors of the present document.

2.1 Surveys and Benchmarks

The literature on the rework mainly comes from the production (in particular in Lean Manufacturing [1]) and in software development (for example in the agile methods [2]). Several benchmarks (with other aeronautical or electronic companies, for example) directed our study of the rework in the processes of development. For most of the improvement initiatives studied, the starting point for rework initiatives is the awareness of a major and sudden shift in costs and delays on one or more projects. However, the diversity of the consequences and the multitude of the causes of these shifts lead to confusion on this waste and misunderstandings on the notions it covers. Hence our need to specify the terms used and to illustrate this waste using a methodology applied to our fields of activity.

2.2 Diagnosis

The numerous interviews conducted with technical leaders and with project managers allowed us to draw up a diagnosis on a phenomenon that is always too important. Here are some reports below:

- Similarity in effects of the phenomenon, irrespective of the discipline involved (hardware, software and system engineering);
- Disparity between causes of the phenomenon depending on the entities involved (influence of organizations, difference in profiles, etc) and depending on the disciplines concerned (differences in the life cycles according to the constituents of a complex system);
- Existence of a visible/ observable part and a hidden / latent part;
- Collective as well as individual contributions to the phenomenon.

> *"how can we explain this cost shift on our project X?", "why has our schedule been delayed?", "what caused this extra cost at the end of our project Y?".*
>
> *(Extracts of external benchmarks to illustrate the occurrence of rework)*

2.3 Stakes

To manage our initiative efficiently, we have determined the in and out of scope of our approach. Here are some details below:

- Define a behavioral model of a general phenomenon despite local specificities;
- Define a method to assess the scale of the phenomenon (relative assessment, if possible absolute);
- Separate the technical causes (usually related to the disciplines and the products) from the non-technical causes (usually related to the organizations, profits of the players, human behaviors);
- Involve all players to reduce/limit the phenomenon.

2.4 Definition of Rework

The definition of the rework formulated in the work of P. Crosby, 'Quality is Free' [3], answered our diagnosis. Our initiative retained:

Rework: "work done to correct defects"

Defect: "failure to conform to requirements"
(even if this requirement has not been explicitly specified).

These definitions are compatible with notions of rework applied in production and wastes identified in the approach Lean Engineering [4].

Examples :

"Incomplete or misinterpreted requirements at the start of a project resulting in rework in cascade through to subcontractors", "Late changes in requirements cause high levels of rework throughout the life cycle of products", "Low defined designs result in expensive reworks to meet the customers' true requirement".

(Extracts of interviews to illustrate the rework in development of systems)

Counter-examples :

1) "Multiple reconfigurations of a test bench due to numerous changes of the type of products tested create additional work. This is not due to a defect but to work organization." ('task-switching' type waste)

2) "A new manager forces his team to use a control system more like the one applied in his previous jobs. The extra work involved in updating is not due to a defect but to a change of method or to relearning team work." ('hands-off' type waste).

(Extracts of interviews to illustrate of 'false-friends' of rework)

The rest of the present document summarizes the scientific approach which we followed to answer the problem of the rework within the scope of the previous definition.

3 Rework Model

This chapter formalizes the behavior of the rework phenomenon: it describes a model and confronts it with the observations described in the presented diagnosis.

3.1 Behavioral Description

The prefix of the word ' re-work ' indicates a looped phenomenon. When developing solutions, the relation between 'work' and 'rework' plays a central role in the generation of delays, extra costs and risks introduced in cascade in the process' downstream activities.

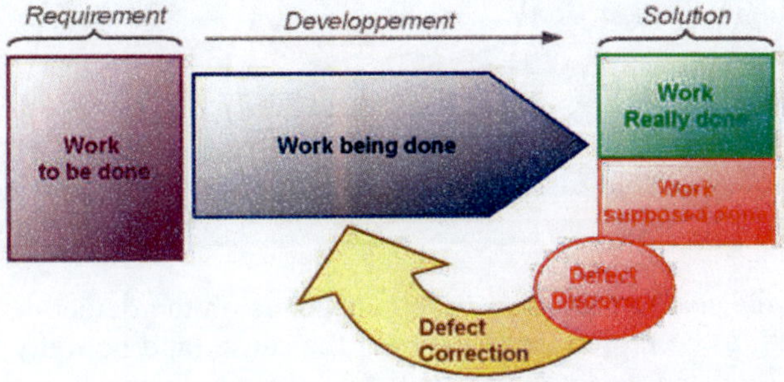

The model proposed, a loop associating work and rework, corresponds to the defect correction process (from discovery of the problem to its complete and definitive resolution [5]).

3.2 Defect Correction Process

The behavioral model of the rework makes the measurable part of the phenomenon visible: the additional cost associated with this waste. Without a correction of the defect, the rework would be null and the defect would cost nothing (apart from dissatisfaction). However, the need to bring the product into conformity requires correction and therefore generates an extra cost so that the solution complies with the requirement.

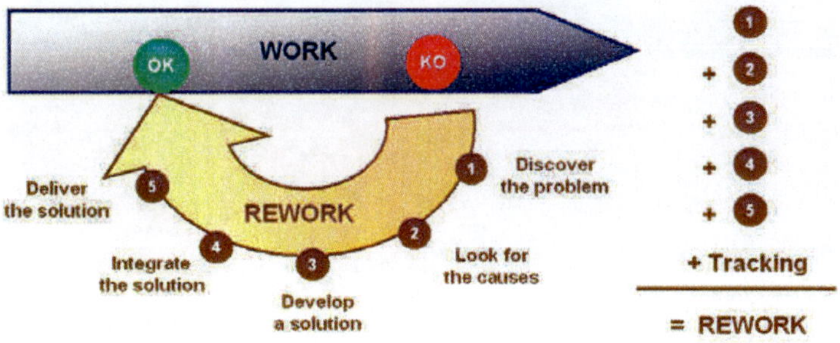

To a first approximation, the measurable part of the rework corresponds to the accumulation of all the impacts of a defect on the costs and/or the delays of a project (impacts made visible by each step of the defect correction process).

3.3 Deduction of Rework

The rework generated by a defect results from the second journey of the phases of a process of development. This journey is defined during a causal analysis led at the discovery of a defect.

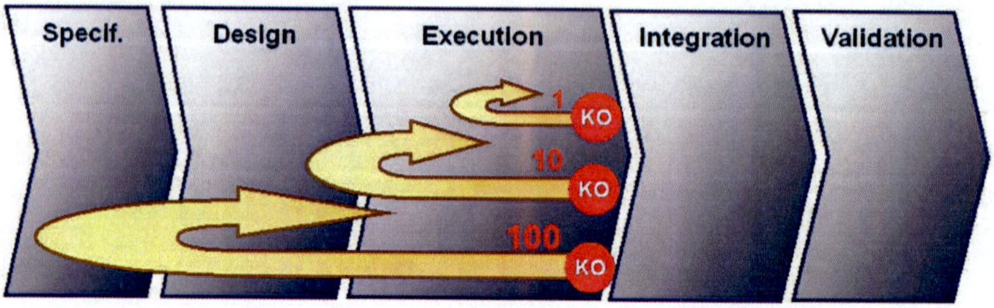

Therefore, the scale of the rework loop depends on the defect discovery phase and on the path to be reproduced to correct the cause (and not only to reduce the effect): *"The later the defect is detected, the larger the impact on the delays and costs of a project: correction within a phase will generate an extra cost of 1 whereas correction of this defect between phases will have an impact of 10 or even 100 on the entire project"*. The faster the second journey, the lower the disturbance on the project: the reactivity of the correction process represents a lever to limit the phenomenon.

3.4 Induction of Rework

"sub-systems partially validated", "choice of immature technologies", "solutions not tested" are defects heard during our interviews. Introduced throughout a development process, they behave as real rework inductors.

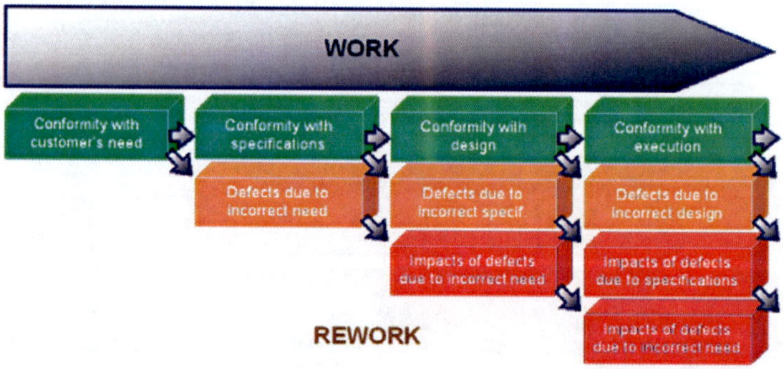

The rework appears from then on as a risk more than a random phenomenon! (Counter) reactions are thus possible to reduce and/or limit its impact, its probability of occurrence or its non-detection. The sooner the defect is detected, the lower its consequences on the project results.

In conclusion, the proposed model supports the fact: "The more defined the engineering upstream, the lower the downstream rework".

4 Quantification of Rework

This chapter implements the model through a mathematical formulation.

4.1 Correction Process Translated into Data

Modelling of rework led to the following observation: "rework is the accumulation of all the impacts of each step in the defect correction process". Without corrective action, no extra cost therefore no rework. The proportion of rework compared with the total work can then be evaluated by observing the impacts of a defect, impacts resulting in additional costs and/or delays on a project.

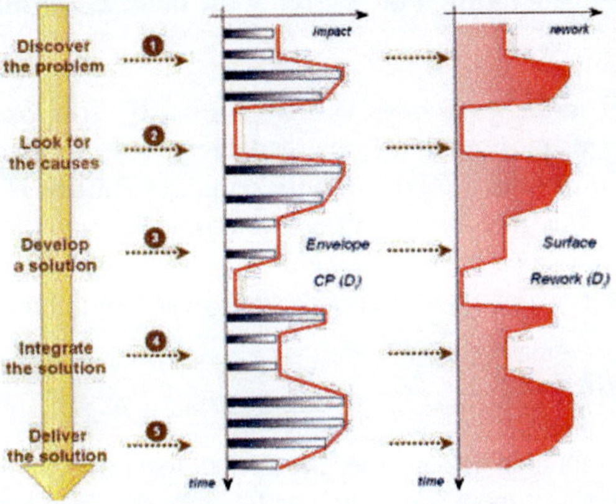

Each step of the correction process is associated with an extra cost or a delay. To a first approximation of the envelope of impacts, the correct process CP of a defect D_i forms a mathematical function CP = f (D_i).

4.2 Mathematical Modeling

If additional costs recorded throughout the process of correction of the defect which generated it forms a function, their accumulation corresponds to a surface enveloped by the function CP. In mathematics, a surface leads to an integral:

$$\text{Rework}(D_i) = \int_{t_{start}}^{t_{stop}} CP(D_i, t)\,.\,dt$$

Advantages

This mathematical model can be easily transpose into an IT code. The formula is an expectation for its implementation in any tool of quantification (any statistical tool, queries in databases, etc). Independent of the measurement units (costs, delays), it can be applied to any organization (only the correction process of an entity is related to the model). Lastly, the model measures $CP(D_i)$ and not directly D_i. Therefore, it measures effects and not causes of the defect, which allows its use in all levels in a complex system (and in any engineering disciple).

User cases

To determine the rework of a particular phase of a V cycle means replacing the t_{start} and t_{stop} in the integral by the dates of the studied phase: the cumulated data comes from a narrower but still homogeneous source or period. Moreover, the value units (€ or $ but also hour, day or week) can be changed to correspond to the local organizations. Lastly, a change in the correction process of an entity does not disturb the model since it evaluates the scale of the phenomenon (i.e. extra cost) and not the way of generating it or resolving the detected problem.

Limitations

The mathematical model possesses its own limits. If it imposes the availability of inputs (such as dates of events, origins of defects, etc), it does not guarantee their reliability. In addition, the model can only compare entities if their defects share a standardized classification. Lastly, since the model is based on a defect correction process, it does not take into account defects that are not recorded, amount of waste before defect discovery or dependencies between defects.

4.3 Data Availability

In aeronautical companies as for other sectors, defects and their resolution need traceability. Defect management tools such as ClearQuest [6] facilitate the recording of events, key parameters or information relative to the process of correction of a defect.

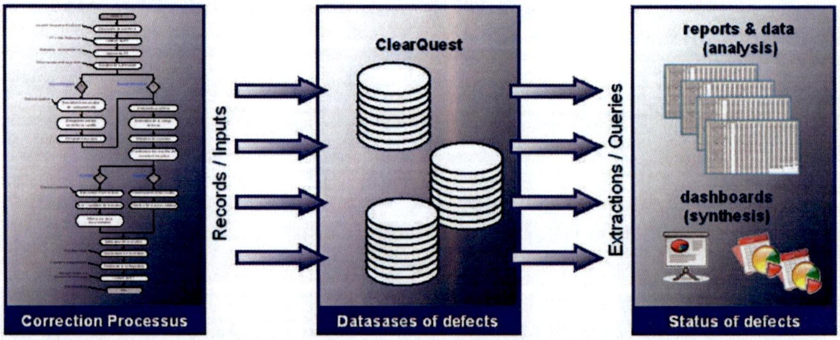

Nota : If for every defect D_i, the recordings insure the existence of data to quantify the rework, they do not insure their reliability, their accessibility or their compatibility. Therefore the confidence level of the calculated quantification depends on the quality of the sources.

Interview of an american expert: "In God, I trust. With rework, I need data!"

In the literature, the rework is usually quantified by a formula $y = f(x_1, x_2, \text{etc})$. Our paper proposes a simpler formulation by introducing an integral $y = \int g(x)$. This mathematical solution reduces the number of variables to be managed. Easy to implement, this integral formula demonstrates that the rework is measurable!

5 Metrics of Rework

This chapter evokes two approaches to exploit the model and the quantification of the rework: a capitalization to prepare a new project and a permanent implication of staff to improve project in progress.

5.1 *Capitalization of a Process*

A virtuous circle must oppose the vicious circle of rework. During the preparation of a new project, a relevant source of information shows to be the analysis of the processes of correction of the defects of previous projects. Applied to the available recordings, the team exploits the presented model to assess the rework and to evaluate the performance of the current process of correction.

Relevance of the correction process

Having determined the status of available data (reliability of the records, the maturity of information fields, etc.), the team guarantees the representativeness of the sample of data for a statistical analysis.

Analysis of 'volume of defects per month' (extract from project 5.A) :

The model filters the monthly data, according to the dates of creation or closing of the correction processes.

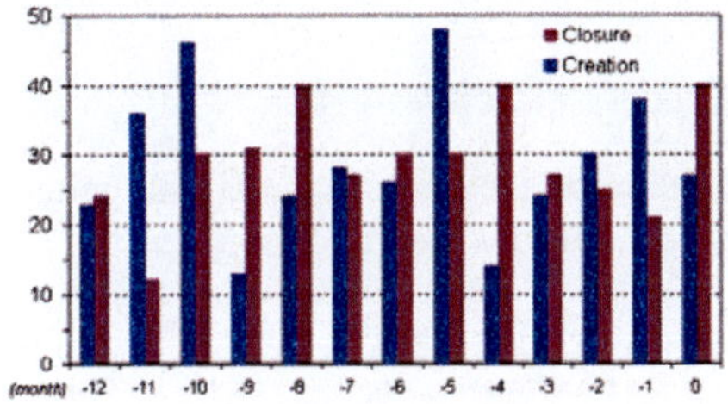

Root Causes of Defects

After establishing groups of the most frequent or the most penalizing defects, the team defines categories and parameters which are the scales of rework in the ended projects: they are the inductors of rework.

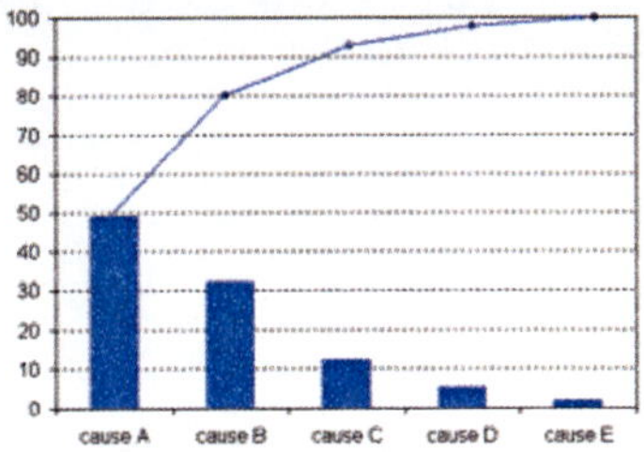

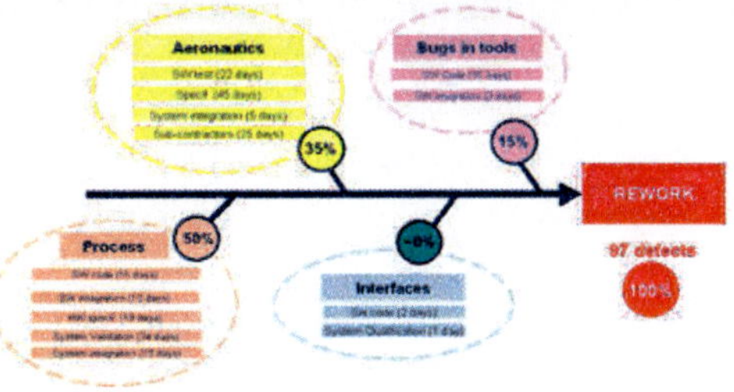

During the new project, when a defect occurs, the team characterizes it and, thanks to the categories and parameters previously identified, determines the risk of potential rework for the project. Charged with costs and delays, the defects are included into the management of the project. They are the indicators of rework. The team assesses the possible disturbance due to new defects and defines its priorities according to its capacities of reaction.

In conclusion, the project integrates the phenomenon of rework through its management decisions and allows the team to anticipate possible difficulties.

Control indicator of corrected defects (extract from project 5.B) :

The model accumulates workloads associated with defects (consumed and estimated) to obtain usual S-shaped curves of project management.

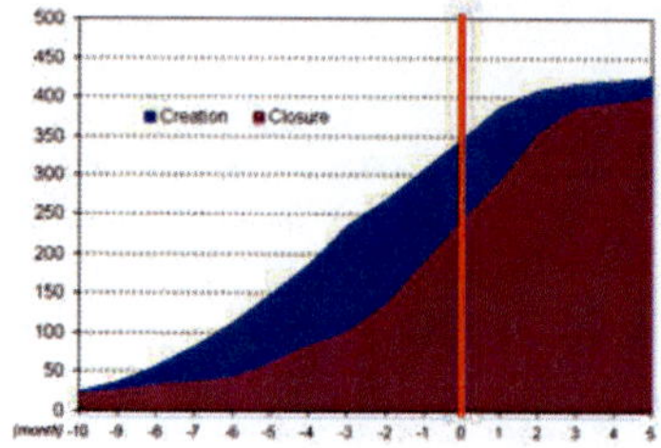

5.2 *Improvement of Processes*

To favor the implication of staffs in the reduction of waste generated by defects, it is essential to associate them with improvement workshops launched locally. By using the proposed quantification, the team measures the influence of their improvements before/after the workshop.

Process Effectiveness

Databases (restricted to a homogeneous perimeter: one product type or a given project) allow the team to assess the real performances of its correction process.

Correction Process:	Discovery	> Creation	> Analyze	> Decision	> Delivery	> Closure	
Defects (80%):	To	+ 0	+ 58	+ 75	+ 94	+ 105	*(days)*
Defects (Average):	To	+ 4	+ 63	+ 80	+ 100	+ 120	

(before)

Observations of the ground offer numerous levers of improvement (e.g. how dynamic is the sequence of process steps to solve a problem). A causal analysis identifies the root causes of weaknesses in the process and suitable countermeasures are implemented by the local team. A regular check with the quantification tool sustains the change.

Correction Process:	Discovery	> Creation	> Analyze	> Decision	> Delivery	> Closure	
Defects (80%):	To	+ 0	+ 16	+ 22	+ 46	+ 51	*(days)*
Defects (Average):	To	+ 1	+ 34	+ 48	+ 83	+ 88	

(after)

Process Efficiency

Similarly, the model can be used to determine the efficiency of the development process by comparing the total workload and the proportion of rework (for example by comparing on a given project and for a given period of time the hours of work and the hours spent to correct defects). To make positive impacts visible involves the team: either by simulating new actions or by pursuing the improvement workshops.

'Rework vs Work' indicator (extract from project 5.D) :

The model compares periodically global workloads with workloads associated with rework. To make wastes visible involve teams.

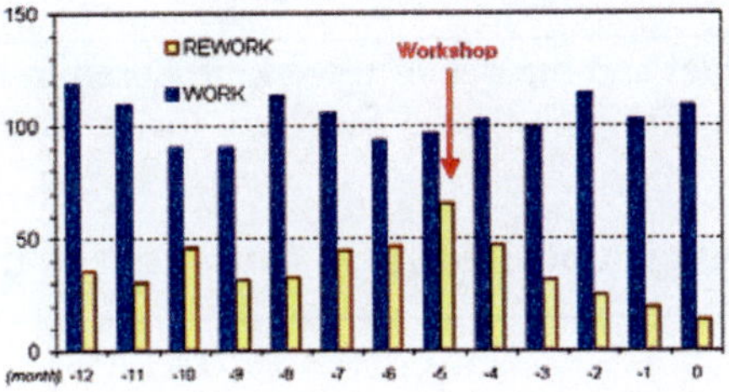

6 Feedback

6.1 Contributions for System Engineering

The model proposed in this document was deployed during local workshops. The teams involved gave feedback on several contributions of our initiative:

- *"Rework is a risk, not a random event! It is not fate!"*
- *"The model describes a behavior and provides keys to observe it. The rework is part of project management"*
- *"The mathematical model allows sharing between entities, avoiding the disadvantage of specific and/or local tools"*
- *"Measures of rework are directly performed by local teams; however, training can be carried out based on a global and common approach"*

6.2 Methodologies

About Systems methodologies (or close), here is some feedback :

- *"governance by facts and the use of existing databases"*
- *"a generic methodology of rework, common to every level of complex systems, applicable to any development in progress"*
- *"An implication of the technical teams through local workshops; an effect confirm by individual but also collective efforts"*
- *"an approach making waste visible to obtain concrete and efficient improvement levers; easy to implement"*

6.3 Deployments

The scientific approach described in the present document as well as the examples to deploy the model and the quantification of rework are applicable in other companies. Here are some clues:

- implementation via the personal initiative of a local manager convinced by this approach (a bottom-up workshop applied on defects of a local project)
- implementation via the major processes described in the Company Reference System (a top-down deployment on defect correction processes)
- implementation via a Lean Engineering approach (identification and elimination of waste in engineering processes [8]).

In conclusion, the model and metrics of rework proposed in this document can be applied to all the engineering processes necessary for the Design of Complex Systems.

"Rework is not 'bad luck' but a risk to manage".

Authors

Edmond TONNELLIER
System Engineering Expert
With many experiences dedicated to the development of complex products, Edmond Tonnellier is currently the System Engineering Expert for Thales Airborne Systems and contributes to the Thales Group System methodology ('SysEM Handbook' and 'Corporate Guide -Value Analysis'). CMMI SEI DEV and ACQ evaluator, he conducts numerous audits inside and outside the Thales group. In addition to his role as SE reference, he contributes to numerous IS training courses at Thales University
AFIS and INCOSE member (CSEP certified, BKCASE reader)
Engineer (Post grad degree), 1977, ISEP, France

Olivier TERRIEN
Lean Engineering Expert
Olivier Terrien is a Thales Lean Expert and is the Thales Airborne System reference for Lean Engineering. He has implemented numerous process improvement workshops based on Lean Manufacturing and/or Lean Engineering approaches (in systems engineering, software development and customer commitment). His background is in the engineering processes (design of microwave components, development of electronic warfare receivers, integration of naval radar systems and airborne electronic warfare suites). He has published more than 250 pages in the worldwide Press.
Engineer (Post grad degree), 1997, ESEO, France
MBA, 2006, IAE Paris-La Sorbonne, France

Acknowledgements. To Harry Gilles, Roger Martin and Michel Baujard for their valuable support during this initiative.

References

[1] Womack, J.P., Jones, D.T.: Lean thinking (1996)
[2] Poppendieck, T., Mary: Lean Software Devpt, Agile toolkit (2003)
[3] Crosby, P.B.: Quality is Free (1979)
[4] Terrien, O.: Incose 2010: The Lean Journey (2010), http://www.thalesgroup.com/Lean_Engineering_Incose2010.aspx
[5] Cooper, K.G., Mullen, T.W.: The Swords & Plowshoes. The Rework Cycles of Defense Software (2007)
[6] Gilles, H.: Thales, Problem & Change Report Management with ClearQuest Tool (2007)
[7] Crosstalk: The journal of defense software engineering 16/3 (2003)
[8] Oppenheim, B.: « Incose 2010, WG Lean for SE (2010), http://cse.lmu.edu/about/graduateeducation/systemsengineering/INCOSE.htm

Chapter 9
Proposal for an Integrated Case Based Project Planning

Thierry Coudert, Elise Vareilles, Laurent Geneste, Michel Aldanondo, and Joël Abeille

Abstract. In previous works [14], models and processes enabling to make interacting a project planning process with the system design one have been presented. Based on the idea that planning and design processes can be guided by consulting past experiences, an integrated case-based approach for coupled planning project and system design processes is proposed in this article. The proposal is firstly based on an ontology of concepts that permits to gather and capitalize knowledge about objects to design, i.e. tasks and systems. Secondly, the integrated case-based process is presented taking into account planning of tasks and tasks of design. For the retrieve task, a compatibility measure between requirements and past cases is calculated. It is based on the semantic similarity between past and required concepts as well as past solutions and new requirements. Finally, integrated adaptation and retention of new solution(s) are done.

1 Introduction

Within a competitive context, system design processes have to interact closely with all other business processes within a company. Based on the idea that the planning of a design process and the design process itself can be guided by consulting past experiences, an integrated case-based approach for coupled planning project and system design processes is proposed. In [14], models and processes enabling to make interacting a project planning process (project of design) with the system design one

Thierry Coudert · Laurent Geneste · Joël Abeille
University of Toulouse, ENIT, LGP, 65016 Tarbes
e-mail: {Thierry.Coudert,Laurent.Geneste}@enit.fr

Elise Vareilles · Michel Aldanondo · Joël Abeille
University of Toulouse, EMAC, CGI, 81000 Albi
e-mail: {Elise.Vareilles,Michel.ALdanondo}@mines-albi.fr

Joël Abeille
Pulsar Innovation, 31000 Toulouse
e-mail: joel.abeille@pulsarinnovation.com

have been presented considering an information viewpoint (part of the french ANR ATLAS Project). In this article, the coupling between both domains is considered taking a knowledge viewpoint. An integrated case-based approach which permits to reuse contextualized knowledge embedded into experiences for solving new design problems (planning tasks of design and design a system) is proposed.

2 Background and Problematic

Design can be seen as: i) a project that aims at creating a new object or transforming an existing one, ii) a knowledge discovery process in which information and knowledge are shared and processed simultaneously by a team of designers involved in the life phases of a product. Among existing design methodologies (see [15, 1, 5]), Axiomatic Design (AD) proposed by Suh [15] is a top down and iterative approach which links requirements or functions to be fulfilled to technical solutions (design parameters and process variables). Pahl and Beitz [1] describe a systematic approach for all design and development tasks. The design process is composed of the following four sequential stages guiding the design of a product from scratch to full specification: requirements clarification (it defines customer or stakeholders requirements), conceptual, embodiment and detailed design (activities that serve to develop products or systems gradually). Taking the viewpoint of System Engineering standards, the product lifecycle is taken into account. In ISO/IEC 15288 standard, activities are: stakeholder requirements definition and analysis, architectural design, implementation, integration, verification, transition, validation, operation, maintenance, and disposal. In INCOSE SE Handbook 3.2, the activities are: Conceptualize, Develop, Transition, Operate, Maintain or Enhance, Replace or Dismantle. These structured activities have to be planned within a project management framework. For the proposal, a simple compatible process is used. It gathers both activities: requirements clarification and solutions developments. In previous works, integrated models and processes have been proposed [14] where each information about a system is linked with a dedicated task within a project. A system is composed of Requirements and of Solutions. A system is developed within its project. The project gather a System Requirements Search task and some Alternative Development tasks (one to each solution).

The Project Time Management (PTM) process defined by the Project Management Institute [6] describes six activities: 1) identification and 2) sequencing of activities, estimation of 3) resources and 4) durations, 5) scheduling of activities, and 6) control of the execution of the schedule. In order to guide design tasks as well as planning tasks of a project, the use of contextual knowledge embedded within experiences is a current and natural practice. In such a context, methodologies enabling to manage these experiences are required. From a new design problem, previous experiences or cases have to be identified in order to be reused. CBR methodologies are well suited for that.

Classical CBR methodologies [7, 2] are based on four tasks: Retrieve, Reuse, Revise and Retain. A fifth steps can be added for problem representation as in the proposal.

CBR has been widely used for system design in many domains (see for instance [11, 8, 9, 3]). Generally, these methods are based on requirements which constitute the new problem to solve. Similar cases are retrieved from a case base which contains some design solutions. Solutions are then reused, revised and retained. In project planning domain, some CBR studies exist also. For instance, in [12], cases which describe software projects are stored and a CBR approach retrieves some prior projects, offering to decision makers to adapt or duplicate them to fulfill a new project planning problem. In [10], a project planning method which is based on experiences and which minimizes the adaptation effort is described.

To carry out efficient CBR process requires a well structured knowledge as well as clear semantic in order to be able to identify prior solutions to new problems. Within a distributed context, designers teams, project planners and managers have to do rapidly their activities with common understandings. So, a semantic interoperability is important to maintain in order to be able to define unambiguous and comprehensive requirements and develop them. In this area, concepts and ontologies bring some interesting solutions [4]. A concept is an organized and structured understanding of a part of the world. It is used in order to think and communicate about this part. The set of attributes or descriptors of a concept (its intention representation) is determined by identifying common properties of individuals which are represented by the concept. Based on concepts, an ontology is an organized and structured conceptualization or interpretation of a particular domain. Then, a set of concepts, gathered into an ontology is a set of terms shared by an expert community in order to conceive a consensual semantic information.

However, there is a lack of integrated processes which take into account simultaneously project planning and system design referring to prior cases or experiences. Therefore, the proposal of this article is concerned with an integrated case-based approach using an ontology of concepts for guiding project planning and system design processes. Firstly, the proposed ontology is described. Then, in section 4, the integrated process that mixes planning of projects tasks and design tasks is defined. The object formalization is proposed in section 4.2. Finally, the case based process is defined in section 5.

3 Ontology for Project Planning and System Design

The proposed ontology is represented by means of a hierarchical structure of concepts. The root of the ontology is the most general concept called `Universal`. The more a concept is far from the root, the more specialized. Concepts are linked by arcs which represent a relation of generalization/specialization. Any concept inherits all the characteristics from its parent. Ontologies are built and maintained by experts of domains. A concept is firstly described by means of a definition or semantic. It is generally sufficient to promote semantic interoperability between teams of designers but not for carrying out an integrated case-based process. Therefore, knowledge about domain is integrated into the ontology for reuse. This knowledge embedded into a concept C is formalized by:

- a set (noted ϑ_C) of **models of conceptual variables**. They can be seen as descriptors of the concept;
- a set (noted c) of **models of domains** for models of variables. A model of domain represents the authorized values of a model of variable. For instance, for a discrete model of variable called *COLOR*, its model of domain can be $\{blue, white, red\}$ i.e., the set of values which have a sense for *COLOR*;
- a set (noted Σ_C) of **models of conceptual constraints** related to some models of variables. A model of constraint links many models of variables giving some authorized (or forbidden) combinations of values;
- a set (noted Ω_C) of **similarity functions**: each model of variable is associated with a function which gives the similarity between two values of the model of a variable([2]). If a model of variable is discrete, a matrix permits to gather the similarities between discrete possible values. Placing Σ_C within the concept permits to specialize similarity functions following models of variable of the concept (for instance, the similarity between two *axis radius* models of variable can be defined with different manner for a `Clock` concept and for a `Motor` concept).

For a concept, some models are inherited from its ancestors and other ones are specific. A concept can be used in order to firstly characterize a system or a solution. Immediately under the `Universal` concept, two concepts appear: `System` and `Task`. The `System` concept is then specialized into different categories of systems. The use of these concepts is described into the next section. An example of ontology is represented in figure 1.

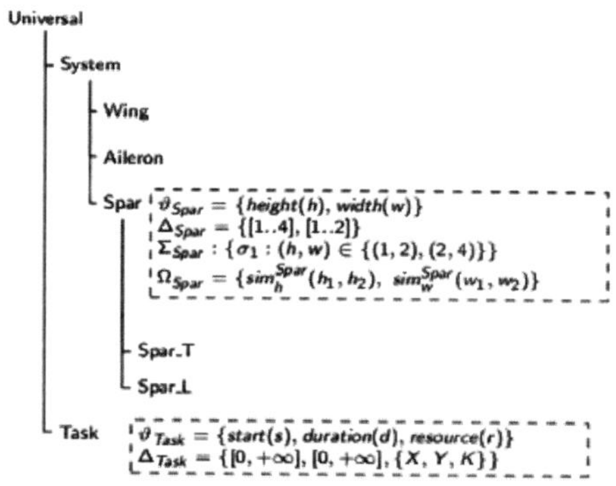

Fig. 1 Example of ontology

4 Integrated Project Planning and Design Process

The proposed integrated process mixes tasks defined by PMI and design tasks composed of requirement definition task and solution definition tasks for a system. The aim is to define, sequence and schedule tasks before launching and controlling them. The difference to a classic approach for project planning is that tasks are defined, sequenced and scheduled only when it is required using available and certain knowledge. When a task is defined, then it is possible to begin immediately its realization even later ones are not totally defined, planned or known. The process is based on `Planning` tasks. A `Planning` task gathers the identification of tasks, the sequencing of tasks, the estimation of resources and durations, the scheduling of tasks. Therefore, any system design task is preceded by a `Planning` task. The outcomes is concerned with a schedule of the identified task(s), its (their) duration(s) and the resources to be used. The `Planning` task is specialized into `Planning of System Requirements Search` task (noted PSRS) and `Planning of Alternative Development` task (PAD). System design tasks are then composed of one `System Requirement Search` task (SRS) followed by one or more `Alternative Development` tasks (AD) which can be realized in parallel.

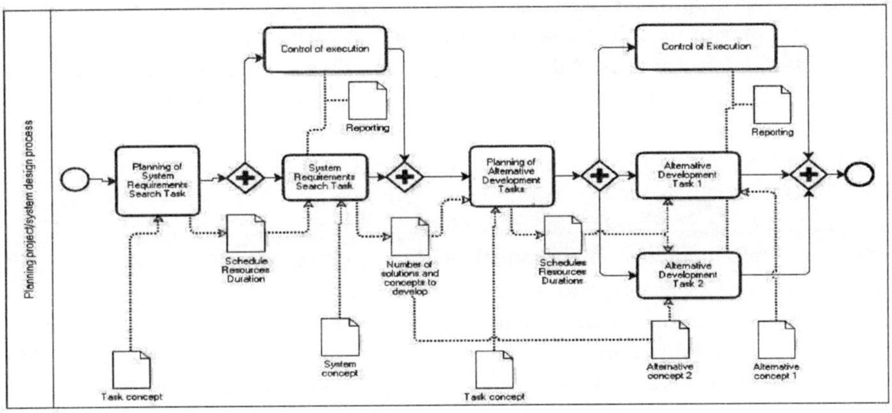

Fig. 2 Integrated Project planning and System Design Process

4.1 Integrated Process Description

The process begin by the PSRS task. The System Requirements Search task is defined using a `Task` concept coming from the ontology. All the variables required for its definition are then copied (resources, durations, release and due dates, etc.) as well as conceptual task constraints from models of the `Task` concept. After planning, according to the schedule, the SRS task is carried out in parallel with a Control of Execution task (*CE* task). During the SRS task, the designer has to choose a specialization of the high level `System` concept corresponding to the system to design and noted System Concept (*SC*). It permits to formalize requirements using copies of conceptual variable models as well as conceptual constraint models which are proper to this concept. The designer can add other variables in order to better characterize the system requirements. It can also add other specific constraints on variables as required. The goal is to formalize all the requirements by means of constraints. This formalization is aided by the ontology. Choosing a concept of the ontology, the designer is guided by: conceptual variables, their domain and conceptual constraints. He can be concentrated only to customer needs formalization defining the appropriate requirement constraints of the system. Finally, the designer has to define, in accordance with the planning manager, the number of solutions to develop as well as the concept associated to each solution. The concept associated to a solution corresponding to a system is either the System Concept *SC* itself or a specialization of *SC*. That permits to define several solutions for a same system. For instance, a `Spar` concept can be specialized into `Spar_L` and `Spar_T` concepts according to their shape. A wing spar system to develop following two solutions will be associated to the `Spar` concept and its solutions respectively to `Spar_L` and `Spar_T` concepts.

The second part of the integrated process is concerned with the Planning of Alternative Development tasks and their execution. When a PAD task is carried out, n Alternative Development tasks (AD tasks) are planned. Each AD task is associated to a `Task` concept in order to guide the definition of its characteristics by means of variables and constraints. All the AD tasks are then scheduled, the required resources are affected and durations are stated. Based on this schedule, each task is carried out and a CE task controls their execution using reporting. The execution of one AD task by a designer (or a designers team) consists in defining one solution that fulfills the requirements (one solution is called *System Alternative*). Firstly, each system alternative *SA* is associated to its own concept *AC* (e.g. $AC =$ `Spar_L`). Conceptual variable models, domain models and conceptual constraint models coming from the concept *AC* are copied into the system alternative. The designer in charge to develop this solution can add variables (with their domains) as well as new constraints corresponding to his own solution. Therefore, the formalization of a viable solution consists in fixing values for variables which satisfy all the constraints or to reduce to a singleton the domain of each variable. Clearly, during execution of tasks and when problems occur, it is possible to ask for re-planning in order to change dates, durations, resources, etc.

4.2 Formal Description of Objects

A system S is associated with Requirements R, a System Concept SC and n System Alternatives SA_i such as $i \in \{1,2,...,n\}$.

1) Requirements: R associated to the system S are described by the following sets:

- V^S: the set of System Variables which characterize the system S to design. It is the union of copies of conceptual variables models (V^S_{SC}) coming from concept SC and of added variables V^S_{add} such as $V^S = V^S_{SC} \cup V^S_{add}$;
- σ^S: the set of System Constraints which formalize the requirements dedicated to the system S. These constraints affect some variables of V^S. It is the union of copies of conceptual constraint models (σ^S_{SC}) coming from the concept SC and of requirements constraints σ^S_{Req} defined in accordance with the customer needs for the system S such as $\sigma^S = \sigma^S_{SC} \cup \sigma^S_{Req}$;

- d^S: the set of domains of variables. Each variable of the set V^S is associated with its own domain. d^S is the union of copies of conceptual variable domain models and of variables domains added by the designer such as $d^S = d^S_{SC} \cup d^S_{add}$. Note that the domain of an added variable is defined by the designer itself (for instance $\mathbb{R}^+$ for an added variable r corresponding to a radius).

2) System Alternatives: all the variables, the domains and the constraints corresponding to the requirements R of the system S are copied into each system alternative. That permits to develop each solution independently. Therefore, a System Alternative *SA* is described by means of the following sets:

- V^{SA}: the set of alternative variables. It is the union of:
  - the set of copies of each system variable (noted $\hat{V}^S$);
  - the set (V^{SA}_{AC}) of copies of conceptual variable models coming from the alternative concept *AC*;
  - the set V^{SA}_{add} of variables added by the designer in order to better characterize the system alternative *SA*;

  such as $V^{SA} = \hat{V}^S \cup V^{SA}_{AC} \cup V^{SA}_{add}$ and $V^{SA} = \{v^{SA}_i\}$,
- σ^{SA}: the set of alternative constraints. It is the union of:
  - the set of copies of each system constraint (noted $\hat{\sigma}^S$);
  - the set of copies of conceptual constraint models (σ^{SA}_{AC}) coming from the alternative concept *AC*;
  - the set of constraints added by the designer in order to define new requirements derived from the solution itself (e.g. the choice of a technology for a solution leads to add a new constraint about security);

  such as $\sigma^{SA} = \hat{\sigma}^S \cup \sigma^{SA}_{AC} \cup \sigma^{SA}_{add}$ and $\sigma^{SA} = \{\sigma^{SA}_i\}$,
- d^{SA}: the set of alternative variable domains. It is the union of:
  - the set of copies of each system variable domain (noted $\hat{d}^S$),
  - the set d^{SA}_{AC} of copies of models of domains for conceptual variables of the set V^{SA}_{AC},
  - the set d^{SA}_{add} of domains corresponding to variables V^{SA}_{add} added by the designer,

  such as $d^{SA} = \hat{d}^S \cup d^{SA}_{AC} \cup d^{SA}_{add}$ and $d^{SA} = \{d(v^{SA}_i)\}, \forall\, v^{SA}_i \in V^{SA}$.

When the design of a system alternative *SA* is finished, it is considered that the domain of each variable v^{SA}_i is reduced to a singleton. The set of these singletons is noted $Val(d^{SA})$.

3) Alternative Development Task: such a task (noted *AD*) is described by means of a set of variables (noted V^{AD}), a set of domains (d^{AD}) and a set of constraints related to these variables. Some variables, domains and constraints are conceptual ones (they are copies of models coming from the Task concept). Others ones are user defined and specific to the task. When task planning is finished, the domain of variables is reduced to a singleton (release and due dates, resources, durations, etc.).

5 Case Based Design and Project Planning Process

The standard process for designing a system is compared to the integrated process presented in the previous section and to the five standard tasks CBR process (see figure 3). Within the standard system design process, the first task corresponds to the requirements definition. With regard to the integrated process presented in section 4, it corresponds to the definition/planning of the task (PSRS) and its execution (SRS) (the control is not represented). Based on the CBR process presented in section 2, the following three sequential tasks for the requirements definition are defined: 1) Planning of SRS and RETRIEVE tasks followed by 2) the execution of the SRS task and finally, by 3) the execution of the RETRIEVE task. At the end of the RETRIEVE task, the requirements are formalized by means of constraints and some retrieved cases are available for potential reusing.

The second part of the standard system design process is concerned with the solutions development. Within the integrated process of section 4, the System Alternatives development tasks are firstly planned and then, executed following their schedule. The proposed case-based process carries out the REUSE and REVISE tasks of CBR for both. Firstly, schedules about system alternatives development tasks are reused (if possible) and then revised. The reuse of schedules provides Suggested Schedules and their revision provides Confirmed Schedules which can be carried out. Secondly, the information about suggested system alternatives are reused (if possible) and then, revised in order to be consistent and compatible with all the system requirements. Finally, the Confirmed Schedules and System Alternatives are Retained. This information constitutes a new learned case which is capitalized into the case base. In the proposed process, the REUSE tasks are entirely done by planning manager or designer teams without automatic adaptation. Either a copy of retrieved information is done, or an adaption to the new problem is realized. In the worst case, there are no compatible retrieved cases and solutions have to be developed from scratch. The REVISE task is also entirely done manually.

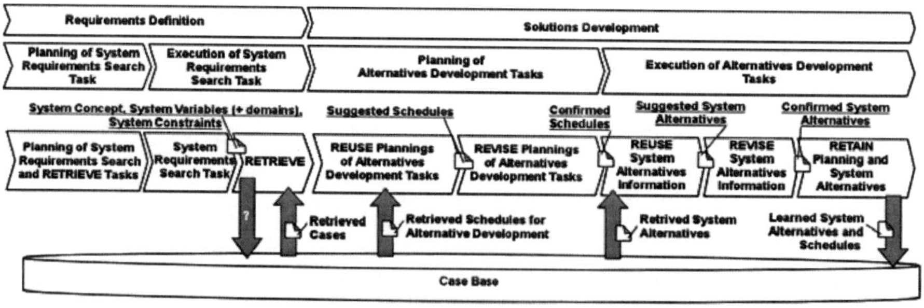

Fig. 3 Case Based Planning and design process

Case Representation. A case is considered as the association of an Alternative Development task with its System Alternative solution. A case gathers: V^{SA}, d^{SA}, σ^{SA}, V^{AD}, d^{AD}, σ^{AD} and its concept AC. Clearly, a case also embeds other information about design solution (files, models, plans, formulas, calculus, etc.).

RETRIEVE Task Description. The input of the RETRIEVE task is defined by: the System Concept SC, the set of System Variables V^S with their domains d^S and the set of System Constraints σ^S (see section 4.2). The proposal is based on the idea that only system requirements can be used for searching and not project planning ones. If the input is used strictly for defining the target case, the risk is to turn down not fully compatible source cases during retrieving phase even these cases could be very helpful to the designer after repair. So, the proposed method introduces flexibility in constraints. Therefore, the aim of the retrieve task is to identify more or less compatible source cases (system alternatives) with regard to requirements and to the system concept. For each ones, two measures indicating how much the source cases are compatible are provided. The identification of cases is done confronting: i) the value of each variable in system alternatives (source cases) to the constraints of the system S to design and ii) the target System Concept to the source Alternative Concepts. More the constraints are satisfied and more SC is near to AC, more the source case is compatible. The degree of constraint satisfaction is defined by a *compatibility function* which returns values between 0 and 1.

The method proposed in this article is restricted to discrete (numeric or symbolic) variables and constraints of σ^S are only discrete. Let σ_i be a constraint (such as $\sigma_i \in \sigma^S$) which defines the authorized associations of values of a set of n discrete variables noted X (such as $X \subseteq V^S$ and $X = \{v_1, v_2, .., v_n\}$). Let X_{autho} be a set of p authorized tuples for X such as $X_{autho} = \{(x_{11}, x_{21}, ..., x_{n1}), (x_{12}, x_{22}, ..., x_{n2}), ..., (x_{1p}, x_{2p}, ..., x_{np})\}$ with x_{ij} a discrete symbolic or numeric value of the variable v_i in the tuple j. The constraint σ_i is defined by $\sigma_i : X \in X_{autho}$. Let Y be a vector of m discrete variables and $Val(Y)$ the vector of their values. The compatibility function of $Val(Y)$ with regard to σ_i (noted $C_{\sigma_i}(Val(Y))$ is defined as follows:

$$C_{\sigma_i}(Val(Y)) = \begin{cases} 1 & \text{if } X \subset Y \text{ and } Val(X) \in X_{autho} \\ \max_j \left[\left(\sum_{i=0}^{n} \omega_i * \left[sim_{v_i}^{AC}(x_i, x_{ij}) \right]^{\beta} \right)^{1/\beta} \right] & \text{if } X \subset Y \text{ and } Val(X) \notin X_{autho} \\ 0 & \text{Otherwise} \end{cases}$$

ω_i: weight given to variable v_i of X such as $\sum_{i=0}^{n} \omega_i = 1$

x_i: value of variable v_i in $Val(Y)$

x_{ij}: authorized value of variable v_i in tuple j of X_{autho}

$sim_{v_i}^{AC}(x_i, x_{ij})$: similarity between x_i and x_{ij} for the variable v_i. This function is defined within the AC concept

β: parameter of the Minkowski aggregation function ([2])

If the set of variables Y includes all variables of X and if the corresponding values $Val(X) \subset Val(Y)$ are authorized by the constraint, then the compatibility is equal to 1. If Y includes all variables of X but their values are not authorized by the constraint, similarities between values are used. For each authorized tuple $X^j_{autho}/j \in \{1,2,...,p\}$, the similarities between values of variables in $Val(Y)$ and X^j_{autho} are evaluated and aggregated by means of a Minkowski function ([2]). Then, the compatibility is equal to the maximum of the aggregated values calculated on the p tuples of X_{autho}. β permits to tune the aggregation: $\beta = 1 \Leftrightarrow$ weighted average, $\beta = 2 \Leftrightarrow$ weighted Euclidean distance, $\beta \rightarrow \infty \Leftrightarrow$ Maximum. If it exists some variables of X not included into Y, the compatibility is equal to 0.

Let SA_k be a System Alternative from the source cases. The compatibility of SA_k with regard to a constraint σ_i (noted $C^{SA_k}_{\sigma_i}$) is given by: $C^{SA_k}_{\sigma_i} = C_{\sigma_i}(Val(d^{SA_k}))$ with $Val(d^{SA_k})$, the vector of values of variables of V^{SA_k} (see section 4.2). When the compatibility of $Val(d^{SA_k})$ has been calculated for each constraint of σ^S, an aggregation has to be performed in order to provide the compatibility of SA_k with regard to requirements (noted $C^{SA_k}_{\sigma^S}$). The compatibilities with regard to each constraint of σ^S are aggregated using a Minkowsky function a second time such as:

$$C^{SA_k}_{\sigma^S} = \left[\sum_{i=O}^{|\sigma^S|} \omega_i * \left(C^{SA_k}_{\sigma_i} \right)^{\beta} \right]^{1/\beta}$$

The second evaluation is concerned with the target System Concept *SC* and the System Alternative Concept *AC*. In order to determine the similarity between *SC* and *AC* (noted *sim*(*SC*,*AC*), a semantic similarity measure is required. The measure of Wu and Palmer [13, 16] is chosen because of its intrinsic simplicity and efficiency:

$$sim(SC,AC) = \frac{2 * depth(C)}{depth(SC) + depth(AC)}$$

- C is the most specialized common parent of *SC* and *AC* in the ontology
- $depth(C)$ is the number of arcs which link the `Universal` concept to C

Then, each system alternative SA_k is associated to a couple $(C^{SA_k}_{\sigma^S}, sim(SC,AC))$ which represents how it is compatible with the new requirements for the system S. From these retrieved cases, the designer can take a decision to reuse or not. The structural coupling between project objects and system objects permits, from a retrieved system alternative to retrieve as well its Alternative Development task.

REUSE and REVISE Plannings Task. If the decision to reuse some retrieved System Alternatives is done at the end of the RETRIEVE task, the new Alternative Development tasks can require less time and resources because of the reuse of information about solutions. Therefore, the REVISE plannings of alternatives development task is necessary and adaptation has to be done. A direct copy of informations is not possible for tasks. Durations have to be adapted (reduced) as well as resources and new current constraints about the new project of design have to be taken into account. The reuse activity (i.e. adaptation) can be aided by a scheduling tool in order to define new dates and resource affectations. A suggested schedule is proposed for each reused alternative. Then, the suggested adapted schedules have to be REVISED in order to test their feasibility and provide confirmed schedules for new Alternative Development tasks to carry out.

REUSE and REVISE System Alternatives. If retrieved System Alternatives have to be reused (decision taken at the end of the RETRIEVE activity), their intrinsic information has to be REUSED (copied or adapted). If the compatibility of a retrieved System Alternative with regards to the new requirements is near to 1, then the adaptation effort will be light. The value of variables of V^{SA} has to be defined from the retrieved case. If the compatibility is low, a lot of adaptations have to be done. The REVISE task has to test the adapted or copied solution in order to provide a solution which fulfills all new requirements. In order to verify the solution, the values of the variables can be confronted to each constraint of σ^{SA}. Each constraint have to be satisfied. Note that in [14], basis of an integrated process which permits to verify and validate associated System Alternatives, AD tasks, Systems and Projects have been proposed.

RETAIN Planning and System Alternatives. When an Alternative Development task has been done, information gathered into the corresponding case has to be capitalized into the case base. If n solutions have been investigated, n new cases have to be capitalized. Each case is capitalized, even it has not been validated. A non-validated case can be useful for designer after completion and/or repair. Therefore, a specific attribute show that the case is not verified [14].

6 Conclusion

The aim of this article was to present a case-based integrated project planning and system design process. The ontology which formalizes knowledge about projects and design has been described and the integrated process has been proposed. Then, the case-based approach has been detailed. Clearly, this content is a preliminary one and many perspectives have to be developed. Firstly, requirements formalized by means of different kind of variables (temporal, continuous, discrete) have to be taken into account and integrated to models. Retrieving mechanisms have also to be developed for such constraints introducing the same kind of flexibility within the search process than into this article. Secondly, the proposed approach has to be extended in order to take into account multilevel decompositions of systems into subsystems and projects into sub-projects. Thirdly, models embedded into concepts of the ontology enabling to better define requirements closer to customer needs have to be defined as well as adapted retrieving mechanisms.

References

1. Pahl, G., Beitz, W.: Engineering Design: a Systematic Approach. Springer, Heidelberg (1996)
2. Bergmann, R.: Experience Management: Foundations, Development methodology and, Internet-based Applications. Springer, Heidelberg (2002)
3. Yang, C.J., Chen, J.L.: Accelerating preliminary eco-innovation design for products that integrates case-based reasoning and triz method. Journal of Cleaner Production (in press 2011)
4. Darlington, M.J., Culley, S.J.: Investigating ontology development for engineering design support. Advanced Engineering Informatics 22(1), 112–134 (2008)
5. Dieter, G.E.: Engineering design - A materials and processing approach, 3rd edn. McGraw-Hill International Editions (2000)
6. PMBOK Guide. A Guide to the Project Management Body of Knowledge, 3rd edn. Project Management Institute (2004)
7. Kolodner, J.: Case-based reasoning. Morgan Kaufman Publishers (1993)
8. Avramenko, Y., Kraslawski, A.: Similarity concept for case-based design in process engineering. Computers and Chemical Engineering 30, 548–557 (2006)
9. Negny, S., Le Lann, J.M.: Case-based reasoning for chemical engineering design. Chemical Engineering Research and Design 86(6), 648–658 (2008)
10. Lee, J.K., Lee, N.: Least modification principle for case-based reasoning: a software project planning experience. Expert Systems with Applications 30(2), 190–202 (2006)
11. Gomez de Silva Garza, A., Maher, M.L.: Case-based reasoning in design. IEEE Expert: Intelligent Systems and Their Applications 12, 34–41 (1997)
12. Grupe, F.H., Urwiler, R., Ramarapu, N.K., Owrang, M.: The application of case-based reasoning to the software development process. Information and Software Technology 40(9), 493–499 (1998)
13. Wu, Z., Palmer, M.: Verb semantics and lexical selection, pp. 133–139 (1994)
14. Abeille, J., Coudert, T., Vareilles, E., Aldanondo, M., Geneste, L., Roux, T.: Formalization of an Integrated System / Project Design Framework: First Models and Processes. In: Aiguier, M., Bretaudeau, F., Krob, D. (eds.) Proceedings of the First International Conference on Complex Systems Design & Management, CSDM 2010, pp. 207–217. Springer, Heidelberg (2010)
15. Suh, N.: Axiomatic Design: Advances and Applications. Oxford Series (2001)
16. Batet, M., Sanchez, D., Valls, A.: An ontology-based measure to compute semantic similarity in biomedicine. Journal of Biomedical Informatics 44(1), 118–125 (2011)

Chapter 10
Requirements Verification in the Industry

Gauthier Fanmuy, Anabel Fraga, and Juan Llorens

Abstract. Requirements Engineering is a discipline that has been promoted, implemented and deployed for more than 20 years through the impulsion of standardization agencies (IEEE, ISO, ECSS,...) and national / international organizations such as AFIS, GfSE, INCOSE. Ever since, despite an increasing maturity, the Requirements Engineering discipline remains unequally understood and implemented, even within one same organization. The challenges faced today by industry include: "How to explain and make understandable the fundamentals of Requirements Engineering", "How to be more effective in Requirements authoring", "How to reach a Lean Requirements Engineering, in particular with improved knowledge management and the extensive use of modeling techniques".

This paper focuses on requirements verification practices in the Industry. It gives some results of a study made end of 2010 about Requirements Engineering practices in different industrial sectors. Twenty-two companies worldwide were involved in this study through interviews and questionnaires. Current requirements verification practices are presented. It gives also some feedbacks of the use of innovative requirements authoring and verification techniques and tools in the industry. In particular, it addresses the use of Natural Language Processing (NLP)

Gauthier Fanmuy
ADN `http://www.adneurope.com`
17 rue Louise Michel, 92300 Levallois Perret - France
`gauthier.fanmuy@adn.fr`

Anabel Fraga
Informatics Dept, Universidad Carlos III de Madrid
Avda Universidad 30, 28911 Leganes, Madrid – Spain
`afraga@ie.inf.uc3m.es`

Juan Llorens
Informatics Dept, Universidad Carlos III de Madrid
Avda Universidad 30, 28911 Leganes, Madrid – Spain
`juan.llorens@uc3m.es`

at the lexical level for correctness verification (on the form, not on the substance) of requirements, the use of Requirements boilerplates controlled by NLP for guiding requirements writing and checking, the use of Ontologies with NLP to verify requirements consistency, and the application of Information Retrieval techniques for requirements overlapping.

1 Introduction

Several studies clearly underlined the importance of requirement management in Systems Engineering [Brooks1987], [Chaos-Report2003], [SWEBOK2004]. Among these studies [NDIA2008], the SEI (Software Engineering Institute) and the NDIA (National Defense Industrial Association) made a study on the efficiency in Systems Engineering. The Systems Engineering Division (SED) of the National Defense Industrial Association (NDIA) established the Systems Engineering Effectiveness Committee (SEEC) to obtain quantitative evidence of the effect of Systems Engineering (SE) best practices on Project Performance. The SEEC developed and executed a survey of contractors for the defense industry (i.e., government suppliers) to identify the SE best practices utilized on defense projects, collect performance data on these projects, and search for relationships between the application of these SE best practices and Project Performance.

The SEEC surveyed a sample of the population of major government contractors and subcontractors represented in the NDIA SED. The survey data was collected by the Carnegie Mellon2 ® Software Engineering Institute (SEI). Project Performance was then assessed based on satisfaction of project cost, schedule, and scope goals. This study showed a strong correlation between project performances and requirements engineering capabilities:

- Organizations with low capabilities in Requirements Engineering are likely to have poor performance projects
- At the opposite, organizations with high capabilities in Requirements Engineering are likely to have good performance projects: over half of the Higher Performing Projects exhibited a higher capability in Requirements Engineering.

Thus, it can be understood that Requirements Engineering is a key success factor for current and future development of complex products. As highlighted in [GAO2004], [SGI2001] the existence of poor requirements, or lack of them, is one of the main causes to project failures. Even more, although there is no complete agreement on the effort and cost distribution of activities for the development process (the requirements phase of software projects estimates between 5% and

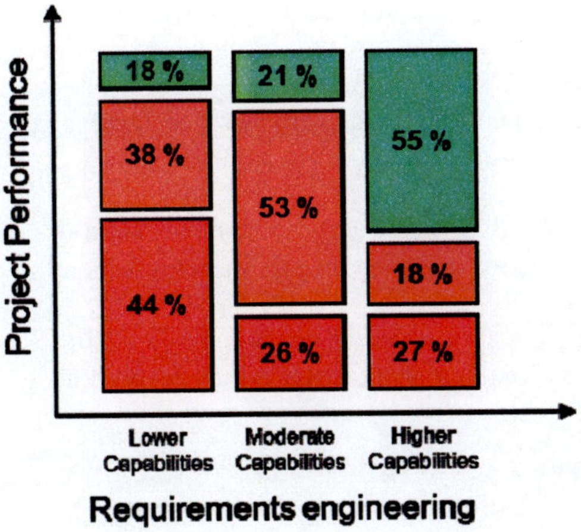

Fig. 1 Relationship between Requirements Capabilities and Project Performance

16% of total effort, with too much variance), empirical studies show that rework for identifying and correcting defects found during testing is very significant (around 40-50%); therefore, more quality control upfront will lead to less rework [Sadraei2007].

Another study [Honour2011] showed that an effective System Engineering is obtained at a level of approximately 15% of a project cost optimizes cost, schedule, and stakeholder success measures.

All the previously mentioned studies suggest a more complex reality: the mere production of huge sets of detailed requirements does not assure effective and quality project performance. To ask for Requirements engineering higher capabilities inside an organization implies much more that strategies for mass production of requirements.

A requirement is "An identifiable element of a function specification that can be **validated** and against which an implementation can be **verified"** [ARP4754].

In the Systems Engineering community, requirements verification and validation are very well described as "Do the right thing" (Validation) together with "Do the thing right" (Verification) [Bohem1981]. To do "*the right thing right*" becomes the essential activity in the Requirements Engineering process. Therefore, the quality factor becomes essential at all levels: the quality of single requirements must be measured, as well as the quality of sets of requirements. The quality assessment task becomes really relevant, and perhaps the main reason to provide this field with a specific engineering discipline.

This issue has raised new needs for improving Requirements quality of complex development products, especially in terms of tools to assist engineers in their Requirements verification or requirements authoring activities.

The first part of this paper describes some results of a study on Requirements Engineering industrial practices, in the framework of the RAMP project (Requirement Analysis and Modeling Process) - a joint initiative undertaken in France between major industrial companies, universities / research labs and Small & Medium Enterprises. Current requirements verification practices and gaps are presented.

The second part of this paper describes how Natural Language Processing (NLP) can help into more efficient requirements verification and Lean requirements authoring.

The third part of the paper gives an overview of existing academic or commercial tools that support Requirements verification and authoring.

2 RAMP Project

The RAMP Project (Requirement Analysis and Modeling Process) started in September 2009 from common needs expressed by 3 industrial companies (EADS, EDF, RENAULT), research studies done in Requirements Engineering (University of Paris 1, ENSTA, INRIA, IRIT) and solutions proposed by SME (ADN, CORTIM) [Fanmuy2010].

In particular, some of the common needs are:

- Requirements are often written in natural language, which is a source of defects in the product development process.
- Obtaining consistency and completeness of requirements remains difficult by the only human-visual review since several thousands of requirements are managed in most of cases.

The objective of the RAMP project is to improve the efficiency and quality of requirements expressed in natural language throughout the life cycle of the system, and thus the performance of projects in design, renovation and operation.

The axes of the Project are:

- Improvement of the practice of Requirements definition written in natural language: assistance in requirements authoring (lexical and syntactical analysis, models…).
- Improvement of Requirements analysis: assistance in the detection of inconsistencies, overlaps, incompleteness (modeling, compliance to an ontology…).
- Assistance in the context analysis: assistance in the identification of data in scenarios enabling requirements elicitation and definition in their context, assistance in identifying impacts in a context of evolution of a legacy system.

3 Industrial Practices in Requirements Engineering

At the initiative of the RAMP project, ADN created and developed a study on industrial Requirements Engineering practices in 2010.

This study was based on:

- Interviews of major companies in different industrial sectors (Aviation, Aerospace, Energy, Automotive...)
- Survey in the Requirements Working Group (RWG) of INCOSE (International Council On Systems Engineering)

Different topics were addressed:

- Needs, requirements: Definition of needs & requirements, Identification and versioning of requirements, Number of requirements, Prioritization of needs & requirements, Elucidation techniques for needs, roles of Stakeholders, Efficiency of the elicitation of needs, Requirements management, Quality rules for Requirements, Specification templates, Formatting of specification documents, Capitalization of the justification towards needs & requirements, Requirement Engineering tools.
- Design of the solution: Derivation of requirements, Systems hierarchy, Requirements allocation, System Analysis.
- Verification and validation of requirements: Most common defects, Verification/validation of requirements by inspections, Verification/validation of requirements by use of models, Traceability of requirements and tests, Integration / verification / validation / qualification improvements.
- Management of changes to requirements: Change management, Change management improvements.
- Configuration management: Configuration management process, Configuration management tools, Improvement of configuration management.
- Risks management
- Customers/suppliers coordination: Maturity of suppliers in requirements engineering, Exchanges/Communication between Customers/Suppliers.
- Inter-project capitalization: Re-use of requirements, Improvement in the reuse of previous requirements.

The results of the Requirements verification and validation section were the following:

- Most common defects in Requirements:
  The most common defects encountered within requirements were the ambiguity and expressing needs in the form of solution.

  Consistency and completeness of requirements repositories are also difficult points to address.

  The input data (needs, scenarios, mission profiles...) of a system are not always complete, precise etc. particularly in the case of new or innovative systems.

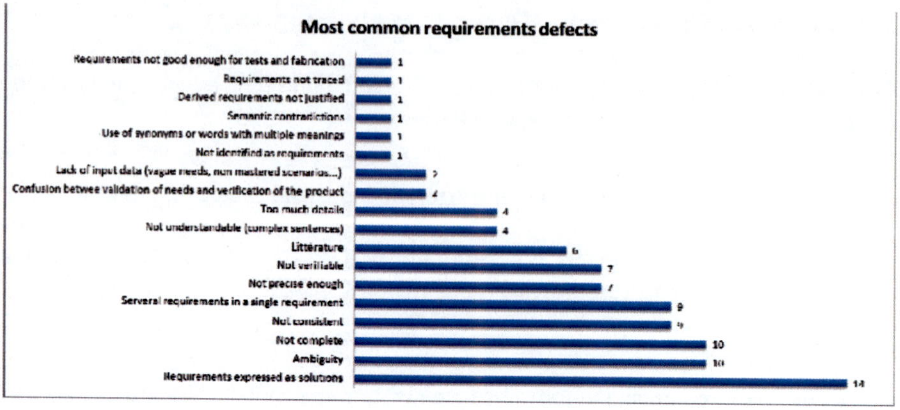

Fig. 2 Most common defects in Requirements

- <u>Verification/validation of requirements by inspections</u>
  Inspections are the most commonly used means to verify/validate requirements. They can take several forms: cross readings, peer reviews, QA inspections with pre-defined criteria etc.

  Most organizations have requirements quality rules at corporate or project level. But most of the time these rules are not correctly applied:

  Correctly applied: 15%
  Not applied : 35%
  Variably applied: 50%

  The review process is burdensome but effective with the intervention of the good specialists / experts. Nevertheless the analysis of needs remains difficult when the final customer does not participate in reviews but is only represented.

  Concerning software, a requirements review process is estimated to be present in about 10% of a global development effort.

  Other means could be used to verify or validate the requirements. For example: the use of executable models to verify the requirements by simulation, the proof of properties (particularly in the case of software).

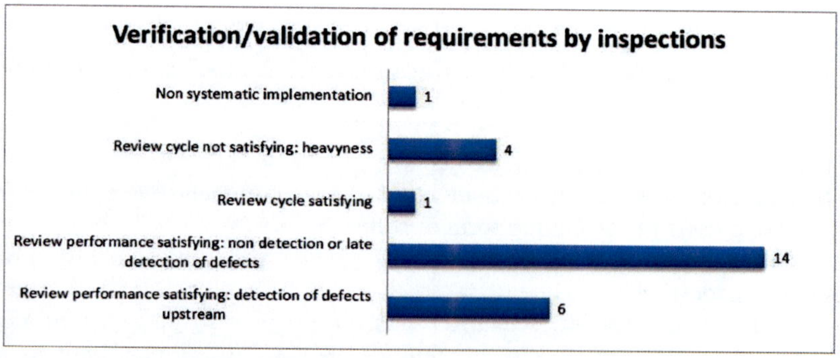

Fig. 3 Verification/validation of requirements by inspections

From a general point of view, in a given context, a ratio time/efficiency could be defined for each method of verification/validation of the requirements.

- Verification/validation of requirements by the use of models
  The use of models for verification/validation of requirements is a practice which is not as common as requirement reviews.
  Nevertheless, this practice is judged as providing real added value in improving engineering projects.
  Examples of the use of models:
  - Support in analyzing for consistency and completeness
  - Evaluation of the impact of requirements and their feasibility level
  - Evaluation of the system behavior within a given architecture

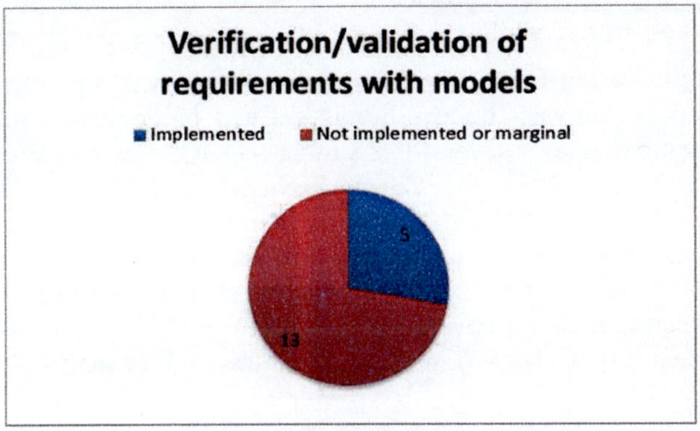

Fig. 4 Verification/validation of requirements by the use of models

The different types of models most often used are, in decreasing order (results from RWG survey):

- UML
- BPMN
- Simulink
- SysML
- Performance models
- Costs models
- CAD Models

- Tools assistance for requirements verification
  Only few organizations use tools to assist engineers or quality assurance in verifying requirements.

  The following practices are encountered from the most to the less frequent:

  - Requirement verification within the RMS[1] tool: requirements are reviewed directly in the RMS tool (e.g. DOORS®, IRQA®). Some attributes or

[1] RMS: Requirements Management System.

discussions facilities are used to collect review comments. Traceability is checked in the tool for incompleteness, inconsistencies between requirements, for tests coverage.

This practice is often limited to the RMS tool because not all tools are always supporting traceability and thus it is difficult to have a global view of requirements connected with other engineering artifacts (e.g. tests, change requests…).

Very often the RMS tool is not deployed in the organization and it is difficult to verify traceability or to make impact studies in case of changes.

In better cases, some add-ons to RMS tools have been developed for identifying semantically-weak words (e.g. detection of words such as fast, quick, at least…).

Typical use cases: compliance to regulation, correctness of requirements, correctness of traceability in design, impact analysis in a change process.

- Requirements verification with specialized tools which perform traceability and impact analysis: traceability is defined in dedicated tools (e.g. Requirements in one or several RMS tools – DOORS®, MS Word®, MS Excel®; Tests in QMS[2] tool; Change Requests in CM[3] tool). Tools are not connected, but specialized tools (e.g. Reqtify® with Reviewer®) capture requirements and other engineering artifacts from any source and generate traceability matrices with rules checker (e.g. reference to missing requirement, requirement not covered with tests…).

  Traceability is checked in the tool and traceability matrices are generated for prove issues.

  In some cases, some scripts have been developed for weak words identification purposes, or for documents verification properties (e.g. a column of a table has/or not content).

  Typical use cases: compliance to regulation, correctness of requirements, correctness of traceability in design to code and tests, impact analysis in a change process.

- Requirements verification with specialized tools which perform lexical and syntactical analysis of requirements

  As requirements in natural languages are imprecise, ambiguous, etc. the use of tools (e.g. Requirements Quality Analyzer®, Lexior®,...) to check requirements regarding SMART (Specific, Measurable, Attainable, Realizable, Traceable) quality rules enable to correct defects in the form before business or project reviews.

  Such tools enable to detects the use of wrong words (e.g. weak words,…), bad grammatical sentences (e.g. passive voice, use of pronouns as a subject,…) and multiple requirements (e.g. sentence length,…)

  This enables to identify critical requirements and correct them before reviews.

[2] QMS: Quality Management System.
[3] CM: Change Management.

Typical use cases: Analysis of Requests for Proposals (Bid), Requirements verification before a business or project review.

4 Towards a Lean Requirements Engineering

One of the conclusions of the survey is that the problems still found, despite the use of a Requirements Engineering approach, concern about the difficulty of understanding the fundamentals of Requirements Engineering.

Teams still face difficulties in the transition from theory to practice. In particular:

- formalization of requirements
- consistency of textual requirements
- requirements that describe solutions
- definition of specification models

One of the identified Industry challenges is to reach a Lean Requirements Engineering: be more efficient in Requirements authoring. This leads to assist practitioners in writing quality (SMART) requirements from the very first attempt and to improve reuse of knowledge.

5 Advanced Requirements Verification Practices

Obtaining the "right-requirements-right" is a complex, difficult and iterative process, in which engineers must respond to the double challenge of *discovering* and *formalizing* the needs and expectations that the clients usually are able to describe using different understanding and representation schemes. This different ways of understanding & representing requirements between clients and engineers leads to real problems at the time of clearly formalizing requirements. To add more complexity to the problem, the clients can several times provide confusing, incomplete and messy information.

The success of this requirements process requires continuous and enhanced collaboration & communication between clients and system engineers: the more complete and unambiguous the requirements specification is, the higher performances will the project have [Kiyavitskaya2008]. We are sure that this need of communication among all stakeholders has been the reason why requirements are mainly expressed in *natural language* for most industrial projects [Kasser2004], [Kasser2006], [Wilson1997] instead of more formal methods. The market supports this idea by in almost all cases offering requirements management tools based on natural language representation [Latimer-Livingston2003].

Due to the reason that industrial system projects can potentially handle thousands of requirements, the human-based verification process becomes extremely expensive for the organizations. Therefore, industrial corporations apply tool-based, semi-automated techniques that reduce the human resources needs of efforts. Usually these techniques are based on the application of transformations to the natural language requirements to somehow represent them in a more formal way [TRC].

It is clearly understood in the market [TRC], that a strong correlation exists between the complexity level of requirement representations and what one can extract out of them. The more semantic knowledge we are interested to validate from requirements the more complex the formal representation must be. The Natural Language Processing discipline organizes the different techniques to be applied to natural text according to the level of semantics [De-Pablo2010]. They are summarized in figure 5.

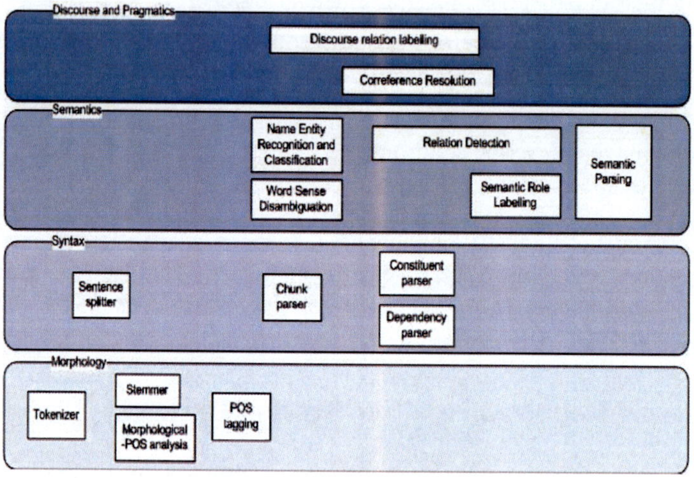

Fig. 5 Natural Language Processing summary of techniques

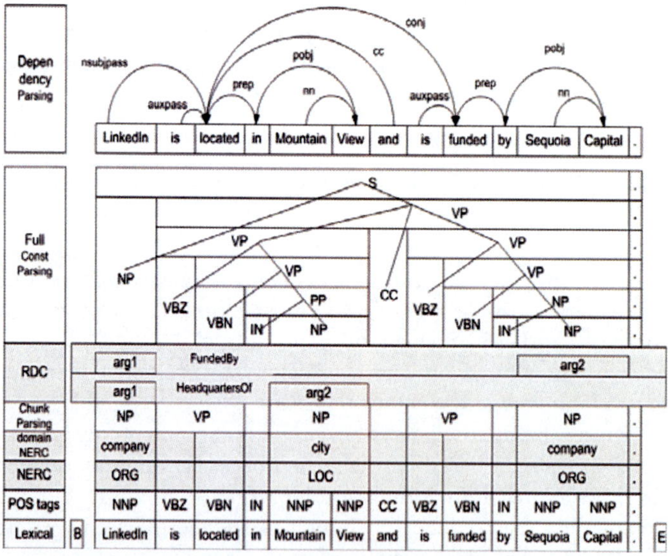

Fig. 6 Application of syntactical techniques to Natural Language

Morpho (lexical)-syntactic techniques can be applied to correctness verification of requirements, as they can somehow produce a grammatical analysis of the sentence. Well-formed sentences will statistically translate into better written requirements. Figure 6 presents an example of syntactical techniques [De-Pablo2010].

An evolution of the syntactical techniques, in the form of identifying sentence patterns (or requirements patterns), is what the industry calls "boilerplates" [Hull2010], [Withall2007]. The identification and application of boilerplates in requirements engineering leads to quality writing and checking and improves authoring.

For example:

UR044: The Radar shall be able to detect hits at a minimum rate of 10 units per second

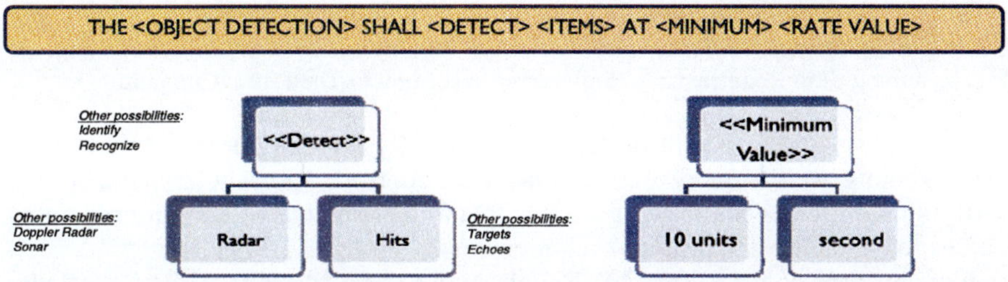

Fig. 7 Requirements modeling with boilerplates and NLP techniques

The application of ontology engineering in requirements engineering and in model-based systems engineering, in general, seems to be a very promising trend [TRC], [Chale2011]. We call it requirements verification based on Knowledge Management, and it deals with assisting System Engineers to get a **complete** and **consistent** set of requirements (e.g. compliance to regulation, business rules, non-redundancy of requirements…) by using Ontologies, which represent the domains of knowledge of an organization.

In its application to requirements verification, extending NLP with inference rules capabilities, taxonomies, thesaurus and semantic groups enable computer tools to enhance requirements consistency. The Reuse Company [TRC] offers a particular vision of ontologies in their application to Requirements Engineering (figure 8).

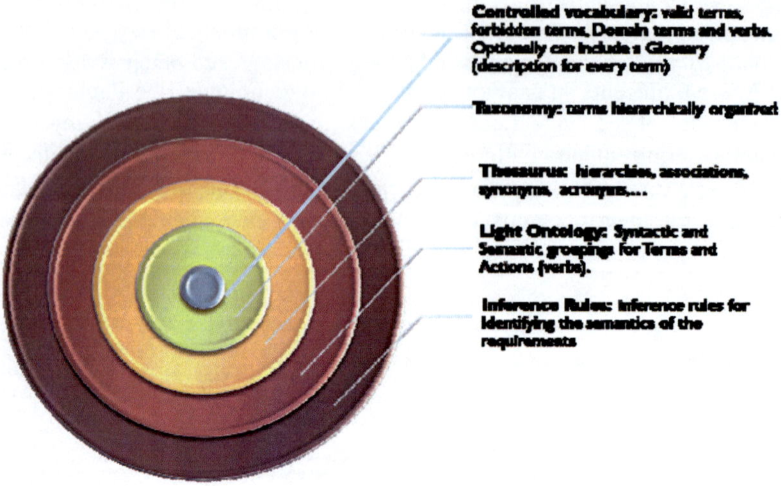

Fig. 8 Ontologies in Requirements Engineering according to The Reuse Company

Ontologies, together with boilerplates, assist System Engineers to write from the beginning well-formed requirements with controlled vocabulary, rather than verifying requirements afterwards. Their relational structure offers multiple possibilities for semantic expansion of requirements.

Finally, the combination of Requirements Engineering with Knowledge Management, by information retrieval from existing sources (business rules, feedback,...), allows the verification process to measure quality of a set of requirements: traceability, consistency/redundancy, completeness and noise. Information retrieval enables also to verify the completeness and of the ontology.

6 Tooling

Different tools are available on the market, from research to commercial tools.
In the scope of its Lean Systems Engineering research, EADS IW, one of the partners of the RAMP Project made in 2010 an assessment of 15 tools [EADSIW2010].

The assessment was made over 3 pilot cases: 2 test cases on AIRBUS (A320NG & A350) and 1 Test case on ASTRIUM (ATV).

8 tools were selected in the short list.

The assessment was based on 74 criteria.

The results are the following:

- One is promising: Requirements Quality Analyzer from The Reuse Company
- One is an important challenger : Lexior® from Cortim
- One is interesting but out of scope: DESIRe (more like a support for requirement authoring)
- The remaining tools are not compliant with industrial expectation: ARM / Requirements Assistant / DXL / QuARS.

Tool	Company	Country	Use	Kind	Result
RQA	The Reuse Company	Spain	Industrial	Commercial	Most promising as assessed so far
LEXIOR	CORTIM	France	Industrial	Commercial	Most important challenger
Requirements Assistant	Sunny Hills	Holland	Industrial	Commercial	Not integrated with DOORS
ARM	NASA	USA	Industrial	Open Source	
DESIRe	HOOD	Germany	Industrial	Open Source	Authoring Support Application
QuARS	University of Pisa	Italy	Academic	Commercial	Academic Tool Service Oriented Business Model
TigerPro	University of South Australia	Australia	Academic	Open Source	
DOORS / DXL	IBM	USA	Industrial	Commercial	Too complex Compliance with COTS Policy Cost of Maintenance

Requirements Quality Analyzer (RQA)

RQA allows defining, measuring, improving and managing the quality of requirements. It is integrated in DOORS®, IRQA® and MS Excel®. It is based on evaluating metrics. These metrics will help the engineers to sort, classify and prioritize the review of their requirements from the most to the least critical defects.

There are non-text-based metrics such as Text length, Readability, Number of punctuation marks, Dependencies …

There are other metrics computed by processing the requirement text with natural languages processing techniques: lack of shall, use of opinions; use of negative sentences, use of implicit and weak words, use of connectors… It enables also to detect the use of design terms in higher-level requirements.

Finally, there are other metrics not based on a single requirement but in sets of requirements:

- Incompatible metric units: inside the same set of requirements it could be a mistake use two different metric units belonging to different metric systems
- Overlapping requirements: this metric computes how similar in meaning are two different requirements.. This allows to delete duplicate requirements or trace them if they belong to a different level in the requirement hierarchy

RQA uses of ontologies and requirements boilerplates for both:

- Analyzing and storing conceptual information that makes it easier for a machine to understand the real meaning of a requirement

- Storing inference rules that allows the tool implement 'artificial intelligence' algorithms to emulate a human reasoning

Lexior

Lexior is a lexical analysis tool based on requirements rules. It is integrated with DOORS®. An XML interface is also available.

It is based on evaluating metrics per requirements writing rule such as: absence of shall, use of negative sentences, use of verbal forms, use of passive voice, use of pronouns as subject, use of weak words…

Scoring of requirements enables to identify the most critical requirements.

It is also possible to generate some RID (Review Item Description) for a review.

7 Conclusion

Assessments results have shown that about 25% of requirements are critical and can grammatically be improved. Typical defects are the following:

- Absence of shall: 8 to 10%
- Forbidden words: 10 to 15%
- Subject, multiple objects, design: 15%
- Incorrect grammar: 50%

"Classical" review process is not performing as expected and is costly and time consuming. Support of tools for lexical, syntactic analysis enables to correct bad requirements writing before business of project reviews.

One of the next challenges in the Industry is to reach a Lean Requirements Engineering process: assist systems engineers to write SMART requirements at a first shot.

The use of ontologies and boilerplates is a promising way of doing better requirements engineering and reuse knowledge:

- Find missing requirements
- Find ambiguous requirements
- Find noise in requirements
- Find inconsistencies

References

[ARP4754] Guidelines for Development of Civil Aircraft and Systems. Society of Automotive Engineers. ARP4754A, `http://standards.sae.org/arp4754a`

[Bohem 1981] Boehm, B.: Software Engineering Economics, p. 37. Prentice-Hall, L/S 004.41 BOE (1981)

[Brooks1987] Brooks, F.P.: No Silver Bullet. Essence and Accidents of Software Engineering. IEEE Computer 20(4), 10–19 (1987); reprinted in: Brooks, F.P.: The Mytical Man-Month, Essays on Software Engineering. Addison-Wesley (20th Anniversary Edition) (1995)

[Chale2011] Chale, H., et al.: Reducing the Gap between Formal and Informal Worlds in Automotive Safety-Critical Systems. In: INCOSE Symposium 2011, Denver (2011)

[Chaos-Report2003] The Standish Group, Chaos Report (2003), `http://www.standishgroup.com/`

[De-Pablo2010] de Pablo, C.: Boostrapping Named Entity resources for adaptive question answering Systems. PHD. Universidad Carlos III de Madrid. Informatics Department, Universidad Carlos III de Madrid (2010), `http://hdl.handle.net/10016/9867`

[EADSIW2010] Requirements Quality Monitoring - Main results and outcomes. In: RAMP Workshop (June 10, 2010)

[Fanmuy2010] Fanmuy, G., et al.: Requirements Analysis and Modeling Process (RAMP) for the Development of Complex Systems. In: INCOSE Symposium 2011, Chicago (2011)

[GAO2004] United States General Accounting Office, Defense Acquisition: Stronger Management practices are needed to improve DOD's software-intensive weapon acquisitions (2004)

[Hull2010] Hull, E., Jackson, K., Dick, J.: Requirements Engineering. Springer, Heidelberg (2010)

[Honour2010] Eric, H.: Systems Engineering return on investment (SE-ROI). In: INCOSE International Symposium, Chicago, IL

[Kasser2004] Kasser, J.E.: The First Requirements Elucidator Demonstration (FRED) Tool. Systems Engineering 7(3), 243–256 (2004)

[Kasser2006] Kasser, J.E., Scott, W., Tran, X.L., Nesterov, S.: A Proposed Research Programme for Determining a Metric for a Good Requirement. In: The Conference on Systems Engineering Research, Los Angeles, California, USA (2006)

[Kiyavitskaya2008] Kiyavitskaya, N., Zeni, N., Mich, L., Berry, D.M.: Requirements for tools for ambiguity identification and measurement in natural language requirements specifications. Requirements Engineering 13(3), 207–239 (2008)

[Latimer-Livingston2003] Latimer-Livingston, N.S.: Market Share: Requirements Management, Worldwide (2003) (Executive Summary); Gartner Research (July 1, 2004), `http://www.gartner.com/DisplayDocument?ref=g_search&id=452522`

[NDIA2008] Report CMU/SEI-2008-SR-034

[Sadraei2007] Sadraei, E., Aurum, A., Beydoun, G., Paech, B.: A field study of the requirements engineering practice in Australian software industry. Requirements Engineering 12(3), 145–162 (2007)

[SGI2001] Standish Group International, Extreme Chaos (2001)

[SWEBOK2004] SWEBOK Guide to the Software Engineering Body of Knowledge. IEEE Computer Society (2004), http://www.swebok.org/ironman/pdf/SWEBOK_Guide_2004.pdf

[TRC] http://www.reusecompany.com (Website last visited July 5, 2011)

[Wilson1997] Wilson, W.M., Rosenberg, L.H., Hyatt, L.E.: Automated Analysis of Requirement Specifications. In: Proceedings of the 19th International Conference on Software Engineering, ICSE 1997, Boston, Massachusetts, USA, May 17-23, pp. 161–171 (1997)

[Withall2007] Withall, J.: Software Requirement Patterns. Microsoft Press (2007)

Chapter 11
No Longer Condemned to Repeat: Turning Lessons Learned into Lessons Remembered

David D. Walden

Abstract. Many organizations have Lessons Learned processes in place, and both positive and negative lessons are dutifully captured in corporate Lessons Learned databases. Even with a well-defined process and an effectively designed database, many of these supposed Lessons Learned are not really learned at all. Subsequent projects often struggle to find lessons that are both accessible and relevant to their current endeavor. As a result, many organizations "relearn" these lessons over and over again.

This paper explores how an organization can transition from Lessons Learned to Lessons Remembered. Instead of having future projects "pull" information from a Lessons Learned database, it describes how Lessons Remembered can be "pushed" to relevant stakeholders via updates to an organization's set of standard process assets. In this manner, the Lessons Learned become Lessons Remembered.

1 What Are Lessons Learned and Why Are They Important?

> "Progress, far from consisting in change, depends on retentiveness. When change is absolute there remains no being to improve and no direction is set for possible improvement: and when experience is not retained, as among savages, infancy is perpetual. **Those who cannot remember the past are condemned to repeat it.**" – George Santayana (Santayana 1905)

The often misquoted phrase above by Santayana is illustrative of why organizations attempt to capture and apply Lessons Learned (LL). By capturing LL, organizations strive to build on past successes and avoid repeating past failures. However, many LL initiatives fail to achieve the retentiveness that Santayana correctly points out as critical to success.

David D. Walden, CSEP
Sysnovation, LLC

A good definition of a LL is:

> "Results from an evaluation or observation of an implemented corrective action that contributed to improved performance or increased capability. A lesson learned also results from an evaluation or observation of a positive finding that did not necessarily require corrective action other than sustainment." (CJCSI 2008).

As can be seen from this definition, LL come in two flavors; negative and positive. Negative LL, sometimes known as *shortcomings*, highlight things that either did not go as well as expected or were unpleasant surprises. Positive LL, sometimes known as *best practices*, highlight things that either went better than expected or were pleasant surprises.

There are benefits to collecting LL for both current and future projects, as well as the organization. For current projects, the LL process allows them to reflect on what went well and what could have been improved during the execution of their effort (Seningen 2005). LLs can also be injected during project execution, thereby allowing for immediate improvements of the current projects. For future projects, existing LL should allow them to avoid some of the more common pitfalls and take advantage of some of the successes of their predecessors. From an organizational perspective, the capturing and application of LL allows more opportunity to improve productivity and increase customer satisfaction. In many organizations, the LL process is just piece of an overall Knowledge Management effort (Meakim and Wilkinson 2002)

2 Typical Approaches to Capturing Lessons Learned

Many organizations have well defined LL processes in place. The process typically specifies the frequency and types of LL sessions that are to take place. Many times, these sessions are tied to project phases or major milestones. An example of a LL process is shown in Figure 1.

A good practice related to LL is to express them in a consistent and actionable form. Each action should be stated in terms of the **result** (positive or negative), the **cause**, and a **recommendation** (how to reinforce or avoid repeating). In general, "result ... **because** ... cause ... **therefore** ... recommendation" is an effective structure. General observations that do not state a clear result and the underlying cause are not valuable and should be avoided. Consistent with the two major types of LL, both things that did not go well (negative LL) and things that went well (positive LL) should be identified. We want the team to propose a recommendation, since they are the ones closest to the LL and many times know what worked, or what could have helped them avoid trouble.

The results of LL sessions are typically captured in corporate LL database. There are numerous examples of LL databases. A search on the web produced numerous examples across many domains, a sampling of which is shown in Table 1. (Note that some of these on-line databases require registration to gain access.) The United States (US) Department of Transportation (DOT) Intelligent Transportation Systems (ITS) is recommended as an example of a well-designed LL database (DOT 2011). A good analysis of the US National Aeronautics and Space Administration's (NASA's) LL process is given in (GAO 2002).

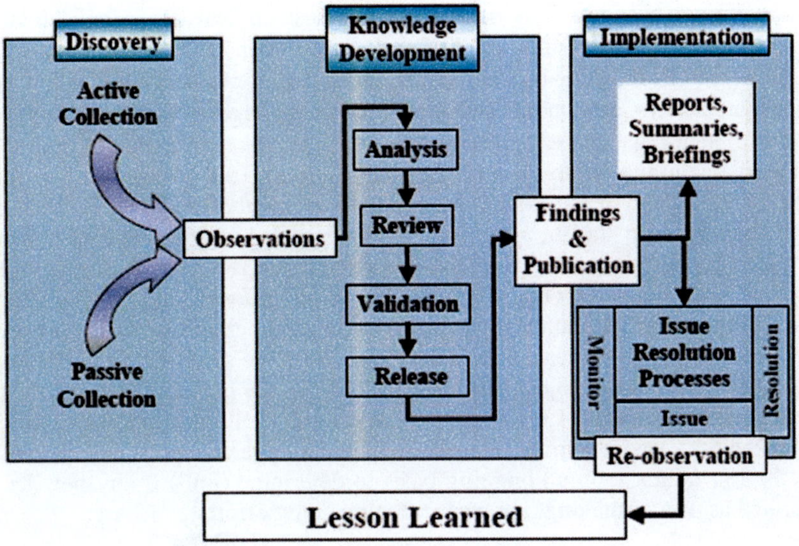

Fig. 1 CJSCI Lessons Learned Process. Source: (CJCSI 2008)

Table 1 Example Lessons Learned Databases

Lessons Learned Database Name	Website and Reference
United States (US) Department of Transportation (DOT) Intelligent Transportation Systems (ITS) Learned Information Database	http://www.itslessons.its.dot.gov/ (DOT 2011)
NATO Joint Analysis and Lessons Learned Centre (JALLC)	http://www.jallc.nato.int/ (JALLC 2011)
US National Aeronautics and Space Administration (NASA) Headquarters (HQ) Lessons Learned Information System (LLIS)	http://llis.nasa.gov/llis/search/home.jsp (NASA 2011)
US Army's Combined Arms Center (CAC) Center for Army Lessons Learned (CALL)	http://usacac.army.mil/cac2/call/index.asp (CAC 2011)
US Department of Homeland Security (DHS) Federal Emergency Management Agency (FEMA) Learned Information System (LLIS)	https://www.llis.dhs.gov/index.do (DHS 2011)
Mine Action Information Center (MAIC) Mine Action Lessons Learned Database	http://maic.jmu.edu/lldb/ (MAIC 2011)

Given that an organization has a well-defined LL process and a well-designed LL database, Figure 2 shows a typical way LL are captured and used. Existing projects hold regular LL sessions to discuss and gather project LL. Many times these sessions are aligned with decision gate reviews or major technical reviews.

Existing projects also typically have a final LL session near the end of the current stage of the effort. The results of all of these LL sessions are captured (pushed) in a LL database. In more mature organizations, existing projects also review (pull) the LL database to determine if any LL can be applied as part of the ongoing project (re)planning efforts.

The organizational-wide Enterprise Process Group (EPG) is responsible for the creation and upkeep of the LL database. They create the LL database and define the LL format and metadata to be supplied with each LL entry. The EPG also reviews the LL database on a periodic basis to determine if certain lessons should be combined or eliminated. The EPG may also add additional metadata to help with the usability of the LL database. The EPG typically performs causal analyses to determine if any root causes can be determined from the set of LL in the database.

Finally, when a new project is initiated, as part of the project planning, the team is expected to review the LL database to determine (pull) what LL are applicable to their project. Again, in more mature organizations, the new project will also review the LL database on an ongoing basis to determine (pull) if any new LL can be applied as part of the ongoing project (re)planning efforts.

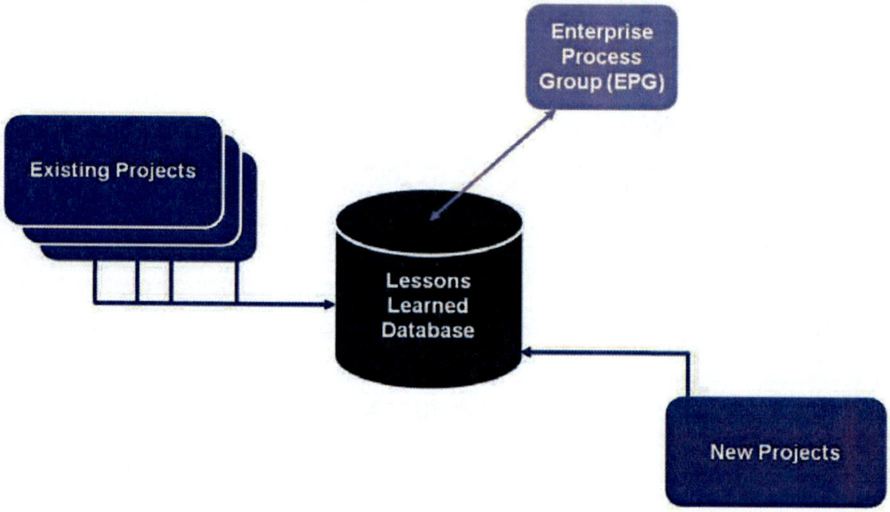

Fig. 2 Typical Lessons Learned Deployment

While there are recognized advantages to deploying a LL systems as described above, there are also numerous issues and challenges that present themselves. For the existing projects, the key issue is determining what is appropriate to capture and how often. Typically, the EPG has already defined the format and metadata to be recorded, so that is not an issue. Many times, determining what to share may be more based on cultural or image factors (e.g., not wanting to share bad results), rather than pure technical factors. Business pressures may also limit the effectiveness of the LL process. For example, the organization may not provide funding to conduct an effect set of LL sessions. In some cases, legal liability concerns may prevent an organization from capturing all relevant LL.

The main challenges for the EPG are to define and maintain the LL database. The challenge in defining the LL database is to ensure there is enough detail to be useful without being too burdensome to the projects (see Egeland 2009). The challenge in maintaining the LL database is determining how to effectively manage the LL entries. Many times, the LL database grows to contain a significant number of LL entries. Members of the EPG may not have the technical or domain experience to effectively combine or eliminate LL entries. These factors are important because if people go to the LL database and are not easily able to locate the relevant information they need, they are not likely to return. Another challenge is to ensure management support is maintained so the necessary LL database activities can be accomplished. The EPG also needs to develop a LL process that can effectively handle both process- and product/system-related LL.

The main challenge for new projects is extracting information out of the LL database that is relevant in their situation. In general, LL efforts do not fail from lack of input, but rather lack of effective utilization. The LL database can help ameliorate some of these issues by including metadata such as project size, project domain, system complexity, and customer. However, sometimes these "aids" can actually exacerbate the situation by obfuscating desired synergistic relationships. Of course, new projects typically need to find and leverage relevant information during the intense project start-up and planning efforts.

As an example, let's assume we are the lead Systems Engineers for a new automobile to be used as part of an Intelligent Transportation System (ITS). We know from Table 1 that the US DOT has an excellent ITS LL database (DOT 2011). We have identified a radar sensor on the automobile system we are responsible for as one of the key risks, specifically related to interference with other ITS radio frequency (RF) sensors. We go to the ITS LL database and search on "sensor" and 105 unique LL entries are returned. In some ways this is comforting to know that there have been that many LL created about sensors, but we wonder, "Which ones are relevant to our automobile?" We look at the first LL that is titled "*Employ sensors that can account for a range of parking lot vehicle movements.*" While interesting, it is not relevant to our risk. Hoping to find fewer, more relevant results, we search on "radar AND sensor" and 14 unique LL entries are returned. This is much more manageable, but then we notice that the second LL is "*Carefully select a project manager to be responsible for deployment and testing of new ITS technology*" and we immediately lose confidence in the LL database.

> Note: The ITS LL database (DOT 2011) is an excellently designed and deployed tool. This example is in no way intended to denigrate this tool, but rather to show some of the limitations upon which most LL tools are developed.

3 How Do We Transition for Lessons Learned to Lessons Remembered?

As we can see from the above example, many of these supposed LL are not really learned at all, as subsequent projects often struggle to find lessons that are both accessible and relevant to their current endeavor. Figure 3 shows a revised deployment approach to turn LL into Lessons Remembered (LR) within an organization.

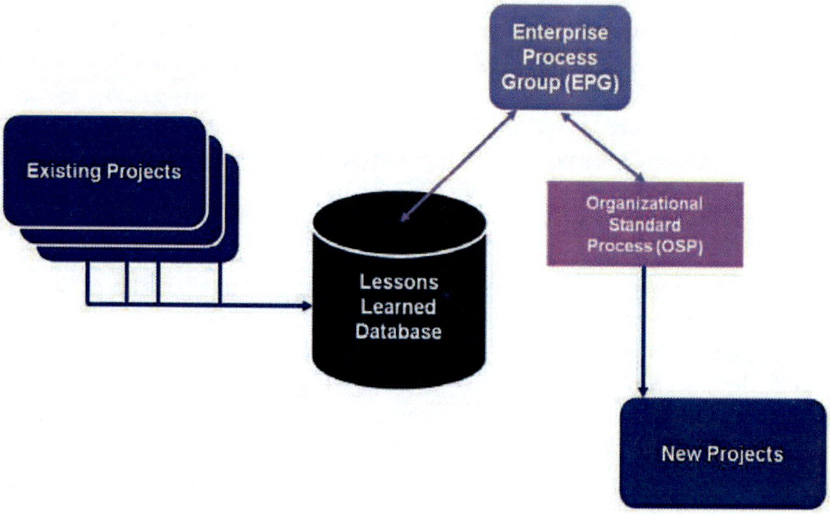

Fig. 3 Approach for Deploying Lessons Remembered

The key differences from Figure 2 are a) now the primary task of the EPG is to convert the LL database entries into updates to the Organizational Standard Process (OSP) and b) the new projects now get these updates pushed to them as part of their tailoring process via the OSP, rather than having to pull them from the LL database. This is consistent with the views of (Knoco 2010):

> "We are beginning to come to the view that a lesson is not an end in itself. A lesson is a temporary step along the way to a process, or to a process improvement. Therefore a lessons database is a holding-place which feeds a library of best practice - something that forms a temporary home in a workflow, rather than a permanent repository."

From an existing project perspective, not much has changed. However, since the primary "reviewers" of the LL database are now the EPG (as opposed to potentially every future project), the existing projects may be more willing to share sensitive LL (if the EPG is proven to be a "trusted agent" of the organization).

From the perspective of the EPG, they still have the same definition and review responsibilities. However, their primary focus shifts from tweaking the LL captured the database to incorporating LR into the OSP that obviate the need to retain the LL. In addition to looking backwards across the prior project's lessons, the EPG can also be forward-looking, accounting for the organization's strategic direction (Kirsch 2008). Also, the EPG can more effectively address the legal liability aspects of the LR by acting as the "gatekeeper" for the LL database.

One of the issues the EPG faces is where in the OSP the LR should be captured. There are several approaches to structuring a set of OSP assets. For the purposes of this paper, we will follow the conventions of (Walden 2007) and define a process hierarchy with the "types" of processes shown in Figure 4.

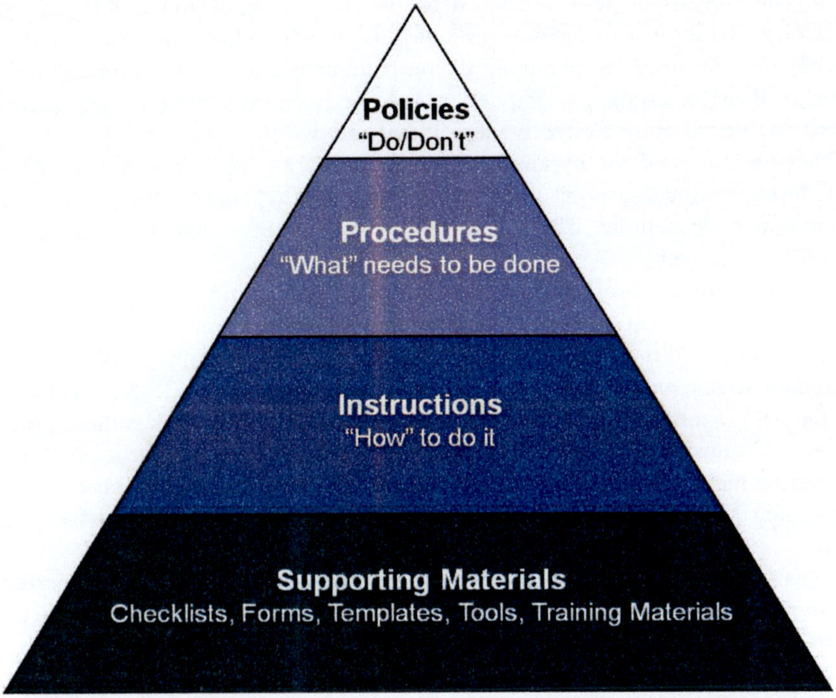

Fig. 4 "Types" of Processes

At the top of the pyramid are policies. Policies document organizational intent. They are usually expressed as a set of "dos" (will be an equal opportunity employer, etc.) and "don'ts" (will not tolerate the use of bribes or kickbacks, etc.). The most effective policies are short and sweet.

Below the policies are the procedures. Procedures capture the "what" needs to be done. Many times, these are considered the requirements of your OSP. As an example, a procedure for risk management may state that the following activities need to be done:

- Risk Identification
- Risk Analysis and Prioritization
- Risk Treatment

- Risk Monitoring
- Risk Reporting
- Risk Retirement

So, for this example, completing the above set of activities at the procedural level and producing the associated set of work products ensures risk management is accomplished.

Below the procedures are the instructions. Instructions capture "how" an activity is to be done. Not all activities need instructions; in fact many should not have them. There needs to be a balance between the skill level and training of your workforce and the particular process areas that require additional detail. Following our risk management example, if you have detected a lack of consistency in the area of risk analysis, you may choose to create an instruction to help increase consistency and improve effectiveness in that activity.

The foundation of the pyramid is the supporting assets. These are the checklists, forms, templates, tools, and training materials that support the policies, procedures, and instructions. These supporting assets can be called out at any level of the process hierarchy.

Regarding process-related LR, they should be primarily incorporated at the procedure and instruction levels. Ideally, your OSP should already contain a significant amount of the material related to the LR; so your primary task should be to update your existing procedures and instructions, extending them as required for the LR. An example would be updating your OSP risk management process for an additional LR category of risk to be considered. Of course, some of the supporting materials may have to be updated, incorporating key aspects of the LR. Significant analysis should be performed before codifying the LR at the policy level.

Regarding product/system-related LR, they should be primarily incorporated at the supporting material level. Most of the product/system-level LR will have to be abstracted to some kind of generic representation. A causal analysis can help determine what OSP changes best accommodate this type of LR. An example would be updating an engineering specialty checklist (e.g., reliability checklist) for an additional LR identified criterion. It should be recognized that some of these product/system-related LR can be very product specific and may not easily be abstracted and be reflected in OSP supporting materials. In these cases, it is recommended that they be communicated by some other knowledge sharing mechanism such as a Community of Practice (CoP) within the organization.

From a new project perspective, the LR are now captured as part of the OSP. Therefore, instead of going to the LL database and attempting to find (pull) relevant LL for their project, the project now gets the LR as part of their tailoring of the (pushed) OSP. In some cases the project will benefit from updated procedures and instructions. In other cases, the project will benefit from updated supporting materials such as checklists and templates. In mature organizations, the replanning process will also point the project back to the OPS on an ongoing basis for the latest set of LR.

4 Experiences with Approach

This approach has been used with success within industrial settings. For example, a US defense firm had just completed a large Preliminary Design Review (PDR) for a software-intensive system that was part of a larger system of systems (SoS). They wanted to capture their LL in such a way that they could become LR. They had a well-defined LL process and post-event analysis was completed in a straightforward manner. Upon identification of the set of LL, the team did a detailed review to determine which ones could be effectively captured into the OSP as LR, and at what level.

Due to the nature of a PDR, many of the LR became codified through updates of the OSP PDR entry and exit criteria. For example, one of the LL dealt with a lack of focus on Total Ownership Cost (TOC), especial in the context of the larger SoS. The LR was captured as an update to the PDR entry criteria and therefore available for all future PDRs.

Some of the LL became LR through updates to the OSP set of procedures. One example was adding activities to the OSP risk procedure to capture the SoS influences now present in this type of program. It was incorporated such that these SoS activities could then be easily tailored out for future programs when not applicable.

However, some of the LL did not lend themselves to becoming LR. The primary LL in this category dealt with the actual system that was under review at the PDR. Technical details that were very specific to the actual system were difficult to capture as updates to the OSP.

5 Limitations and Considerations

No single approach is ideal for every situation, including this proposed LR approach. This paper describes some of the key aspects and anticipated benefits of transitioning from LL to LR. This section outlines some of the limitations of the LR approach and lists some additional considerations.

Usage of this approach does require a level of organizational maturity. Obviously, it is difficult to update an OSP with LR if such an OSP and its corresponding infrastructure do not exist. Organizations should not even consider moving to a LR approach unless they are at least CMMI® (SEI 2010) Level 3 or higher (or some equivalent measure of organization maturity). In addition to having a well-defined OSP, there needs to be an effective process improvement system in place and the EPG has to be staffed with competent people who are empowered to execute the LR activities.

It is important to note that the LR process is more applicable to processes rather than products/systems. As mentioned in the example above, many times the LL related to details of the product/system do not easily translate into changes to the OSP. This is because the OSP, by its nature, is focused on the generic process of how to develop any product/system. As mentioned previously, these types of LL should typically be communicated by some other knowledge sharing mechanism, such as a CoP.

There is a risk that any scheme that attempts to provide a single systematic framework for LR capture will also end up being inflexible. There is a related risk that the scheme will inevitably end up removing critical information in some cases. A mature EPG is necessary to minimize information loss as the LR is incorporated into the OSP. The EPG should leverage the inputs of the LL originators and clarify any questions that arise. From the usage end, the ability to tailor the OSP is key. Not every process is applicable in every situation. Things that are not relevant should be tailored out of the OSP as part of the projects initial and on-going tailoring and planning activities (INCOSE 2011).

Finally, it is important to note that moving to a LR approach will not eliminate the tendency of people to hide or paint over negative results. This is especially true if the EPG is perceived as a policing entity. It is critical that the EPG understand its role and be a "trusted agent" of the organization, not just become a bureaucracy that ends up having too much power. In addition to addressing the above concerns with the EPG, every organization has to embrace a culture where true LL (both positive and negative) are welcomed and appropriately turned into LR.

6 Summary and Conclusions

This paper describes an approach for transitioning from LL to LR. The key concept is to transition from new projects "pulling" relevant information out of a LL database to being "pushed" LR as part of the OSP. Several ways that this process should improve both the submission and application of lessons were elaborated. For this process to work, an OSP needs to exist. The EPG also needs to be empowered to perform continual process improvement on the OSP.

Using this process should result in fewer resources being utilized. In terms of human resources, the slight increase in the EPG efforts should be more than offset by reductions in each of the new projects. In terms of information technology resources, the incorporation of LR into the OSP, and the subsequent deletion of LL for the database should reduce storage requirements. Changes to the OSP should also lead to increased productivity and customer satisfaction.

References

CAC, Center for Army Lessons Learned (CALL), US Army's Combined Arms Center, http://usacac.army.mil/cac2/call/index.asp (accessed April 2011)

CJCSI, Chairman of the Joint Chiefs of Staff Instruction, 3150.25D, Joint Lessons Learned Program (October 10, 2008), http://www.dtic.mil/cjcs_directives/cdata/unlimit/3150_25.pdf (accessed April 2011)

DHS, Federal Emergency Management Agency (FEMA) Learned Information System (LLIS), US Department of Homeland Security, https://www.llis.dhs.gov/index.do (accessed April 2011)

DOT, Intelligent Transportation Systems (ITS) Learned Information Database, US Department of Transportation, http://www.itslessons.its.dot.gov/ (accessed April 2011)
Egeland, B.: A Lessons Learned Template. Project Management Tips (November 2009), http://pmtips.net/lessons-learned-template/ (accessed April 2011)
GAO, NASA: Better Mechanisms Needed for Sharing Lessons Learned, GAO-02-195, United States General Accounting Office (2002)
INCOSE, INCOSE-TP-2003-002-03-03.2.1, Systems Engineering Handbook: A Guide for Systems Life Cycle Processes and Activities, Version 3.2.1. The International Council on Systems Engineering (INCOSE) (January 2011)
Kirsch, D.: Learning Lessons Learned About Lessons Learned? Dr. Dan's Daily Dose (May 16, 2008), http://it.toolbox.com/blogs/dr-dan/learning-lessons-learned-about-lessons-learned-24710 (accessed April 2011)
Knoco Ltd., Knowledge Management Processes, Lessons Learned, Knoco Ltd. Reference Section (April 2010), http://www.knoco.com/lessons-learned-page.htm (accessed April 2011)
MAIC, Mine Action Lessons Learned Database, Mine Action Information Center, http://maic.jmu.edu/lldb/ (accessed April 2011)
Meakin, B., Wilkinson, B.: The 'Learn from Experience' (LfE) journey in Systems Engineering. In: Proceedings of the 12th Annual International Symposium of the International Council on Systems Engineering. The International Council on Systems Engineering, INCOSE (2002)
NASA, NASA Headquarters (HQ) Lessons Learned Information System (LLIS), US National Aeronautics and Space Administration, http://llis.nasa.gov/llis/search/home.jsp (accessed April 2011)
NATO, Joint Analysis and Lessons Learned Centre (JALLC), North Atlantic Treaty Organization, http://www.jallc.nato.int/ (accessed April 2011)
Santayana, G.: Life of Reason. Reason in Common Sense, vol. 1. Scribner's (1905)
SEI, CMU/SEI-2010-TR-033, CMMI® for Development (CMMI-Dev), Version 1.3. Software Engineering Institute (SEI), Carnegie Mellon University (November 2010)
Seningen, S.: Learn the Value of Lessons-Learned. The Project Perfect White Paper Collection (2005), http://www.projectperfect.com.au/downloads/Info/info_lessons_learned.pdf (accessed April 2011)
Walden, D.: YADSES: Yet Another Darn Systems Engineering Standard. In: Proceedings of the Seventeenth Annual International Symposium of the International Council on Systems Engineering. The International Council on Systems Engineering, INCOSE (2007)

Chapter 12
Applicability of SysML to the Early Definition Phase of Space Missions in a Concurrent Environment

Dorus de Lange, Jian Guo, and Hans-Peter de Koning

Abstract. One of latest trends in the field of Systems Engineering (SE) is increased use of models next to and in stead of documents to capture and control information on the system-of-interest under development. This approach is generally called Model Based Systems Engineering (MBSE). An important element of the MBSE approach is the language that is used. The Systems Modeling Language, SysML, is often seen as most promising language by the SE community.

The Concurrent Design Facility (CDF) is a state-of-the-art facility in the field of concurrent engineering and systems engineering research, used to perform feasibility studies for potential future space mission of the European Space Agency (ESA). Currently the CDF uses a simple Excel-based modelling approach. As not much experience with SysML is present at the CDF, research was performed to investigate the potential added value of applying MBSE using SysML to the work performed in the CDF. A methodology and a standard model were developed, followed by its application in a real CDF study to investigate the added value.

From the evaluation of the research, it is concluded that MBSE using SysML is in line with the concurrent engineering approach but is only partially applicable to early design studies. Many of the activities performed during the case study required

Dorus de Lange
Delft University of Technology (TU Delft), Faculty of Aerospace Engineering,
Chair of Space Systems Engineering
e-mail: dorus.de.lange@gmail.com

Jian Guo
Delft University of Technology (TU Delft), Faculty of Aerospace Engineering,
Chair of Space Systems Engineering
e-mail: j.guo@tudelft.nl

Hans-Peter de Koning
European Space Agency (ESA), Systems and Concurrent Engineering Section (TEC-SYE section)
e-mail: Hans-Peter.de.Koning@esa.int

a significant time effort with too little added value. Also, the tool used proved limiting in some of activities. Added value was observed in the capturing of the trade options, decisions and trade rationale in combination with the requirements. It is proposed to integrate these aspects into the CDF studies as starting point for the transition phase to a more model-based approach using SysML.

1 Introduction

Over the last few decades projects have become more and more complex, challenging project teams to keep their grip on the design process. The practice of Systems Engineering (SE) was introduced in the late 1950's to guide the technical effort of projects by enhancing the communication between team members, ensuring consistency in the design to finally enhance the overall product quality.

One of latest trends in the field of SE is the use of models to capture and control the systems information. This new SE approach is called Model Based Systems Engineering (MBSE) and is a next step in managing complexity. Gathering all of the systems properties and information in one central model is to increase consistency and enhance the quality of the SE tasks and thus the final product. The Systems Modeling Language (SysML) is a standardized graphical notation language developed by the OMG (Object Management Group) in support of MBSE.

This paper presents the research on the applicability of SysML, performed at the European Space Agency (ESA) in the Systems and Concurrent Engineering Section which includes the Concurrent Design Facility (CDF). The CDF is a state-of-the-art facility in the field of concurrent and systems engineering research. As the use of SysML is gaining interest within the global SE community, more experience with the language in the CDF is desired. In particular, can SysML add value to the studies performed at the CDF? The focus of this paper is on SysML and not so much on the broader field of MBSE.

2 ESA CDF

The ESTEC CDF (Concurrent Design Facility), established in 1998, is an integrated concurrent design environment available to all ESA programmes for interdisciplinary and inter-directorate applications.

2.1 CDF Activities and Achievements

The CDF performs mostly internal Phase 0 feasibility studies which are to provide ESA with a "faster, cheaper and better" way to thoroughly evaluate mission concepts before selecting Phase A candidate missions. Studies performed at the CDF include [1]: studies of future missions (130+), new launcher concepts (7), complex payload instruments including platform (11), reviews of Industrial Phase A studies (25), joint studies with external parties like NASA/JPL/PDC-Team X or CNES,

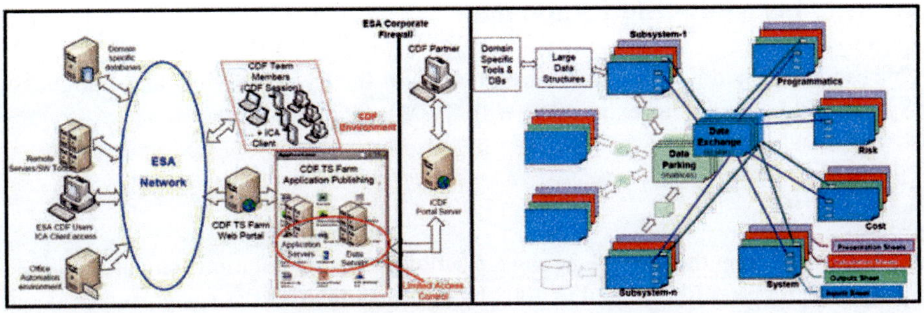

Fig. 1 Overview of the CDF hardware infrastructure (left) and the IDM (right) [1].

anomaly investigation of later project phases as well as educational and training activities. Examples of studies for which Phase-A has been initiated include: ExoMars, Mars Sample Return, Solar Orbiter, PROBA-3, Small Geo and ESMO [1].

2.2 *Study Work Logic*

For the CDF team the starting point is the preparation phase during which the customer, team leader and systems engineers define the study and mission objectives and start defining the mission requirements. The preparation phase ends with the study Kick-Off (KO) where the whole study team (typically 10 to 20 domain experts) get together for the first time. In the KO the study is explained to the team after which the team comes back together for typically 6 working sessions. A working session is half of a working day, and typically there are one or two working sessions per week. In between the sessions the experts work on the design of their subsystem or domain. The study ends with an Internal Final Presentation (IFP).

2.3 *Infrastructure*

The CDF environment is comprised of hard- and soft-ware, an Integrated Design Model (IDM) and the domain experts. The IDM is an Excel-Based model used for the data exchange between domain expert and is essentially a first step towards a more model-centred design approach. All experts have their own Excel workbooks, linked via a data exchange workbook to exchange the necessary parameters. The IDM is only used for data-exchange and the domain experts have their own domain specific tools to help in the design process (e.g. Matlab, STK, ThermXL, CATIA, NASTRAN). An overview of the CDF infrastructure is shown in figure 1.

The CDF systems engineer often makes use of MS Excel for general calculations and parametric analysis, MS PowerPoint to communicate to the team and requirements engineering tools such as DOORS to manage requirements. Up till now, no central systems model is made; systems information is distributed over several files.

3 Systems Modeling Language

This section provides a summary of the basics of the Systems Modeling Language (SysML) and of previous experience with SysML from literature.

3.1 Basics

SysML is based on the UML language, a set of graphical notation standards to create models of systems, initiated in 2001 by OMG and INCOSE. The current version of SysML is version 1.2 [2]. SysML defines 9 diagram types: Activity diagram, Sequence diagram, State machine diagram, Use case diagram, Requirement diagram, Block definition diagram, Internal block diagram, Package diagram, Parametric diagram. These diagrams are grouped in the four pillars that make up the SysML language, shown in figure 2.

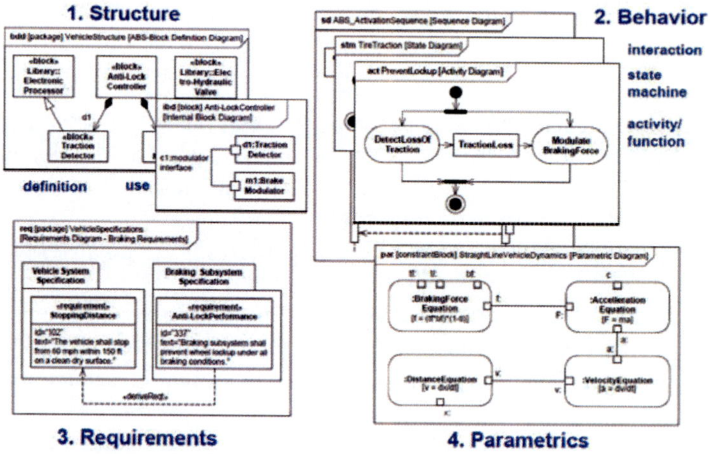

Fig. 2 SysML example model of an automatic braking system, showing the four pillars of the SysML language [11].

3.2 Review of Use in Space Projects

In investigating the use of SysML in the CDF, first a literature survey was performed. Some conclusions and quotes are shown here. Although there has been scepticism that SysML is not yet mature enough, JPL concluded the opposite after a pilot study where the Jupiter Europa Orbiter Mission was modelled using MagicDraw [8], as can be seen from the following quote [3]:

> However, the pilot also found that tools for specifying systems, such as SysML, are mostly ready to do their jobs.

Another example is the "Modeling Pilot for Early Design Space Missions" study performed by JPL [5]. In this study the goal was to investigate the use of MBSE with SysML for early mission phases. They concluded the following:

> The pilot illustrated some of the utility and benefits of the MBSE/SE-CIM (Systems Engineering Conceptual Information Model) approach in the early stages, preformulation/formulation of a space mission. The pilot intended to show that the MBSE approach and in particular the SE-CIM may offer significant benefits for specifying and designing complex mission system in addition to and enhancing the current, more traditional approaches.

Other examples of studies on the use of SysML can be found in [4] [6] [7]. All-in-all, experiments have been performed in the recent past, but up till now no evidence was found of any institute, company or agency to have adapted the use of SysML as a new standard, as opposed to the traditional document based approach [12].

4 MBSE Methodologies

At the start of any project, a methodology is to be selected. It prescribes what to do, how to do it and in what order.

4.1 Existing Methodologies

Many different methodologies exist and a survey of MBSE methodologies was performed by Jeff Estefan, documented in [9], in which he described the current leading MBSE methodologies. Apart from these methodologies, Intecs, ThalesAleniaSpace and ESA developed together an MBSE methodology called the Model-Based methodology to support the Space System Engineering (MBSSE) [10]. The MBSSE is a result of a search for a model based systems engineering methodology, compliant with the relevant ECSS and ISO standards.

4.2 Used Methodology

Before selecting a methodology it is important to understand the purpose of the methodology. Criteria for the methodology were that it should be: "as close as possible" to the current CDF methodology and be "as simple as possible" without leaving out valuable activities and artifacts in making it "simple".

After a critical assessment of the methodologies described by Estefan [9], most methodologies were discarded as they were either too much software oriented or contained too many activities which were not applicable to the CDF environment. The Object-Oriented Systems Engineering Method (OOSEM) and the MBSSE were analysed in detail and the relevant activities from both methodologies were combined with newly defined activities into a more streamlined, better implementable and more applicable methodology to the CDF. The methodology is shown in the work flow diagram in figure 3. The main activity flow (red) contains the main study activities. The middle flow (yellow) contains model maintenance activities and the blue flow contains methodology maintenance activities.

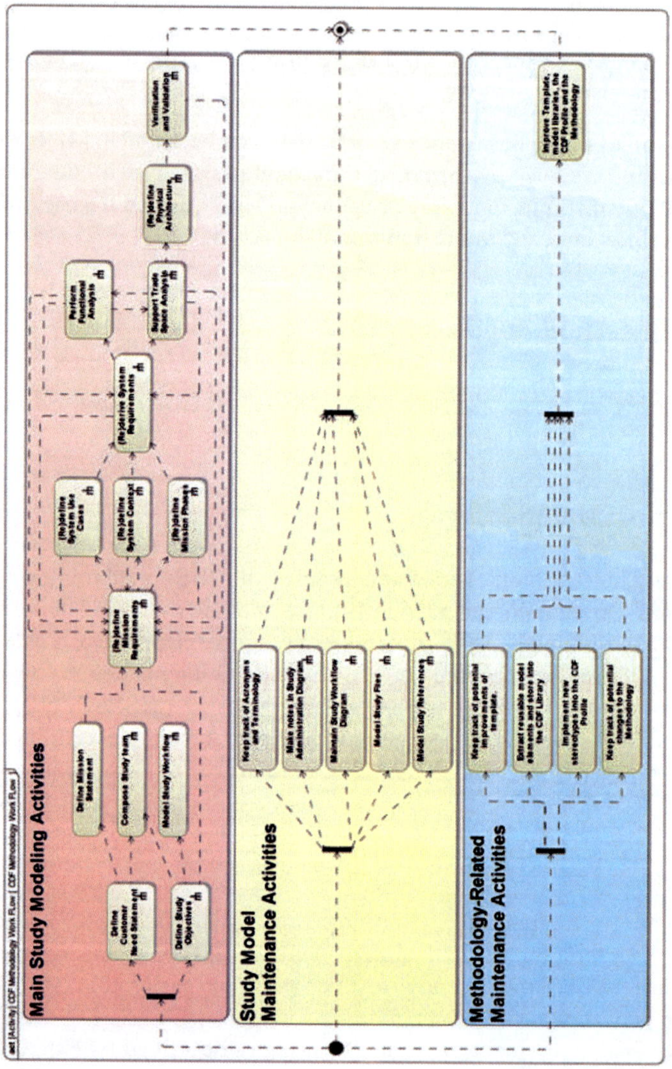

Fig. 3 Simplified work flow diagram (SysML activity diagram) of the proposed methodology to be used in the ESA CDF studies.

5 Case Study

Part of the research included the development of a SysML model that can be used in future CDF studies, called the CMSSM (CDF MBSE SysML Study Model). The CMSSM contains four modules: the CDF profile, the CDF library, the CDF methodology and finally a template. Instead of describing the model it is decided to go to directly to the case study as this shows directly its application.

5.1 NEMS CDF Study Background

The CDF study selected for the case study is the Near-Earth Exploration Minimum System (NEMS) study which started with the Kick-Off on 2011-03-23 and the Internal Final Presentation on 2011-04-20. The NEMS mission statement was defined as follows:

> Transfer a crew of 3 astronauts from the Earth to an accessible Near-Earth target beyond the Earth-Moon system (e.g. NEO), perform a rendezvous with the target, stay for 10 days to perform EVAs and safely return the crew back to the Earth before the end of 2030.

Note that ESA is not designing a human mission to go to a Near Earth Asteroid, but performed the study to gain a better understanding of the implications as a means to identify critical technologies and development activities.

5.2 Model Structure

The drivers for the definition of the modules and the structure of the model were identified as follows:

- the organisation of the model shall support the CDF methodology,
- the packaging of the model shall be "simple" and easy to understand for the CDF system engineers,
- the structure shall be flexible in the sense that it should support any study type,
- the structure shall support multiple mission and system options,
- the model shall be use-able after the CDF study, in next project phases,
- the structure shall encourage/ensure re-use of model elements using libraries,
- a profile shall be set up for CDF specific stereotypes,

The structure of the model was derived iteratively from the drivers listed above, based on new experience gained during the case study. The highest level package diagram is shown in figure 5, showing the four modules: "NEMS Study Model", the "CDF Profile", the "CDF Library" and the "CDF Methodology". The CDF library contains re-useable elements and the CDF profile contains 20 model element stereotypes and three relation stereotypes, specific for the CDF environment, of which an example is shown in figure 4. The CDF methodology contains the definition of the CDF methodology (section 4.2), which is also integrated into the template (and thus study model) by means of hyperlinks to detailed activity diagrams to support the modeller in the modelling process. E.g, the functional analysis overview diagram contains a link to the methodology activity diagram containing instructions on the functional analysis activity.

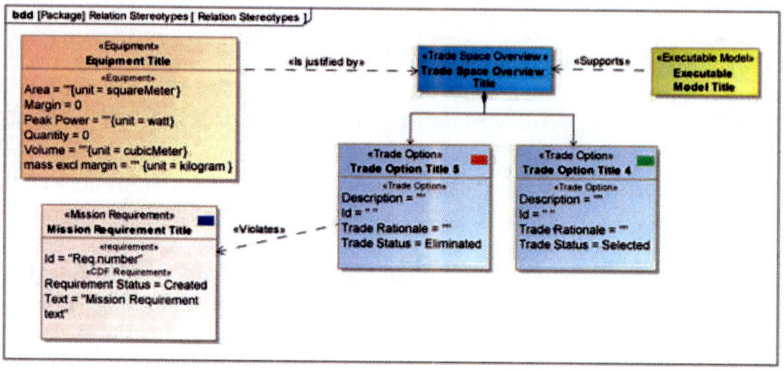

Fig. 4 Example of the relation stereotypes and some of the model element stereotypes as defined in the CDF profile.

5.3 *NEMS Model in Brief*

Figure 5 shows the content overview of the NEMS study model with its main packages. As two mission options were designed during the study, the intention was to model both options. As the options were at least 90 percent the same and variant modelling is not properly supported in SysML or the used tool, it was decided not to model the second design option as no added value to the research would be gained.

The study information is stored in the "CDF Study Information" package. It serves as a starting point to the study, containing most of the background information such as the study objectives, study workflow and references. Requirements management becomes difficult when dealing with multiple mission options in a single SysML model. After trying several methods it was decided to have a single core requirements set for all mission options and add the mission option specific requirements as "Mission Option Requirement" stereotypes, with a different requirement ID prefix as compared to the shared requirements. A new requirement stereotype was introduced to indicate the requirement status by means of Enumeration Literals: created, TBD (To Be Determined), TBC (To Be Confirmed), Fixed and Eliminated (to maintain the ID and history of removed requirements).

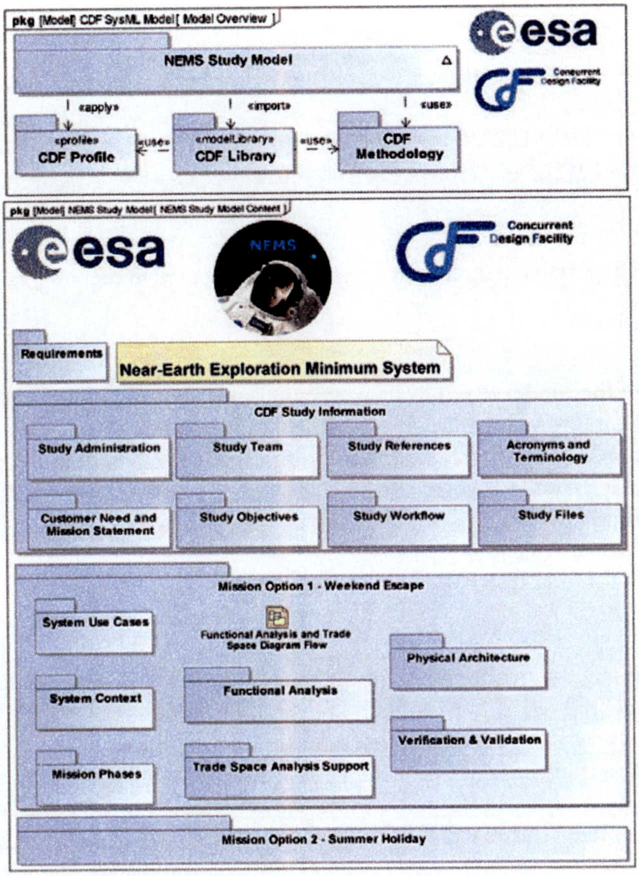

Fig. 5 Highest level package diagram (top) showing the four modules and the model content diagram (bottom) of the NEMS study SysML model.

For the functional analysis the stereotype "Function" of the activity was introduced to capture functions. The support to trade analysis will strongly depend on the type of study. It could vary from support by means of executable SysML in analysis and simulation, to capturing of trade-options and trade rationale. For the NEMS study the latter was the case as the study remained on a relatively high mission level and contained many mission level trades where the identification and rationale of trade-offs stood central. This support was done by introducing the "Trade Option" stereotype of the SysML Block with "Trade Rationale" and "Trade Option Status" as main attributes such that trade option trees can be carefully captured and traced. Trade options can be linked to requirements by means of the "Satisfy" or the "Violates" (stereotype of satisfy, to indicate the opposite) relations. The status indicator is used as is shown in figure 6 and can be set to "Identified" (blue), "Not preferred" (pink), "To be Traded" (yellow), "Selected" (green) and "Eliminated" (red).

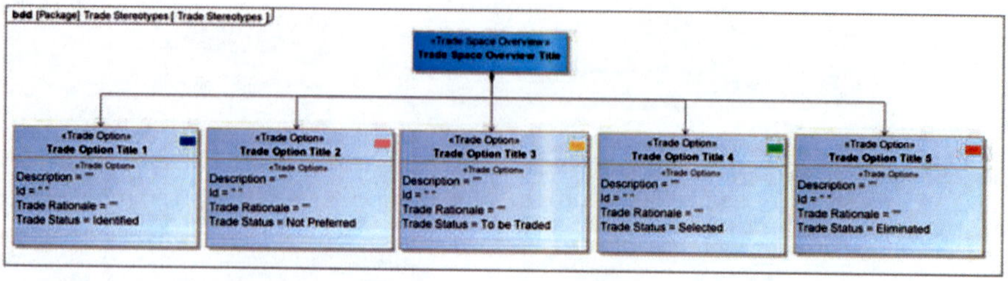

Fig. 6 Trade option stereotype, defined in the CDF profile, used to capture trade options and rationale.

The flow in the model is not everywhere the same, but in general: functions satisfy requirements, functions are allocated to "Trade Space Overviews" which are the trade option discovery trees containing the trade options that are to fulfil the functions and satisfy the requirements up the flow. A trade space results finally in physical architecture elements which make up the system.

6 Evaluation

Four methods for the evaluation of the research are used. The modelling history and model statistics were tracked and analysed, difficulties in the modelling process were analysed, an evaluation based on the modelling experience was carried out and finally a questionnaire containing 19 questions was distributed through the CDF team. The questionnaire guided interviewees through the case study model and seven completed forms were received and analysed. In this section the observed strengths and weaknesses of SysML, MagicDraw (used tool) and the use of SysML in the CDF are summarized.

6.1 SysML

For the SysML language, the following main points are identified:

- SysML has a steep learning curve, slowing down the transition to MBSE,
- Language is flexible and allows to model all the systems aspects,
- Language is standardized, resulting in an increase of consistency in complex projects and better communication,
- Requirements diagram quickly becomes unreadable when dealing with 100+ requirements, and is not adding value when dealing with only a few requirements when the relations are obvious.
- Relations could be made more readable. A relation such as "Allocate" could be renamed "Is allocated to" to make the relations more self-explanatory for non-SysML experts. The "Satisfy" relation could also be interpreted in two directions, namely as "Satisfies" and "Should Satisfy".

Based on the performed research, it can be concluded that SysML is a good step towards the standardization of MBSE, but it for sure is not the last step. Its flexibility has advantages but makes the language difficult to learn, severely slowing down the transition phase. For ESA it is important to be involved in SysML as much as possible, as the language is likely going to be become the standard and now is the time to shape it. The more the language is formally standardized and used in time, the less it can be changed later on.

6.2 Executable SysML

The modelling of the case study concerned a static model, with no use of executable SysML. For the CDF, two potential applications are identified for executable SysML. Firstly, executable SysML could support in trade-off processes by guiding for example the performance calculations. The second form of executable SysML that could add value is the exchange of information with the IDM. After a brief analysis it was concluded that the performance calculations take a significant effort to model, with only limited re-usability due to the high variance in the CDF studies. Integrating the SysML model with the CDF IDM requires a significant development effort with limited added value due to the overlap.

Based on the limited analysis performed in the thesis work, it is concluded that executable SysML is likely not adding value to the studies performed in the CDF due to the variance in the studies and the limited re-use opportunities. To be more certain of this statement, a dedicated study on executable SysML is highly recommended.

6.3 MagicDraw

For the MagicDraw tool, the following main points are identified:

- Tables and matrices need to be improved by adding more flexibility by adding more options for the user. Sorting, zooming, fixing the export function of the generic table which sometimes leaves out some of the attributes,
- Reasonably stable tool with in general an acceptable speed,
- Overall graphics need to be improved, especially the rendering of text in symbols,
- Introduce more basic drawing skills, starting with functionality similar to the PowerPoint "AutoShapes". This allows for making better drawings and thus being able to use the model in communicating to the team,
- Quick layout functions needs to include more user options (for example, distance between symbols) to provide for faster modelling,

Although there are still some weaknesses, MagicDraw is a mature tool for SysML modelling. However, the identified weaknesses need to be diminished before it can be used to its full potential for SysML modelling in the CDF.

6.4 Use of SysML in the CDF

Even though a lot of information is gathered concerning the strengths and weaknesses of SysML, MagicDraw and the used methodology, it is not easy to draw a definite conclusion regarding the applicability of SysML in the CDF. The following concluding statements on MBSE using SysML in the CDF are derived:

- Trade-offs, study assumptions and decisions can be captured in the model to provide the relation to the requirements and **increase traceability**, thus adding significant value,
- Modelling the study/mission/system provides the modeller with an overall **better understanding** due to structured and systematic approach,
- Using MagicDraw (v17), the model is **less effective in communication** to the team as compared to PowerPoint due to limiting drawing capabilities and poor graphical quality. This drawback can be solved by improving the tool. When the diagrams look nicer and clearer, they can be used to communicate to the team, thus removing the need to maintain the usage of both PowerPoint and the model.
- As the IDM is present in the CDF, there is **no use in manually modelling the physical architecture** properties, as long as no automatic link is present,
- **Transition** to SysML will be difficult as SysML is a tough language to learn,
- The significant effort of introducing **executable SysML** likely will not be justified by the resulting added value for the Phase 0 studies performed in the CDF due to limited re-usability opportunities,
- SysML does not provide sufficient support in **variant modelling**.

For now, MBSE using SysML can add value by using it next to the IDM to model the requirements and trade-offs. When the MagicDraw tool improves graphically, slowly it could be used in communicating to the team. Finally, the modelling activity itself provides the modeller with a better understanding of the system.

7 Conclusions

At ESA, attempts have been made before to evaluate the use of SysML in (CDF) studies, but all of these attempts were passive in the sense that whatever was modelled was not used actively during the study itself. Attempts included the modelling of ExoMars during the development of the MBSSE [10] and the modelling of the Mars Atmosphere Sample Return (MREP) CDF Study by a MSc student at the end of 2010. Actively using modelling with SysML results in more direct and practical conclusions related to the strengths, weaknesses and applicability.

Only a small part of the SysML language was used during the case study. It might be that for the Phase-0 studies a kind of "SysML-light" version would be more suitable, after which the full SysML language would be used from Phase-A onwards. The used SysML tool is mature enough to perform valuable work but requires improvements mostly in terms of graphical representation in order to replace the more traditional methods of communication to the CDF team.

From the research, it is concluded that MBSE using SysML is in line with the concurrent engineering approach but only partially applicable to Phase-0 studies. For now, it is recommended to the CDF to introduce the use of SysML next to the IDM to model the requirements and trade-offs. This would be a good starting point for the transition phase to a more model-driven design approach using SysML. Also, a dedicated study on the applicability of executable SysML is recommended.

Acknowledgements. The authors would like to express acknowledgement to ESA ESTEC and TU Delft for providing support to the thesis work in the CDF.

References

1. ESA CDF Team. The ESA Concurrent Design Facility: Concurrent Engineering applied to space mission assessments, look for CDF Presentation, `www.esa.int/cdf` (cited April 12, 2011)
2. OMG SysML Partners. OMG Systems Modeling Language (OMG SysML) Version 1.2, `http://www.rsc.org/dose/title` of Subordinate document (cited April 12, 2011)
3. Cole, B., Delp, C., Donahue, K.: Piloting Model Based Engineering Techniques for Spacecraft Concepts in Early Formulation. In: JPL and INCOSE. July 12-15, INCOSE, Chicago (2010)
4. Alameda, T., Trisch, T.: Deployment of MBSE processes using SysML. In: 13th Annual Systems Engineering Conference U.S. Army Research, Development and Engineering Command, San Diego, October 25-28 (2010)
5. Mandutianu, S.: Modeling Pilot for Early Design Space Missions. In: 7th Annual Conference on Systems Engineering Research (CSER), April 20-23. Loughborough University (2009)
6. Tactical Science Solutions. Inc. Quicklook Final Report V1.19, in support of the Tactical Satellite-3 design effort (2007)
7. Karban, R., Zamparelli, M., Bauvir, B., Koehler, B., Noethe, L., Balestra, A.: Exploring Model Based Engineering for Large Telescopes - Getting started with descriptive models. Society of Photo-Optical Instrumentation Engineers, SPIE (2008)
8. No Magic, Inc. MagicDraw modeling tool, `www.magicdraw.com` (cited April 12, 2011)
9. Estefan, J.: INCOSE-TD-2007-003-02: Survey of Model-Based Systems Engineering (MBSE) Methodologies Rev. B. (2008)
10. Mazzini, S.: Model Based Space System Engineering (MBSSE). In: System Design & Simulation Workshop, Rome (October 21, 2010)
11. Friedenthal, S., Moore, A., Steiner, R.: OMG Systems Modeling Language(OMG SysML) Tutorial. In: INCOSE (2006)
12. Eisenmann, H., Miro, J., de Koning, H.P.: MBSE for European Space-Systems Development. INCOSE Insight 12(4), 47–53 (2009)

Chapter 13
Requirements, Traceability and DSLs in Eclipse with the Requirements Interchange Format (ReqIF)

Andreas Graf, Nirmal Sasidharan, and Ömer Gürsoy

Abstract. Requirements engineering (RE) is a crucial aspect in systems development and is the area of ongoing research and process improvement. However, unlike in modeling, there has been no established standard that activities could converge on. In recent years, the emerging Requirements Interchange Format (RIF/ReqIF) gained more and more visibility in the industry, and research projects started to investigate these standards. To avoid redundant efforts in implementing the standard, the VERDE and Deploy projects cooperate to provide a stable common basis for ReqIF implementation that could be leveraged by other research projects as well. In this paper, we present an Eclipse-based extensible implementation of a RIF/ReqIF-based requirements editing platform. In addition, we also investigate two related aspects of RE that take advantage of the common platform. Firstly, how can the quality of requirements be improved by replacing or complementing natural language requirements with formal approaches such as domain specific languages or models. Secondly, how can we establish a robust traceability mechanism that links different artifacts of a development process like requirements, design etc.

1 Motivation

The increasing complexity of embedded systems presents ever-growing challenges for projects. The implementation of functions demand more than merely designing

Andreas Graf
Name, itemis GmbH
e-mail: andreas.graf@itemis.de

Nirmal Sasidharan
itemis GmbH
e-mail: nirmal.sasidharan@itemis.de

Ömer Gürsoy
itemis GmbH
e-mail: oemer.guersoy@itemis.de

and implementing individual components. The interplay of several components is almost always necessary to provide the required functionality. The resulting communication networks and their relations can become complex and unclear. Considering that such systems are mostly implemented as a collaborative effort of several organizational units, this could become technically challenging. According to Conway, this also influences technical characteristics and processes. Additionally, a Product Line Approach (PLA) of systems development requires development, validation and maintenance of different software variants. This underlines the need of capturing and tracing of system requirements to ensure the quality of deliverables In the research project "VERDE", itemis uses an Eclipse-based tool for capturing and tracing of requirements. The tool is intended to be used as a platform to provide solutions for several problems:

- Requirements Exchange: To enable exchange of requirements with other tools, ReqIF has been implemented as the meta-model. The implementation based on Eclipse platform enables tighter integration of the tool chain.
- Requirements Capturing: In co-operation with the project "ProR", a graphical user interface for the capturing of requirements is provided as open source.
- Formal notations: Requirements have been traditionally created in natural language. This has often led to ambiguity and redundancy. We borrow the trend of 'Domain Specific Languages' from software engineering, which enables creation of descriptive languages for a specific purpose, and integrate the concept and related tools with requirements. This permits users to adopt flexible formal notations.
- Integration with modeling: Analysis and design is often supported by models such as UML, state machines etc. We show how the use of the Eclipse platform supports a tight integration with requirements.
- Traceability: The traceability of requirements is often hindered due to gaps in tool chains. Eclipse, as a platform, offers a solution to alleviate this problem. For this purpose, a tracing model is being implemented, which facilitates tracing of requirements, model elements, source code etc. without having to modify the metamodels and/or its tooling.

2 The ITEA2 VERDE Research Project

The Research Project VERDE, which is carried out under the umbrella of European ITEA2 projects and sponsored by the German BMBF, aims at providing a integrated Eclipse-based tool platform for the verification- and validation-orientated development of embedded systems.

VERDE is the abbreviation for VERification-oriented and component-based model Driven Engineering for real-time embedded systems. The project focuses on the integration of the tools which are already in use by the industrial partners. VERDE develops new tools and methods in the areas where there are gaps in existing tool-chains and procedures. itemis GmbH, as one of the industry partners in the cross-industry project, deals with the aspect of requirements and traceability.

3 The Target Platform Eclipse

Until a few years ago defining a common/homogeneous platform for a research project would have been quite difficult. Tools are based on different programming languages (Java, C or C++) and employ a variety of different libraries. So there is no seamless interplay between them. Many companies experience this problem in the process chain. This has led to the establishment of Eclipse as an open tool platform in recent years. It was originally developed as a Java development environment but has now evolved into a powerful infrastructure for tool programming.

All of the VERDE industrial partners are currently using third-party and internally developed tools which are based on Eclipse. The success and the high suitability for the integration process is due to the architecture and the extensive basic functionality of Eclipse. From data storage and modeling to development environments for different programming languages through to web applications Eclipse projects provide a host of components which would otherwise have to be developed from scratch in a time-consuming and cost-inefficient process. New functionalities could be easily added by developing "plug-ins", the basic building block of the Eclipse framework. Plug-ins define extension points so they in turn can be modified and extended easily. The users only need to start one application where they can find numerous different tools integrated with a common interface. Tedious switching between different programs is no longer necessary. Eclipse supports the functional connection between different tools with a common mechanism for defining and storing application data, the Eclipse Modeling Framework (EMF). This de-facto standard simplifies the integration of data from various tools. An extensive functionality for the different phases of the development process already exists in the form of tools: architecture (SysML), software-Design (UML), AUTOSAR (Artop) and implementation (CDT). An integrated tool chain can build on these tools and on commercial/third party offerings. The economic and organisational advantage for the companies lies in the fact that there is a wide range of know-how available on the market. Open source licenses are gaining an increasingly good reputation as a strategic guarantor for long-term platform availability and maintenance.

SPONSORED BY THE
Federal Ministry
of Education
and Research

Fig. 1 Sponsored by the BMBF

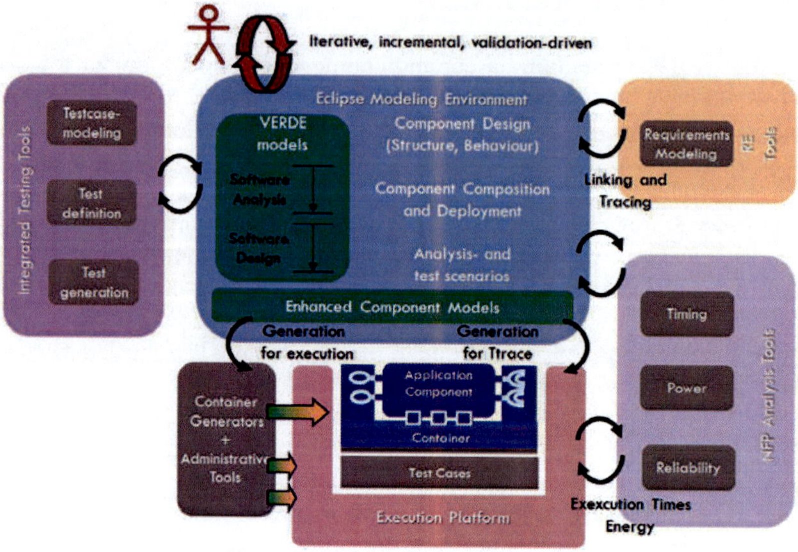

Fig. 2 The scope of the VERDE research project

4 Requirements Exchange

Companies are often faced with a dilemma regarding requirements engineering (RE). It is expensive to establish mature processes, and implementing those processes requires expensive tools. As a result, we often see decent processes (due to continuous improvement) that are held back by inadequate tools like Word or Excel. Also, due to the lack of standards, the development of those processes was usually expensive.

We see parallels in this to the state that modeling was in during the late 1980s and early 1990s. During that time, there were dozens of incompatible modeling methods and notations. Eventually they converged into UML. This paved the way for interoperable tools and vastly improved communication.

We currently see the effects of the Requirements Interchange Format (RIF/ReqIF). RIF/ReqIF [1] is an emerging standard for requirements exchange, driven by the German automotive industry. It consists of a data model and an XML-based format for persistence. We believe that RIF may have a profound impact on the RE tool landscape and on how organizations work with requirements. While there are many differences to the history of modeling, we still hope to gain some insights from that event.

Finally, we are currently building a RIF-based open source platform for requirements engineering in the joint efforts of VERDE and ProR.

4.1 *The Requirements Interchange Format (RIF/ReqIF)*

RIF/ReqIF [1] is an emerging standard for requirements exchange, driven by the German automotive industry. It consists of a data model and an XML-based format for persistence.

4.1.1 History of the RIF/ReqIF Standard

RIF was created in 2004 by the "Herstellerinitiative Software", a body of the German automotive industry that oversees vendor-independent collaboration. Within a few years, it evolved from version 1.0 to the current version 1.2. The format gained traction in the industry, and a number of commercial tools support it.

In 2010, the Object Management Group took over the standardization process and released the ReqIF 1.0 RFC (Request For Comments). The name was changed to prevent confusion with the Rule Interchange Format, another OMG standard.

4.1.2 The Structure of a RIF Model

In ReqIF, a *SpecObject* represents a requirement. A SpecObject has a number of *AttributeValues*, which hold the actual content of the SpecObject. SpecObjects are organized in *SpecHierarchyRoots*, which are hierarchical structures holding *SpecHierarchy* elements. Each SpecHierarchy refers to exactly one SpecObject. This way, the same SpecObject can be referenced from various SpecHierarchies.

RIF contains a sophisticated data model for *Datatypes*, support for permission management, facilities for grouping data and hooks for tool extensions. The details can be found in the ReqIF specification [1].

4.1.3 RIF Tool Support Today

A handful of tools in the market claim to support RIF, the most prominent being Rational DOORS[1]. There are third parties offering RIF-based tools and solutions, including Atego[2] and Enso Manager[3]. While we did not have a chance to try out these products ourselves, we informally talked to attendants at SEISCONF 2010 and GI Fachgruppentreffen RE 2010. The comments we got from users were mixed at best. We got the impression that RIF was merely used as another marketing item on the tool's feature list. We also got comments indicating that the ReqIF 1.0 release is highly anticipated.

---

[1] http://www.ibm.com/common/ssi/fcgi-bin/ssialias?infotype=PM&subtype=SP&appname=SWGE_RA_RA_USEN&htmlfid=RAD14037USEN&attachment=RAD14037USEN.PDF

[2] http://www.atego.com/products/atego-exerpt-synchronizer/

[3] http://reqif.de/

4.1.4 Lessons from the Impact of UML on the Modeling

Our expectations on the impact of ReqIF on requirements engineering practices in companies are derived from experiences with the development of model driven software development after the publication of the UML standard. Before UML was published, a number of incompatible modeling approaches where on the market (Grady, Booch, ROOM).

After the introduction of UML, most modeling books, tutorials and tools flocked around that standard. This made building qualified teams easier, since the common notation removed the strong qualification and the rich availability of information lowered the entry barrier. On the tooling side, the common notation increased the market for low-price tooling and it removed a barrier that hindered open source tooling: The agreement on a modeling notation and thereby increasing the community for open source tools. Tools like Papyrus and Topcased have already been used in commercial projects Although newest trends in modeling like domain specific languages are moving away from the UML, it remains a strong candidate for a lot of modeling projects.

Right now, the low-cost and open source tool landscape for requirements tooling does not over a rich selection for SMEs. The situation is similar to the time before the advent of UML: No established standard of requirements exists. The closest approach might be the requirements specification within SysML. However, this is closely related to the use of SysML as a modeling method, which might not be suitable for all the projects.

The ReqIF standard seems flexible enough to accommodate several approaches on requirements engineering while being specific enough to allow for tool implementation. Since the standardizing body for both UML and ReqIF is the OMG, the chances for industry-wide acceptance seem rather good.

4.2 Model Driven Tool Development

The implementation of paper-based standards is a tedious and error-prone thing. For RIF, however, the standardization partnerships provides an UML-based meta-model for requirements, which is available for a low-cost commercial tool, Enterprise Architect. Based on this model, we used tools for model driven software development to reduce manual implementation.

The foundation of Eclipse-based tooling is the Eclipse Modeling Framework (EMF), which has its own definition format for models. We used the open source tooling Xpand to generate the EMF definition for RIF 1.2 out of the UML-model. This is the basis for several other requirements tools. itemis is currently preparing a project proposal to publish the implementation as an official open source project under the Eclipse Foundation.

5 Requirements Capturing

During the course of the project, we found another project dealing with RIF/ReqIF and started a cooperation. The tool from the Deploy project is called ProR[5] and available for download[4]. It is a full-featured editor for RIF-Documents. To leverage synergies, itemis will maintain the core (requirements meta-model) as the standard evolves and will extend the GUI of ProR. The GUI of ProR is shown in Figure 3. The left pane shows projects that can be expanded to show their content. The view in the middle shows the RIF Specifications, which can be customized to show only selected Attributes in a table view. For the selected SpecObject (the selected row), all Attributes are shown in the Properties View on the bottom. Values can be edited in the Property View or directly in the main view. On the right side is an outline view that allows an alternative navigation of the model.

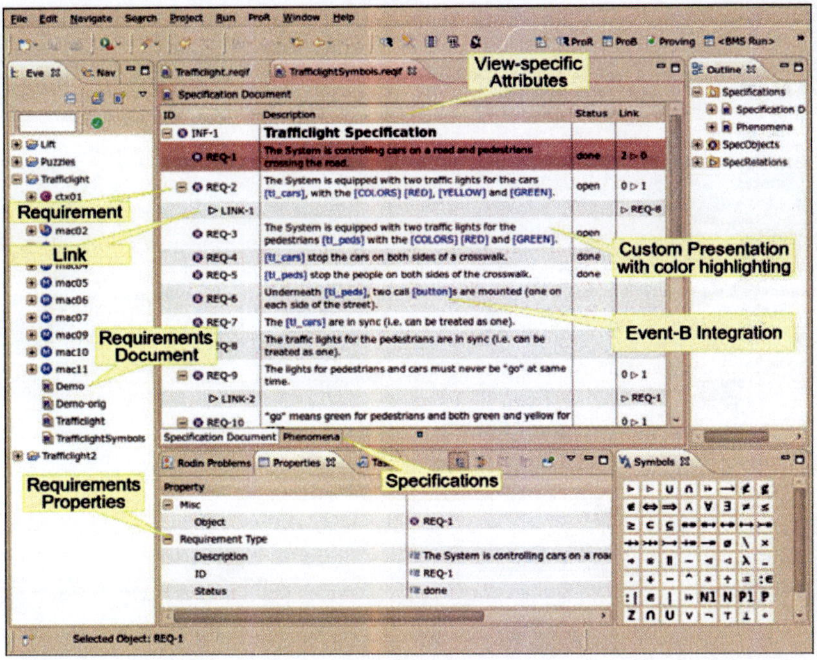

Fig. 3 The ProR GUI (here shown running inside the Rodin Platform)

6 Formal Notations

Requirements are usually expressed in natural language. This however suffers from the disadvantage of imprecision and ambiguity. In addition, it is still very difficult to

[4] http://pror.org

process natural language by software. Accordingly, requirements engineers recommend approaches to increase the structure and degree of formality of requirements. A well known approach is the usage of sentence patterns. A higher degree of formality is accomplished by using "Controlled Language", which is the first step in bridging the gap between being "readable" and "processable by software". This can be improved by textual domain specific languages. Figure 4 shows an example DSL for use case scenarios.

```
role Commander
role CommandModulePilot
role LunarModulePilot

abstract actor Astronaut {
    states:
        alive
        sick
        dead
    verbs:
        speak
        enter
}

actor Armstrong is an Astronaut roles=Commander
actor Collins is an Astronaut roles=CommandModulePilot
actor Aldrin is an Astronaut roles=LunarModulePilot

requisite Speech description="One small step for a man, but one giant leap for mankind"
requisite CommandModule description="The command module carries the astronauts." {
    states:
        ok
        damaged
    verbs:
        separate
}
requisite LunarModule description="The lunar module was used to descent to the moon" {
    states:
        docked
        landed
    verbs:
        land
        launch
}

story LandingOnTheMoon description="How to land on the moon" {
    as a LunarModulePilot
    given
        LunarModule is docked
        Aldrin is alive
    objectives
        Aldrin wants to land LunarModule
    so that
```

Fig. 4 A DSL for expressing use case scenarios

6.1 Domain-Specific Languages

A domain specific language (DSL) is a machine-processable language which is designed to express specific aspects of a system in a specific domain. The concepts and notations used correspond to concepts used by the stakeholders of the system. DSLs are a well established approach in software engineering and have found wide acceptance to due increasing tool support.

It is now very easy to build DSLs with "language workbenches" that take a language specification and generate rich development environments, including syntax-highlighting, navigation, auto completion etc. We use the open source tooling Xtext, which is provided by itemis as a open source tooling and has won several awards. Figure 5 shows a grammar definition for a requirements language that allows both free text, that must use the "Shall"-form, and also a formal notation for conditions.

```
grammar de.itemis.xtext.Oclnl with org.eclipse.xtext.common.Terminals

generate oclnl "http://www.itemis.de/xtext/Oclnl"

Ocl:
(statement+=Statement)*;

Statement:
(subject1=ID) (logicalRule+=LogicalRule)* (predicate=Predicate) |
(subject1=ID) "SHALL" (nllPredicate+=ID)+;

LogicalRule:
(logicalop=LogicalOp) (subject2=ID);

enum LogicalOp:
 and|or|xor;

Predicate:
(condition=('is equal to' | 'is greater than' | 'is less than' |
    'is not equal to' | 'is enabled' | 'is disabled')) (value=STRING)?;
```

Fig. 5 DSL definition with Xtext

6.2 *Integrated Tooling*

Figure 6 illustrates how the editor for the DSL integrates into the requirements tooling. The left pane shows the requirements structure. The content of one of the requirements is displayed on the right. The editor for the content is fully generated from the grammar in 6.1. The fourth line does not conform to the grammar ("rpm" is not followed by a valid keyword) and is marked accordingly, including content assist for a valid choice of keywords.

7 Integration with Models

The support for DSLs does not end here. Requirements will often refer to parts of models, like state-machines, sequence- or class diagrams. With Xtext, both the requirements and the referenced models are available within the same tooling, allowing for a tighter integration. From the requirements editor, the user has full access to other models and references can be fine-grained. Usually, the granularity would be from a requirement to a model or model element (Figure 7 top). It is now possible, to create references from within a requirement (Figure 7 bottom).

8 Traceability

To assure that the requirements are fully implemented in the system, it is necessary to trace them during the whole development process [2]. The effect of a requested

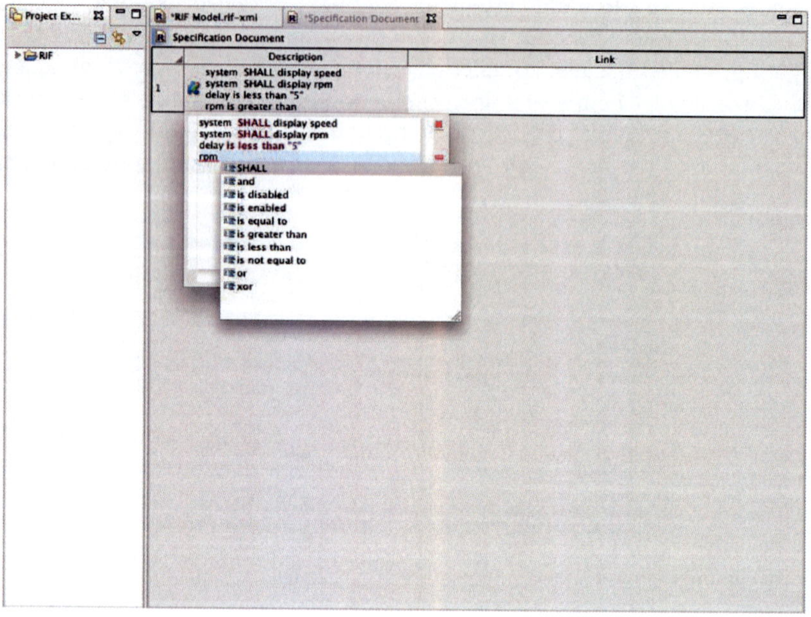

Fig. 6 Requirements and DSL

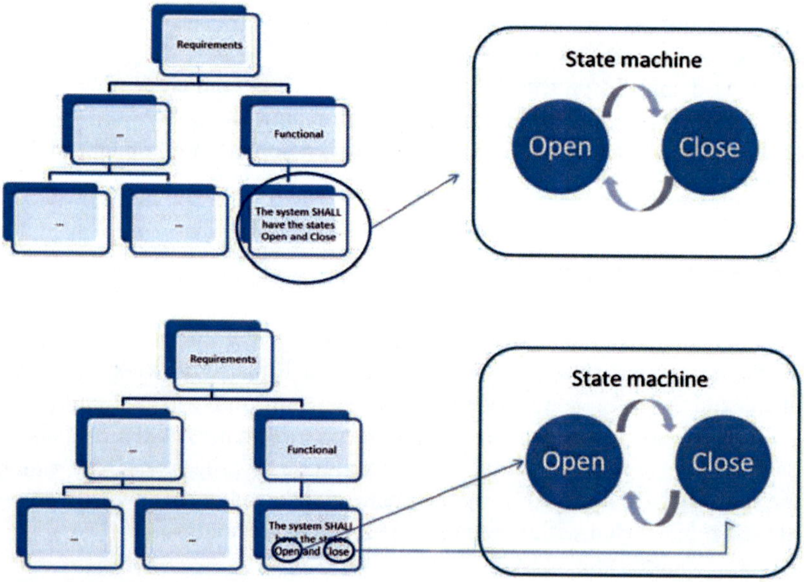

Fig. 7 Fine-grained references

change in any of the software artifacts can be assessed by following the traces up and down the process. Therefore it is necessary not only to assign requirements to artifacts but also to relate artifacts of earlier phases with artifacts of later phases.

8.1 Tracepoint Approach

This general concept of traceability in VERDE led to the decision of the implementation of a traceability that is independent of the types of artifacts that are involved. Since Eclipse-based models are usually built on the Eclipse meta-model Ecore/EMF[3], VERDE implements a generic solution for the traceability of EMF-based elements (see Figure 8). The core of the implemented solution is a mapping table with three elements: source element, target element and additional information like description etcetera. The elements are identified by a data structure, the so called "Tracepoint". The inner structure of a tracepoint depends on the structure of the meta-model that is being traced, but is hidden from the traceability infrastructure. E.g., a UML model element is identified by an unique ID, a C source file in contrast is identified by file path and name. The basic implementation supplies an "assignment window" that integrates in to the existing editors (like UML, C-IDE) to provide a user interface for traceability editing and querying an navigation between linked elements.

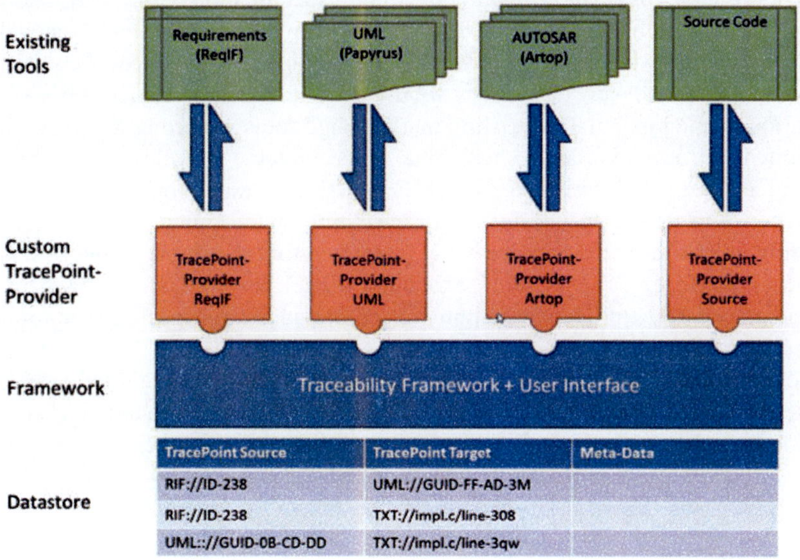

Fig. 8 Traceability Framework

Technical solutions of other tools are often based on the import of requirements into the target model. In this the requirements are imported with the stereotype "Requirement" similar to the UML and traceability is enabled by means of modeling. However, this has the drawback that not all models support extensions[4]. AUTOSAR for example does not have such an extension concept and also the traceability to code is not considered.

Beside this traceability alongside the development artifacts there is the possibility to correlate project management data with the artifacts. In the development process the software is implemented in different release states/integration levels with growing functionality. For coordination and project planning it is crucial to trace what function is implemented in which milestone. If the milestones also are included in EMF, this assignment is traceable with the same mechanisms as the traceability to other artifacts.

9 Future Work

Work on requirements support using domain-specific languages and on traceability mark the beginning of the VERDE process chain. Using the integration of DSLs and models the project partners such as the FZI Karlsruhe can put new concepts for formulating requirements quickly and iteratively to the test, especially non-functional requirements such as memory usage and timing. There has been some preliminary work in this area, for example making the timing specification of the UML real-time profile MARTE into a convenient editor. Changes in the process can be carried out and tested quickly due to the lightweight tool support. These concepts can also be used on an abstract level. This is why we use domain-specific languages to describe scenarios in one project. The existing methods and tools are brought together in an integrated platform. In order to check statements about the non-functional characteristics of a system at an early stage VERDE is analysing execution platforms already in existence to define an interface with execution and communication semantics. Building on this idea, adapters are being integrated for the project partners' existing execution platforms. The SystemC execution model, the AUTOSAR standard and the integration of heterogeneous simulator compounds are especially important in this context.

Not included are the topics access rights, versioning, large models and distributed work. A working group Eclipse Modeling Platform has been founded to create a generic framework which deals with these topics. As soon as there are any results these will be integrated into the VERDE requirements.

References

1. OMG: Requirements Interchange Format 1.0 Beta 1, http://www.omg.org/spec/ReqIF/
2. Gotel, O., Finkelstein, A.: An analysis of the requirements traceability problem. In: Proceedings of the First International Conference on Requirements Engineering, pp. 94–101 (1994)

3. Steinberg, D., Budinsky, F., Peternostro, M., Merks, E.: EMF Eclipse Modeling Framework. Addison-Wesley (2009)
4. Casse, O., Monteiro, M.R.: A Reference Model for Requirements and Specifications A ReqIF/SysML profile example - Requirements exchange and roundtrip. In: ERTS2 2010 (2010)
5. Jastram, M., Graf, A.: Requirements, Traceability and DSLs in Eclipse with the Requirements Interchange Format (RIF/ReqIF). In: Tagungsband des Dagstuhl-Workshop MBEES: Modellbasierte Entwicklung Eingebetteter Systeme VII (2011)

Chapter 14
Mixing Systems Engineering and Enterprise Modelling Principles to Formalize a SE Processes Deployment Approach in Industry

Clémentine Cornu, Vincent Chapurlat, Bernard Chiavassa, and François Irigoin

Abstract. Systems Engineering (SE) is a tried and tested methodological approach to design and test new products. It acts as a model-based engineering approach and promotes for this purpose a set of standardized collaborative processes, modelling languages and frameworks. In a complementary way, Enterprise Modelling (EM) provides concepts, techniques and means to model businesses along with their processes. The purpose of this paper is to provide a method for the deployment of SE processes considering interoperability and building bridges between SE and EM. An application case is given illustrating the definition of the stakeholder requirements definition process defined in the ISO 15288:2008.

Keywords: Systems Engineering, Enterprise modelling, Deployment of processes, Interdisciplinary design, Interoperability, Requirements definition.

1 Introduction

The Systems Engineering (SE) approach is considered today as an efficient methodological and interdisciplinary approach which promotes a set of processes

Clémentine Cornu · Bernard Chiavassa
Eurocopter, ETZP, Aéroport International Marseille Provence,
13725 Marignane Cedex – France
email: {Clementine.Cornu,Bernard.Chiavassa}@eurocopter.com

Vincent Chapurlat
LGI2P - Site de l'Ecole des Mines d'Alès, Parc Scientifique George Besse,
30035 Nîmes Cedex 1 - France
email: Vincent.Chapurlat@mines-ales.fr

François Irigoin
Mines ParisTech-CRI, 35 rue Saint Honoré, 77305 Fontainebleau Cedex-France
email: Francois.Irigoin@mines-paristech.fr

[13] [3] particularly relevant for large businesses providing complex technical products or open-ended services. From a theoretical point of view, to apply SE principles companies must first define SE processes, then put them into practice and finally continually improve them. Unfortunately, from a pragmatic point of view, many obstacles prevent companies from easily deploying SE processes. For instance:

- Reference documents about SE promote various definitions of its processes but they are often inconsistent and defined from such a high-level of abstraction that they are really subject to interpretation. Thus, to clearly define what the company should deploy is a first difficulty.
- There is no generic deployment method guiding companies from the expression of their needs to the physical implementation, so they are forced to develop a method from scratch. In addition to its difficulty, this task requires people mastering SE with an excellent knowledge of the company's organization.

Moreover, research works about interoperability question [2] have underlined for a long time [5] [6] that the latter is a key factor when companies have to lead changes. But, modelling approaches used to describe the company's organisation, behaviour, and constraints are not really suitable to describe conditions under which a couple of resources are interoperable (*i.e. they are able to correctly communicate and use information, and to perform resulting appropriate action [9]*) or not, nor to evaluate them. So, it seems relevant to consider interoperability as a guide for a successful deployment.

The purpose of the research work presented here is to help companies to get over all these obstacles by providing them with a deployment method based on both top-down and bottom-up approaches. Its fundamental strength is to draw its principles from two research fields both based on systematism which do not overlap but could bring a lot of advantages when used together: Systems Engineering (SE) and Enterprise Modelling (EM).

2 Merging SE and EM Principles

Let us highlight the relevant principles of both fields to point out how complementary they are and how interrelated they can be.

Among others, SE relies on two key concepts: the System-of-Interest (SOI) and the System-Used–To-Design (SUTD). On one hand, a SOI can be defined as *any product or service that a company has to provide in order to meet the needs of its market*. For instance, a helicopter manufacturer provides helicopters to its customers but also services like maintenance or supplying of spare parts. On the other hand, a SUTD is *a system in charge of organizing, executing and coordinating all activities required to design a given SOI that, from market's needs, provides an economical and competitive solution*. Its expected output is a completely defined "virtual" SOI taken as input by the "production system". Descriptions of the generic activities the SUTD should perform are provided by SE reference documents in the form of processes. Companies must tailor them to their needs and business areas, and define means and methods to control and improve them continuously.

In addition, as interoperability remains absent or underestimated, companies must also answer these questions: *What existing components of the SUTD could be reengineered to gain efficiency? What new resources or activities must be introduced? How? Who are their stakeholders? What are their needs? How to assess and improve their interoperability?*

Thus, to prepare the deployment, the team must first model the considered SUTD and then appraise its performances especially concerning interoperability. To this end, many developments done in the field of EM are helpful. Among them, we can mention the various:

- Standardised modelling frameworks, languages and methods now widely acknowledged in industries (e.g. [19][20][10][1][7]).
- Techniques and tools to support process control and monitoring (e.g. dashboards, workflows, reporting and information systems).
- Techniques and tools for the verification and the validation (V&V) of processes models (e.g. support to expertise, processes simulation)
- Techniques and tools for interoperability assessment (e.g. [12][4][11][15])

Thus, it appears relevant to use together solutions provided by EM and SE, but the question of the "mixed" language (syntax and semantics) needed to support the deployment must be considered. From one hand, in the frame of EM, the need for a shared constrained language has been established [19]. It must include not only the modelling syntax that can be used (i.e. graphical elements) but also all the semantics. This language is the key that distinguish modelling and drawing [19]. From the other hand, in the frame of SE, many efforts have been done to formalize concepts and their relationships useful to design a System-of-Interest and to manage a System-Used–To-Design (SUTD). The key idea is to encourage the use of a unique (or at least unifying) language which enables the communication between resources involved in SE processes. Consequently, a mixed "deployment language" must be defined. It aims to unify all concepts and relationships required to design the SUTD i.e. to model and adapt the ideal vision of SE processes regarding company's available resources, skills, organisation, constraints, needs, etc. This language must be independent from the process which is the target of the deployment but must be semantically compatible with the SE language used in the company. It plays a major role in the deployment since it facilitates exchanges and interoperability between actors in charge of the deployment. Last but not least, this language, just like all the models built in the context of the SE deployment, are a contribution to the "shared language" enabling the discussion, the description and the management of the enterprise so important to the enterprise architecture vision [14].

3 The Deployment Approach

This section aims to show how to practically merge relevant principles taken from SE and EM fields by presenting the proposed deployment approach and its outcome: a deployment guide equipped with software tools.

3.1 SE Processes Deployment Language

Section 2 highlighted the need for a deployment language. To this end, a conceptual meta-model has been defined (see Appendix). It includes a set of classes and relationships which enables to represent all SE and EM concepts necessary for the deployment. The meaning of classes and relationships is defined by adding to this meta-model a set of textual descriptions, not presented here for space reasons.

3.2 SE Processes Deployment Activities

First of all, a deployment team including experts from SE and EM domains must be created. Once done, the constraints, objectives, needs, expectations, and requirements of the deployment effort must be clearly expressed and shared among all deployment stakeholders. Then, a deployment referential must be defined. It includes both a SE referential (i.e. the selection of reference standards providing the theoretical basis about SE) and a modelling referential (i.e. the selection of modelling language and methods providing a frame to modelling activities). Once all these preliminary activities have been achieved, the following four main stages are proposed. They take as inputs all the deployment choices made until then.

Stage 1 - Model Ideal Processes to Deploy and their Relationships
To deploy SE processes appropriate for its own use, the company must adapt those described in the SE referential to its own context. This is started during this stage since the deployment team models its ideal vision of the activities and flows (of artefacts/information) characterising the processes to deploy, but also all roles needed for their optimal execution. We call this set of models the **IDEAL model** (for an example, see Section 3). "IDEAL" means that the deployment team must not curb its creativity and ideas about company's improvement just because they seem utopian confronted to the current organisation or daily operational difficulties.

Stage 2 - Model the current processes/activities and their relationships when they exist
In order to know where improvements can be done, the way the enterprise design new products or services must be characterized and analysed. To that end, existing activities, roles, actors and resources involved or that could be involved in the design of products and services are characterized and modelled. The resulting set of models constitutes the **AS-IS model** [8]. During this stage, operational needs, expectation and constraints are collected, analysed and translated into requirements which are used as in inputs in Stage 3.

Stage 3 - Specify the processes to deploy
Comparing the IDEAL and AS-IS models, the deployment team can now perceive significant gaps and thus factually highlight ways of improving the current organization. A **TO-BE model** [8] is then proposed, mixing and merging the IDEAL and AS-IS models. It aims to share the trade-offs found between the ideal and current organizations taking into consideration all requirements expressed until then. At

this stage, human expertise and EM reference frameworks such as [20] are used to define and decide what improvement must be done.

Stage 4 - Define practical implementation
Finally, the deployment team defines an action plan for the deployment and provides all its practical details.

3.3 SE Processes Deployment Resulting Guide

In addition to the deployment language and the stages presented above, the equipped methodological guide for the SE process deployment aims to include all elements illustrated in Fig. 1. This guide is currently under development.

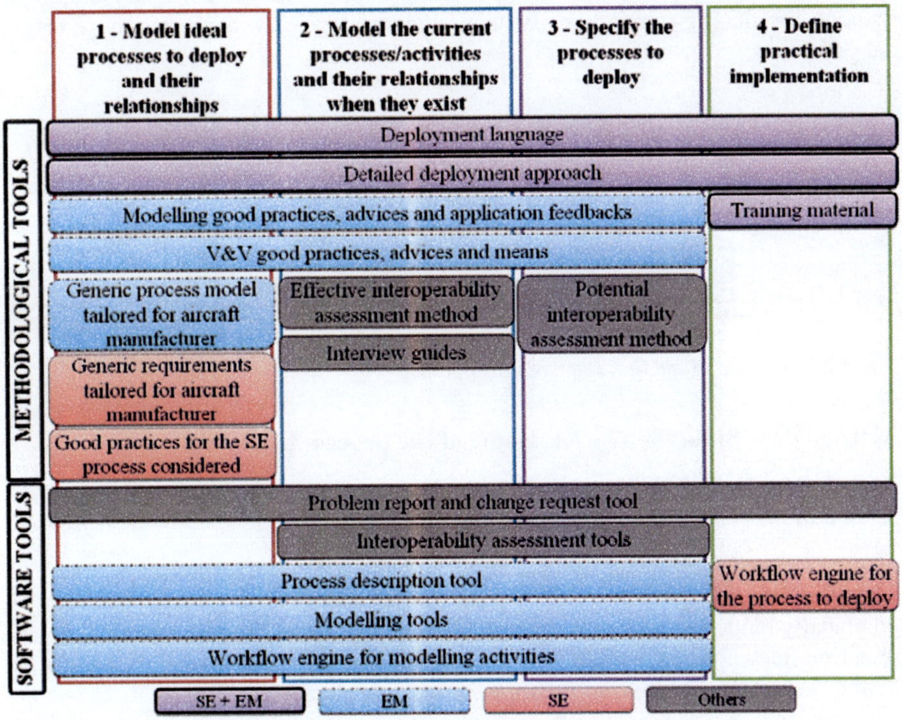

Fig. 1 Main elements of the equipped methodological guide

4 Application: Ideal Definition of the "Stakeholder Requirements Definition Process"

The purpose of this section is to show how to execute the first stage of the approach that is broken down into activities as shown in Fig. 2 and which has already been tested in industry. The process which is modelled is the "Stakeholder

Requirements Definition" process described in [13]. It has been chosen since, according to SE principles, it is the first process that must be deployed. Moreover, [13] has been selected as SE referential since it is a generic standard widely-acknowledged in the SE community and in industry. Regarding now the modelling referential, we have selected the Business Process Modelling Notation (BPMN) [17] for the modelling language since it is a widely-acknowledge standard. Nevertheless, it appears that BPMN suffers from semantic gaps when compared to other languages like the Event-driven Process Chains (EPC) from the ARIS method [18] which, for instance, clearly describes the notion of "Role". This is due to its purpose: the first goal of BPMN is not to build conceptual model but to enable process model execution especially thanks to the Business Process Execution Language (BPEL)[16]. Thus, even if BPMN appears to be particularly interesting in our context, it requires enriching it conceptually with all necessary elements without introducing semantic or syntactic inconsistencies with the language original definition.

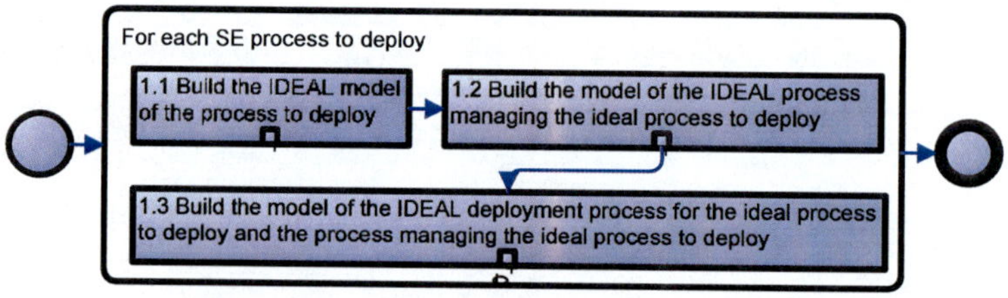

Fig. 2 Break-down of the first stage of the approach

Activity 1.1 – Build the IDEAL model of the process to deploy

Task 1.1.1.- *Collect theoretical activities from the SE referential to have a first version of the process functional architecture*

The SE referential gives us a first version of the process break-down which can be graphically modelled as shown on Fig. 3. However, as the process graphical representation does not describe all attributes defined in the meta-model, they must be defined separately (to save space, they are not introduced here).

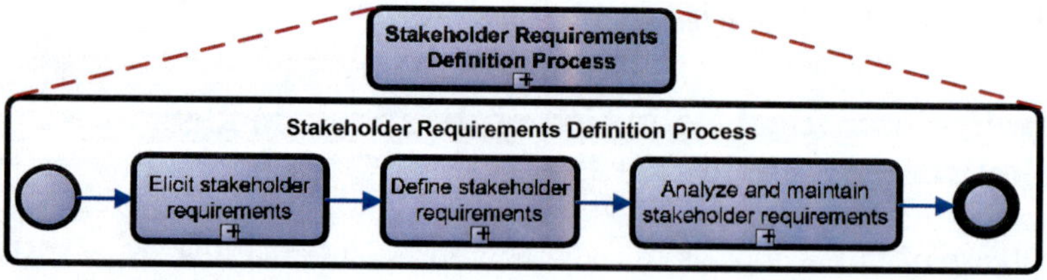

Fig. 3 Illustration of Task 1.1.1

Task 1.1.2. *Identify the inputs and outputs of each activity*
This task consists in the identification of activities' inputs, which are roles that resources have to play, and outputs, which are flows or services. Unfortunately, BPMN does not provide notation for the concept of "role" and its related relationships. Furthermore, the notions of pools and lanes do not totally meet the needs for resource's description since it requires clearly associating resources to the roles they play. That is why an extension of BPMN is required and is currently under development. An example of this task is shown in Fig. 4 with the break-down of activity "Elicit stakeholder requirements". The concept of "role" is represented with ovals (with dotted line for company internal roles) and added relationships are explicitly named.

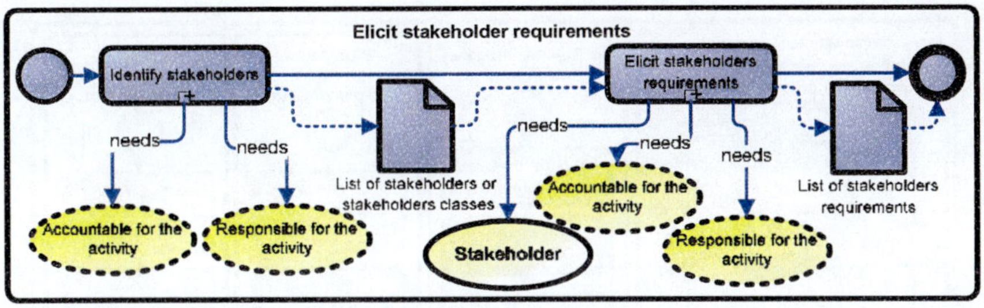

Fig 4. Illustration of Task 1.1.2

Task 1.1.3. Identify who/what will perform the needed roles of identified activities
If needed, roles are broken-down and classes of resources that can play the roles required by the activity are identified. If they must be played only by resources internal to the company, then "internal" roles are identified. Fig. 5 shows the example of the break-down and allocation of the "stakeholder" role. "Inherits from" and "plays" relationships have been added to this end, and classes of resources are symbolised with an icon of man.

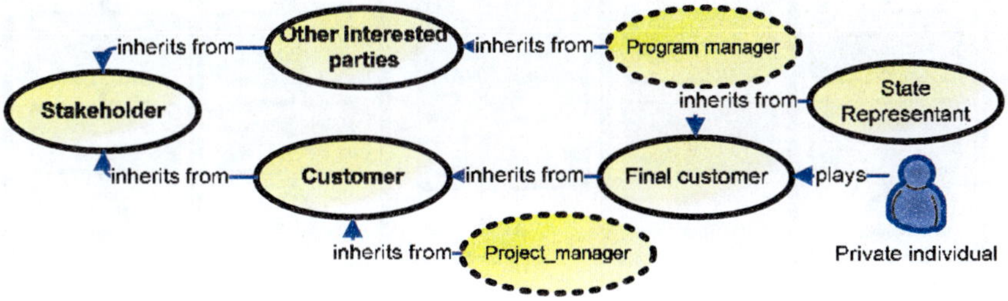

Fig. 5 Illustration of Task 1.1.3 (simplified list of stakeholders)

All these tasks are then performed again but at a lower level of abstraction: the break-down of each activity is performed. During this re-execution, if elements previously defined must be improved or completed, then they are.

Activity 1.2. – Build the model of the IDEAL process managing the ideal process to deploy and **Activity 1.3 – Build the model of the IDEAL deployment process for the ideal process to deploy and the process managing the ideal process to deploy**

These two activities perform again all tasks presented in Activity 1.1 respectively based on the generic processes proposed in Fig. 6 and Fig. 7 which must be instantiated according to the target SE process and the specificities of the company's business area.

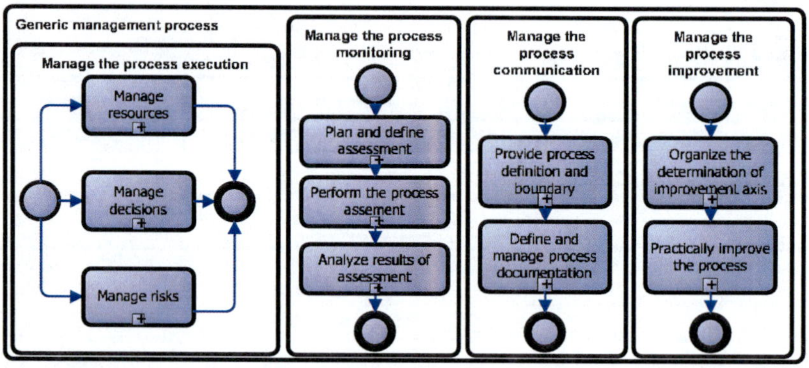

Fig. 6 Management process

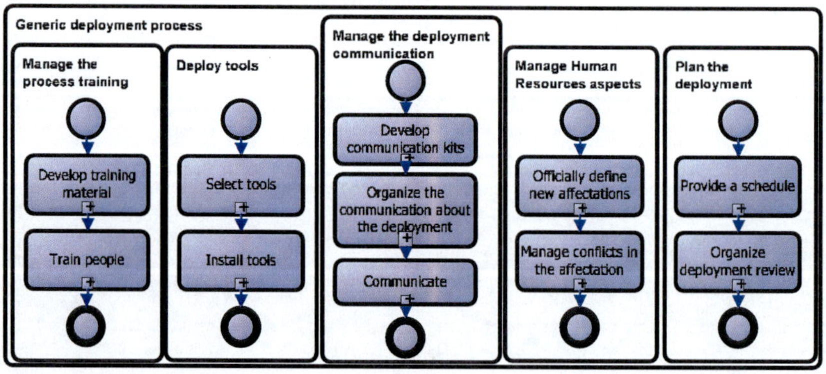

Fig. 7 Deployment process

5 Conclusion

This paper presents and illustrates a guide aiming to help companies in their deployment of SE processes. This guide's main strength is to map, mix and merge concepts taken from SE and EM domains. In addition to a deployment language and approach currently tested by a helicopter manufacturer; a set of support tools including software and methodological tools such as interoperability assessment methods or interview guides is being specified and developed.

References

[1] GERAM: Generalised Enterprise Reference Architecture and Methodology - version 1.6.3, IFIP-IFAC Task Force (1999)

[2] ISO/DIS 11354-1 - Advanced automation technologies and their applications - Part 1: Framework for enterprise interoperability, ISO (2010)

[3] Systems engineering handbook - A guide for system life cycle processes and activities - v3.2. INCOSE (January 2010)

[4] ATHENA project. Deliverable D.A1.4.1 -Framework for the establishment and management methodology (2005)

[5] ATHENA project, Deliverable D.A4.2 - Specification of interoperability framework and profiles, guidelines and best practices (2007)

[6] C4ISR, Final report. C4ISR Architecture Working Group (1998)

[7] Chapurlat, V., Braesch, C.: Verification, validation, qualification and certification of enterprise models: Statements and opportunities. Computers in Industry 59(7), 711–721 (2008); Enterprise Integration and Interoperability in Manufacturing Systems

[8] Chapurlat, V., Vallespir, B., Pingaud, H.: An approach for evaluating enterprise organizational interoperability based on enterprise model checking techniques. In: Proceedings of the 17th IFAC World Congress (2008)

[9] Daclin, N.: Contribution au développement d'une méthodologie pour l'interopérabilité des entreprises, Ph.D. thesis, Université Boredeaux 1 (2007)

[10] DoD, DoD architecture framework - version 1.5 (2007)

[11] Ford, T., Colombi, J., Graham, S., Jacques, D.: The interoperability score. In: CSER - Conference on Systems Engineering Research (2007)

[12] Guédria, W., Chen, D., Naudet, Y.: A Maturity Model for Enterprise Interoperability. In: Meersman, R., Herrero, P., Dillon, T. (eds.) OTM 2009 Workshops. LNCS, vol. 5872, pp. 216–225. Springer, Heidelberg (2009)

[13] ISO/IEC, ISO/IEC 15288:2008 - systems engineering - system life cycle processes (2008)

[14] Kappelman, L.A.: Brindging the chasm. Architecture Governance 3(2), 28 (2007)

[15] Leite, M.J.: Interoperability assessment (1998)

[16] OASIS, Web services business process execution language version 2.0 (2007)

[17] OMG, Business process model and notation version 2.0 (2011)

[18] Scheer, A.-W.: Architecture of integrated information systems. Springer, Heidelberg (1992)

[19] Simons, G.F., Kappelman, L.A., Zachman, J.A.: Enterprise architecture as language. Complex Systems Design & Management (2010)

[20] The Open Group, TOGAF version 9 (2009)

Appendix: The Proposed Meta-Model

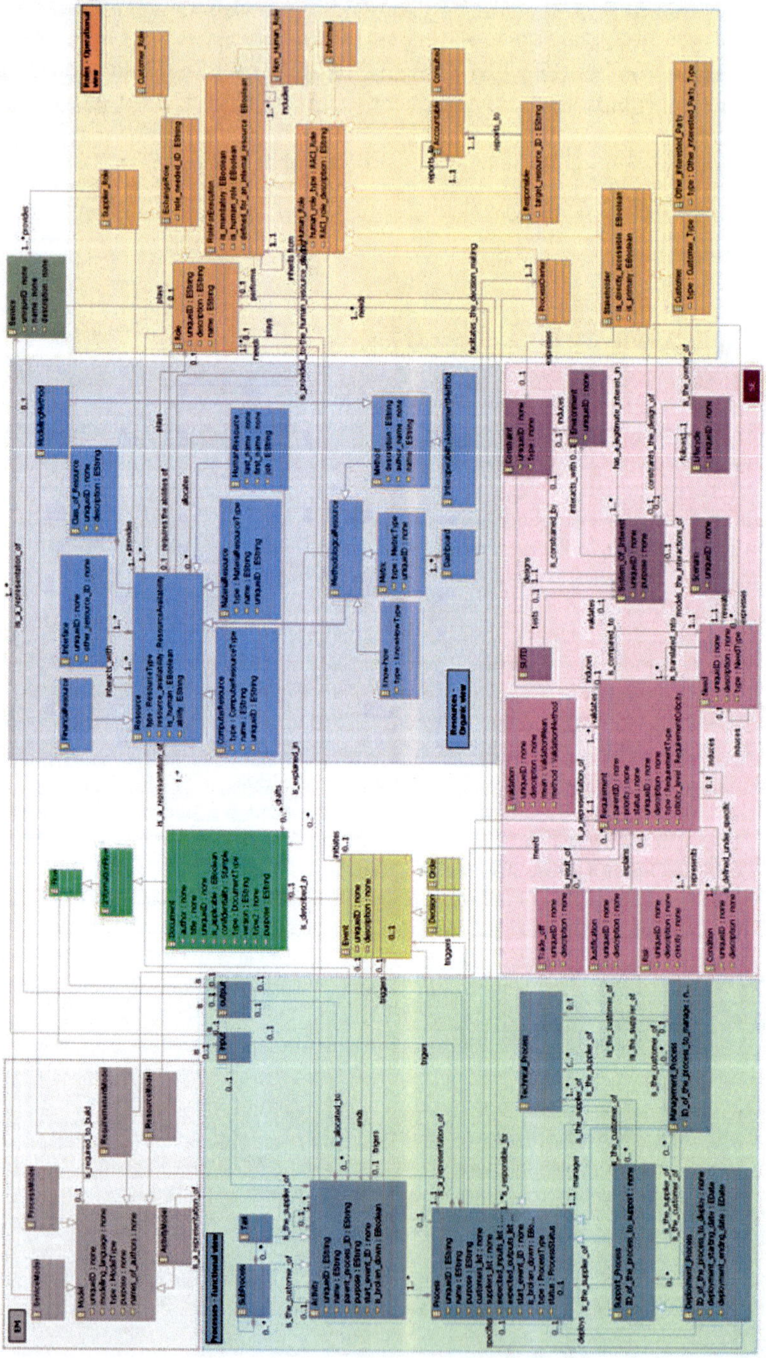

Chapter 15
Enabling Modular Design Platforms for Complex Systems*

Saurabh Mahapatra, Jason Ghidella, and Ascension Vizinho-Coutry

Abstract. In recent times, an emerging trend in several industries that have adopted Model-Based Design has been holistic product platforms where a single systems design is reused and customized to meet diverse customer requirements such as application, cost, and operational considerations. Many of these dynamic changes in nature have required system design component variations referred to as "variants" on top of a fixed master design. One approach to realize this is to create copies of the original design for each variant combination. Additionally, this requires a sophisticated traceability mechanism to propagate any changes in the design to the various implementations. An alternative approach is to design a modular architecture that can reference all the product variations within a single file. Different implementations can then be realized by selecting different system components through a scripting language. This approach promotes design reuse and provides a powerful mechanism to implement traceability. However, such a paradigm requires core tool functionality similar to those available in various UML/SysML implementations before being applied to a systems development process. In this paper, we introduce variant semantics for complex systems design for use within the Simulink modeling environment. We discuss their attributes which can be parametric or structural that can be used throughout the development process. In addition to improving the efficiency and development of product variations, variants present a variety of uses in the context of systems engineering workflows. For example, design exploration, where several alternatives exist for a

Saurabh Mahapatra · Jason Ghidella
MathWorks Inc., Natick, MA, 01760

Ascension Vizinho-Coutry
MathWorks Inc (France), 92190 Meudon

component, can now be managed efficiently to simulate every design possibility in a combinatorial fashion for a given test suite. For large-scale problems, these simulations could be distributed to a high performance computing cluster for overall speedup through a scripting methodology. Design elaboration and integration is a challenging activity that can also be improved through the use of variants, where low fidelity components are replaced by more specialized one's going from mathematical equations to physical or software elements. Since the order in which these components are integrated influence design quality and subsequent iterations, it is possible to carry out several separate integrations that increase confidence. Since there are a number of ways to modularize a design, we also outline a set of best practices for partitioning the design variations for scalability and maintainability. Using Simulink-based examples, we illustrate the above scenarios and outline strategies on how organizations can leverage these possibilities to reuse while enhancing their existing knowledge to meet system design challenges of the future.

1 Introduction

In recent times, the reuse of software and hardware components has been a central design theme in various organizations. With increasing product complexity, high specificity of customer requirements and cost pressures have made it imperative for engineers to take a design reuse-centric point of view. For example, the Joint Strike Fighter (JSF™) program was instituted to design next-generation aircraft that meet US forces' diverse requirements of cost and technology. To lower economic costs, a master design was used as a common design platform with three variants that incorporated special features- F-35A CTOL (Conventional Take-Off and Landing), F-35B STOVL (Short Take-Off and Vertical Landing), and F-35C (Carrier) variants. As a basic example, observe the common theme seen across the designs in the parameter similarities [1] but functional differences due to the special features as shown in Fig. 1.

Modular product architectures and production have long been used in the civil aircraft industry. The variants in Boeing's 747 product family show the same reuse pattern with very similar design parameters but serving different requirements [2]. The 747-200B was a pioneer in low-cost air travel for the masses. The 747-200F was a freighter version with an upward hinging nose. The 747 SP was an extra-long-range-variant featuring a taller tail and short fuselage. The E-4B Command Post variant was equipped to become the wartime emergency base for the US President and his advisors.

Variant	Span (ft)	Length (ft)	Wing Area (ft^2)	Weight (lb)	Special Features
F-35A CTOL	35	50.5	460	18,498	Matches F-16 performance, exceeds stealth, range, avionics
F-35B STOVL	35	50.5	460	13,326	Thrust vectoring, lift fan, smaller weapons bay
F-35C Carrier	43	50.8	620	19,624	Landing gear has high load capacity, larger wing for control on landing on low speed approaches

Fig. 1 Basic JSF design parameters with special features of the variants

Model-Based Design [3,4] has been successfully adopted and deployed across organizations in various industries as an algorithm design platform for complex control, communications, computer vision, signal and image processing applications. The behavioral modeling of the system is represented as a set of interdependent computational functions that typically involve time as an independent parameter. Since non-linear relationships can make closed-form solutions hard to come by, computer-based numerical simulations are used to solve for the system dynamics. Such a paradigm offers advantages of using simulation for early verification of algorithms before deployment on the final embedded hardware. By increasing model fidelity and using powerful computers, the cost of development is significantly reduced as the algorithmic verification activities can be moved upstream, thereby preventing detection of design bugs late in the development process.

Simulink® [5] is a commercial block diagram environment that enables Model-Based Design. The system behavior can be modeled as differential and difference equations by using graphical block constructs and signal line interconnections to represent interrelationships among them. Control logic behavior can be visualized by using statecharts. By using graphical abstractions, a complex functional hierarchical decomposition of the overall system can be obtained which can be numerically simulated to verify and validate text-based algorithm requirements. Thus, the Simulink model forms an executable specification of those requirements. Automatic code generation transforms this specification to C/C++, VHDL (Verilog Hardware Description Language) or PLC (Programmable Logic Controller) code that can be deployed on the hardware for rapid prototyping or hardware-in-the-loop simulations for testing real-time requirements of the algorithm. Since the graphical model forms the basis of the system design, collaboration is enhanced across organizational teams which can rapidly iterate through their designs or refine requirements. It is seldom possible to gain a holistic understanding of a complex system by using the functional decomposition methodology presented above. Other approaches [6] based on formalisms such as UML (Unified Modeling Language) or SysML (Systems Modeling Language) also aim to achieve similar goals by providing complementary modeling views of the system. This paper does not seek to advance ideological arguments [7] that advocate modeling of functional requirements by procedure-based vis-à-vis object-oriented paradigms. An object-oriented modeling systems (OOMS) approach, [8], combines both perspectives to model dynamic systems in Simulink. Approaches such as Computer Automated Multi-Paradigm Modeling (CAMPaM), [9], seek to use multiple paradigms to get a full understanding of the system under study.

Effectiveness of modular design platforms within the context of Model-Based Design will depend not only on topological considerations of the design but also on the tool support for handling variants. In this paper, we use Simulink as the basis of our examples where modular functional design platforms can be achieved through variant handling mechanisms. This restriction, however, should not limit the reader to translate these core ideas to other formalisms or modeling perspectives whenever it is applicable to do so.

2 The Power of Modular Design Platforms

In this section, we offer a definition of modular design platforms and explain the basis of their architectures. We also present a discussion on the challenges with traditional approaches that engineers can use to implement design variants without tool support.

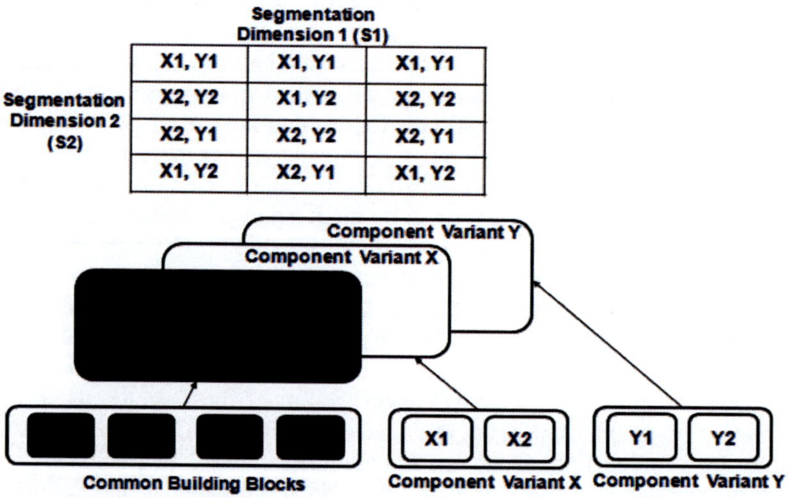

Fig. 2 Redundant design derivatives created from variant choices mapped to market segments

2.1 *Definition and Economic Considerations*

A modular design platform is a finite set of components and their associated interfaces that can form a fixed common structure while allowing for some variability.

These variable components are called as variant components. The goal is to make the architecture common across many designs. Such platforms capable of accommodating new component technologies and variations make it possible for firms to create design derivatives or product variants at incremental cost relative to initial investments in the platform itself. Since the costs associated with the fixed components carried forward are sunk costs, only the incremental costs of creating variations to them accrue to the derivative designs. Typically, these incremental costs are a small fraction of the cost of developing the original design, providing platform leverage [10]. However, if the variability is significantly high across derivative designs, such lowering of costs through economies of scale cannot be attained. Modular design platforms can also improve development cycles of derivative designs by facilitating a more streamlined development process and more frequent model changes [11].

Successful application of modular design platforms is primarily dependent on the market-driven segmentation [10]. For example, Fig. 2 shows a 2-dimensional user-specified market segmentation model containing 12 segments onto which are mapped 2 component variants each containing 2 choices. Since only 4 design derivatives are possible, the redundancy across various segments can potentially result in higher economies of scale thereby reducing both sunk and incremental costs. For the sake of demonstration, we have assumed that each segment represented by *(S1, S2)* has only one design derivative associated with it. This is based on the assumption that sound segmentation schemes, possibly higher dimensional will yield this mapping.

2.2 Traditional Approaches to Handling Variants

The simplest approach to managing variants shown in Fig. 2 is to create a separate file for each of the design derivatives. This would also require the generation of unique file names and the manual creation of associated metadata. The latter would map the relationships between file names, variant information, and market segments. Checks would have to be incorporated to determine the integrity of the mapping. As changes are made to the fixed components in the design, it becomes expensive to propagate those changes without error across each of the separate files in an ordered fashion. The inherent complexity of the architecture would require that such changes be made manually.

A better approach to the above problem is to use a revision control system to create 4 branches corresponding to each variant choice implemented as a file-based component. By locking the common files that constitute the building blocks from changes and using metadata, a more robust file management of variants can be implemented. Any changes made to the common files can then be propagated forward across all branches efficiently. The approach suffers from maintaining a large number of branches corresponding when the number of variants and/or their associated choices increases. Also, the associated metadata would have to be managed inside the revision control system and may require knowledge of its API. If correctly implemented, the above two approaches do provide enough flexibility in managing the variant metadata.

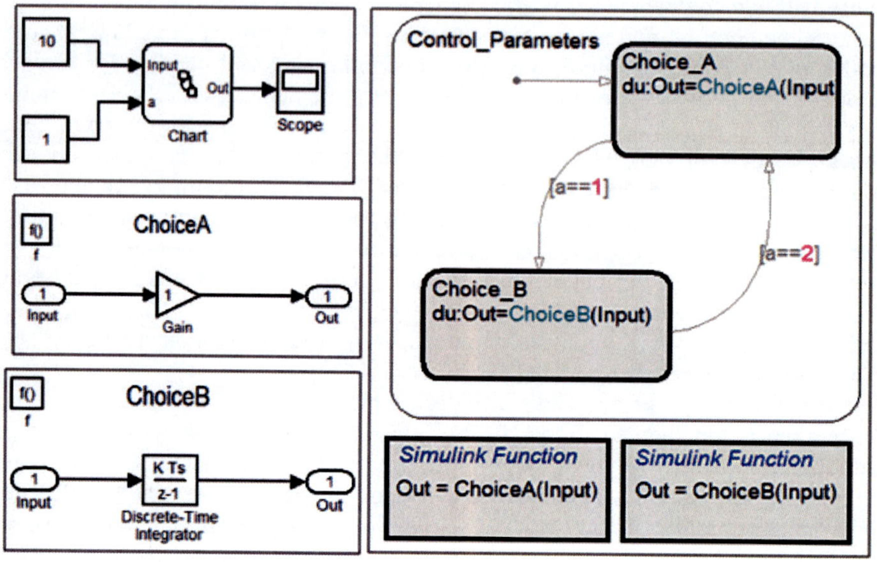

Fig. 3 Implementation of simulation variants which are dependent on control parameter *a*

Yet another approach is to implement the variants as functional components in the model itself that are chosen during the simulation based on certain control parameters. For example, an implementation of such a scheme using a statechart implemented in Stateflow® [12] is shown in Fig. 3. The control parameter *a* is an input parameter that causes the statechart to switch between the two states and trigger the execution of the corresponding Simulink function call subsystem. Observe that the subsystems (*ChoiceA, ChoiceB*) have identical interfaces but different mathematical implementations resulting in functional polymorphism. This is similar to run-time polymorphism in object-oriented programming paradigms where the behavior of the virtual function in the superclass is defined by an instance of the inherited subclass. The disadvantage of such an approach is fundamental- the notion of a design derivative exists outside the context of simulation. Furthermore, an unintended change in the control parameters during the simulation can cause the model to alternate between design derivatives in time. Such a scheme is harder to develop as the number of design derivatives increase or variants are distributed across the model hierarchy. The flexibility of managing the variant metadata is difficult as this information is present inside the model itself. However, such an approach is most applicable when variability in simulation behavior is indeed intended. Since neither of the above is perfect, we outline a balanced approach for handling variants in Simulink that combines the best elements of all the approaches.

3 A Variant Implementation in Simulink

As early as 1970's, Parnas [13,14] proposed modularization, information hiding and separation of concerns principles for handling variants in a product family. In this section, we discuss the framework used for implementing the notion variants within the Simulink environment and compare it with some other implementations.

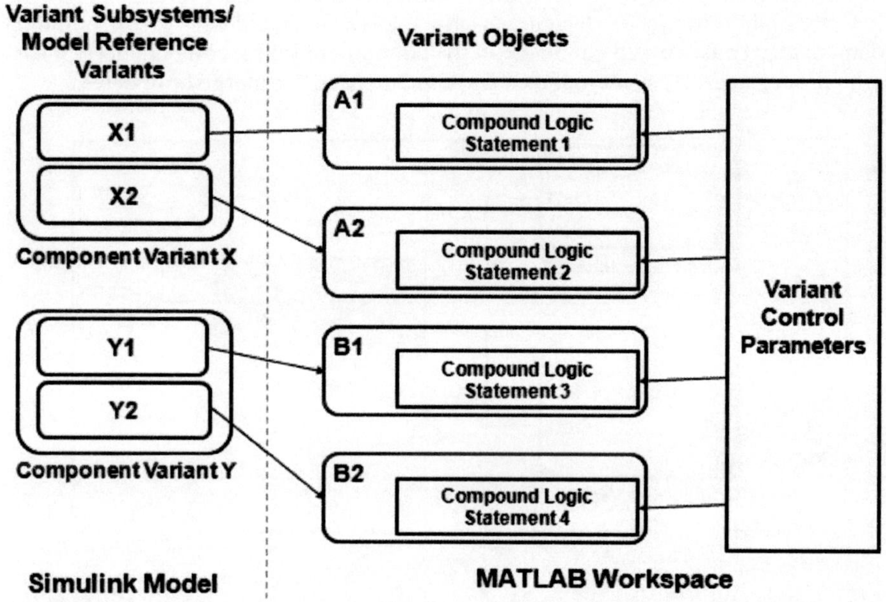

Fig. 4 Design variants in a Simulink model are associated with variant objects that encapsulate atomic Boolean statements based on variant control parameters defined in the MATLAB Workspace

3.1 Understanding the Framework

In Simulink, the approach taken to represent variable components is either through the use of variant subsystems or model reference variants. The former provides a mechanism in which design derivatives can be incorporated within a single model file while the latter allows for them to be implemented in separate files. For large-scale modular platforms, the model reference variant approach provides better performance due to componentization at the file level. A schematic of the implementation is shown in Fig. 4. The variant choices can have a many-to-one mapping onto a variant object which encapsulates a compound logical statement based on the variant control parameters. This mapping is defined at the model-level and is independent of the placement of the variant in the hierarchy. The variant choices have the same interface to ensure consistency during module switching. The encapsulation is atomic in nature and only allows the association of a Boolean logical statement with a variant object. This is a matter of convenience as several logical statements can be combined into a single compound statement thus resulting in a one-to-one mapping between variant objects and the statements. The variant object and the corresponding component in the Simulink model are activated at compile time prior to simulation when the encapsulated Boolean statement evaluates to TRUE. This 2-tiered variant handling approach where compound logical statements are created from variant control parameters and then associated with variant objects which map to the modules offers a lot of flexibility in creating associated metadata. Additionally, variant subsystems and model reference variants plug in as modules whose design can now evolve independently of the common components. Thus, any update made to the common building components can easily be propagated across all modules without any need for merge or updates.

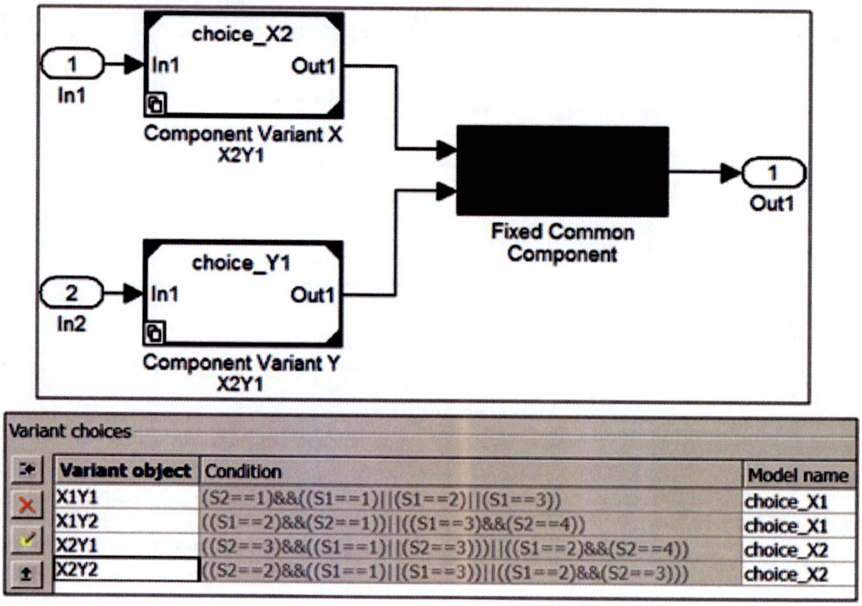

Fig. 5 Implementation of the design derivatives outlined in Figure 2 using model reference variants in Simulink.

An implementation of the design variants is shown in Fig. 5 that can be used to represent the four design derivatives shown in Fig. 2 which are mapped onto 12 segments represented by *(S1, S2)* market segment coordinates. There are four variant objects named *X1Y1, X1Y2, X2Y1,* and *X2Y2*. Each of the variant objects is mapped to its corresponding files by instantiating them in the MATLAB Workspace. The variant objects are shared across the two model reference variants. For example, X1Y1 is mapped to the *choice_X1* file in component variant X and is also mapped to the *choice_Y1* file in component variant Y. Observe that the market segment coordinates *(S1, S2)* are the variant control parameters and are mapped to the variant objects by compound logical statement construction. By using Boolean algebraic theorems, simplifications can be carried that result in compact representation and better readability.

However, such flexibility does come up with the associated risks of creating unsound compound logical statements that may erroneously activate multiple variant choices within a single variant or choose the wrong variant. Checks may need to be incorporated to determine if the variant control parameters are mapped correctly to the logical statements. We address these concerns in the next section.

3.2 Variant Handling in other Domain-Specific Tools

The simplest example of variant implementation is the static polymorphism seen with overloaded member functions and operators in various object-oriented programming paradigms. For example, C++ templates can be used to create a binding between an abstract and a concrete type at compile time.

An XML-based language called XVCL [15] (XML-based Variant Configuration Language) is based on Frame Technology [17] to represent a piece of a reusable asset, which is also instrumented marked-up with XVCL commands to accommodate variants. The frame concept was first introduced by Minsky [16] who introduced it as a static data structure to represent well-understood stereotyped situations. Newer situations could be adjusted to by calling up past experiences captured as frames. Details of these past experiences could be used to represent the individual differences for the new situation. Paul Bassett[20] applied this in the context of software engineering where frames were reusable pieces of program code adapted to meet specific needs.

Software Product Line engineering [18] is a frequently used term among UML users whose goals are similar to that of engineering product platforms. It aims at improving productivity and decreasing realization times by gathering the analysis, design, and implementation activities of a family of systems. A UML *Profile* contains stereotypes, tagged values and constraints that can be used to extend the UML metamodel. Several UML profiles have been proposed that implement concepts similar to ours but within the UML context. For example, Possompès [19] proposes a metamodel that describes feature diagram semantics with a tool implementation to create a UML 2 profile which can link it with various UML artifacts. Another example is that of *SMarty* [20] (Stereotype-based Management of Variability) is composed of a UML 2 profile, a *SmartyProcess* process, and a *SMarty* profile that contains a set of stereotypes and tagged values to represent variability.

4 Scripting Approaches for Handling Variants

In this section, we cover some scripting approaches that can be used for efficient handling of variants. We also discuss the applicability of each method.

4.1 Encapsulation of Variant Metadata

As the variant control parameters and the associated variant objects that constitute the variant metadata are present in the MATLAB Workspace which is a central repository for data across all models, there is a risk of unintentional data tampering. To reduce the probability of this occurring, it is possible to encapsulate the variant object and the variant control parameter data within a MATLAB script that also incorporates suitable checks. Since the script has access to the MATLAB Workspace, this can be used as a utility script within the development environment. A better encapsulation strategy is to create a MATLAB class where the variant-related information is declared as private members and appropriate method is used to activate the desired variants that constitute the design derivative. However, such an approach is fraught with readability issues at the model level where an object's members will have to be accessed using the dot notation. A compact naming scheme may alleviate the issue but does not scale well as the number of variants increase. As our primary goal was to keep variant naming and association simple, we adopted the script-based approach for our implementation at the cost of better encapsulated structures for such information.

4.2 Best Practices for Variant Representations

One of the challenges in the scripting implementation is the optimal representation of the design derivatives which depends on the choice, number and association of variant objects with control parameters. Combined with the number of parameter control values, they also influence the compound logical statement complexity. In the case of large-scale systems, naming plays a critical role in readability.

In Fig. 2, there are 4 design derivatives that are mapped to the 12 customer-specified segments. It is possible to choose between 4 and 12 variant objects to map onto the 12 segment coordinate space. Observe that the lower bound is given by the number of design derivatives while the upper bound is given by the total number of market segments. For example, a 12 variant object representation will have repeated definitions with different names but will have a one-one mapping onto the segment coordinates. Choosing such an implementation also reduces the compound logical statement complexity as shown in Fig. 6(a).

(a)

Variant Object	Compound Logical Statement	Design Derivative
XY_11	S=1	X1Y1
XY_12	S=2	X2Y2
XY_13	S=3	X2Y1
XY_14	S=4	X1Y2
XY_21	S=5	X1Y1
XY_22	S=6	X1Y2
XY_23	S=7	X2Y2
XY_24	S=8	X2Y1
XY_31	S=9	X1Y1
XY_32	S=10	X2Y2
XY_33	S=11	X2Y1
XY_34	S=12	X1Y2

(b)

Variant Object	Compound Logical Statement	Design Derivative
XY_11	S1=1 && S2==1	X1Y1
XY_12	S1=2 && S2==2	X2Y2
XY_13	S1=2 && S2==1	X2Y1
XY_14	S1=1 && S2==2	X1Y2
XY_21	S1=1 && S2==1	X1Y1
XY_22	S1=1 && S2==2	X1Y2
XY_23	S1=2 && S2==2	X2Y2
XY_24	S1=2 && S2==1	X2Y1
XY_31	S1=1 && S2==1	X1Y1
XY_32	S1=2 && S2==2	X2Y2
XY_33	S1=2 && S2==1	X2Y1
XY_34	S1=1 && S2==2	X1Y2

Fig. 6 Two implementations where the number of control parameters impacts the compound logical statement

Two control parameters can now be used that map them to the segment coordinates in the grid which results in simpler logical statements shown in Fig. 6(b). A further reduction in the complexity can be obtained by a single control parameter could be used that takes on 12 values but this captures no information about the variant components in the model as shown in Fig. 6(a) i.e. the model has two variant components *X* and *Y*. The values that each control parameter can take depend on the number of choices available within the variant component. Thus, the minimal number of control parameters has an inverse relationship with the number of values used. Two control parameters can represent 4 design derivatives if each takes 2 values. If maintaining the contextual information at a model level is important, it is a good practice to have the same number of control parameters match the variant components in the model. Also, if retaining contextual information at the component level is important, it is a good practice to map the control parameter values to each of the variant choices. In such a case, a redundant variant object scheme such as that shown in Fig. 6(b) would need to be devised where each variant object maps to a market segment coordinate.

If the contextual information is not important, a compact representation of the variant objects can be used where the market segment dimensions are mapped onto the control parameters. The values that each control parameter can assume are based on the number of segments along that dimension. Such a scheme is shown in Fig. 5 which results in complex logical statements but has fewer variant objects. In this case, the market segmentation information is represented by the control parameters.

The observed inverse relationship between number of variants and compound logical statement complexity is due to the flexibility of the 2-tier variant handling framework where their effects are multiplicative. For a similar reason, the number of control parameters and the number of values also have an inverse relationship. Note that the representation shown in can be derived from Fig. 6(b) by recursively removing each row from the table that has a common design derivative while using a logical OR to combine it with the logical statement in the undeleted rows.

The association of variant objects with a variant component's choices also plays an important role in the implementation. In Fig. 7, the number of variant objects represented by the gray ovals is the same as the 4 possible design derivatives. The first implementation in (a) associates the variant object with a variant choice. Additional scripting is required to create the association between variant choices in *X* with that in *Y*. The second implementation in (b) circumvents this limitation by associating the same variant object with a variant choice in *X* and *Y*. Following similar grouping schemes is advantageous, as the variant objects truly represent the design derivative context.

The set of best practices can be summarized as:

1) The maximal number of design derivatives possible is equal to all the possible combinations of all the variant choices available in the system. The minimal number of variant objects required is equal to this number. The maximal number is dependent on the total number of market segments addressed.
2) If the goal is to maintain contextual information on the variants at the model and component levels, the number of control parameters chosen should correspond to the number of variant components. The number of values each control parameter takes should correspond to the number of choices available within a variant component. The variant objects can capture the market segmentation information. However, if a more compact variant object representation is desired then the control parameters should be mapped to the market segment coordinates.
3) Each variant object should be associated with a grouping of variant choices that represents the design derivative.
4) For readability, use compact but sufficiently descriptive names for control parameters and variant objects. Use enumerated constants for representing values that a control parameter can assume. Use Boolean algebraic theorems to simply the compound logical statements or use redundant variant object schemes.

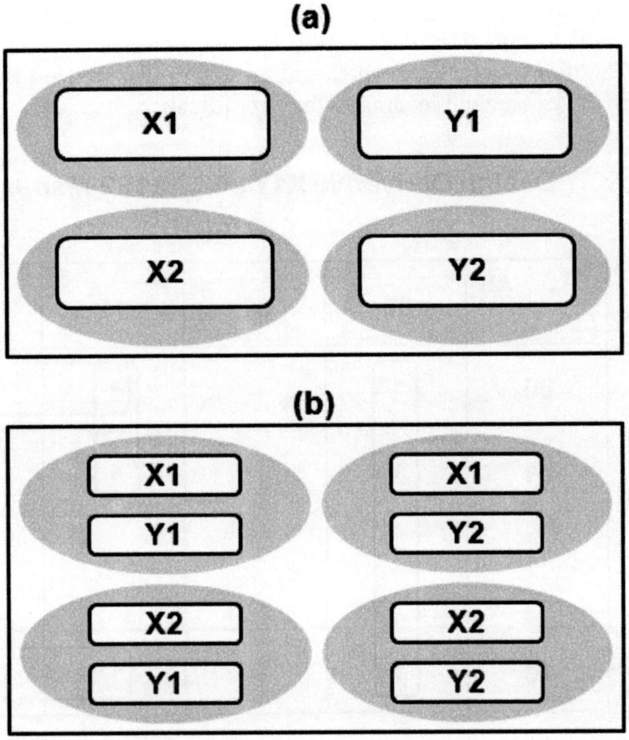

Fig. 7 Two variant object associations with the same control parameter set and values

4.3 Simplifying Compound Logic Using Karnaugh Maps

The Karnaugh map (K map) is a popular technique used to optimize logical expressions in digital systems [21]. Each compound logical statement can be represented as a logical function of the control parameter values. Furthermore, each of the values can be mapped to a binary representation having a suitable number of bits. A logical function of binary-valued variables can either be represented in a sum-of-products form with each term referred to as a *minterm* or an equivalent product-of-sums form with each term referred to as a *maxterm*. The map is a diagram which provides an area to represent logical function output as a function of the input variables. The essential feature of the K map is that the adjoining boxes, horizontally and vertically (but not diagonally) correspond to *minterms* or *maxterms*, which differ only in a single variable, this variable appearing complemented in one term and uncomplemented in the other.

Consider an example of a market segmentation mapping of the design derivative *X1Y1* as shown in Fig. 8 . A presence of a **1** indicates the activation of the variant for that segment. Each of the control parameters *S1* and *S2* are mapped to the segment dimensions can take 4 binary values- **00**, **01**, **10**, and **11**. *S1* is represented by the 2-bit *AB* and *S2* by *CD*. Thus, each segment can be represented as a 4-bit binary string *ABCD*. Using the Karnaugh map, we can realize a compact representation of the logic function by grouping **1**s both along vertical and horizontal directions. The dimension along which the grouping takes place loses either 1 or 2 variables in the expression leading to the simplification.

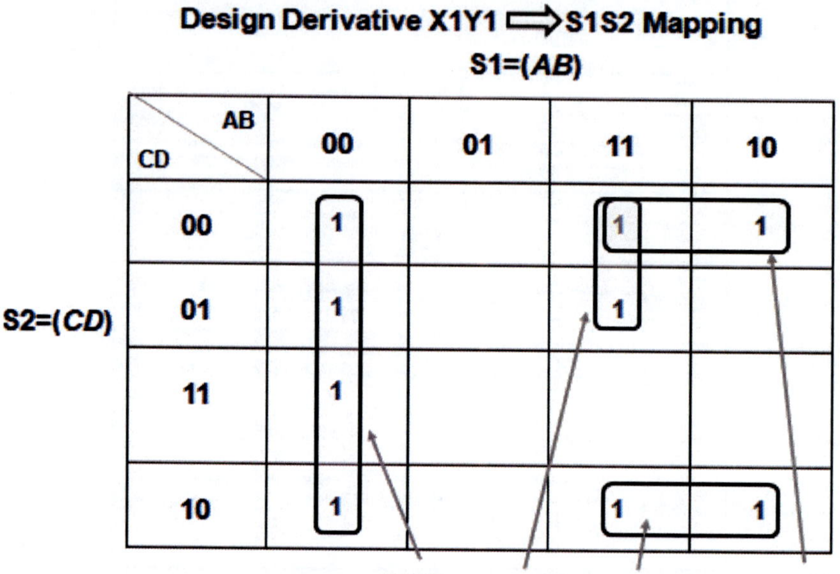

$$f(A,B,C,D) = \overline{A}\,\overline{B} + AB\,\overline{C} + AC\,\overline{D} + A\,\overline{C}\,\overline{D}$$

Fig. 8 Visual simplification of the compound logic statement as sum-of-products expression using Karnaugh map

5 Opportunities for Using Variants in Model-Based Design

Variant semantics support in subsystem and model reference blocks provides some interesting opportunities in Model-Based Design. Since these are the foundational blocks used for abstracting [22] computational elements, the ability to choose their behavior from different implementations at compile time provides a new dimension for well-established concepts and workflows.

The Signal Builder block can be used in a Simulink model for creating a test suite [23] comprising of groups of test vectors. With variant semantics, it is now possible to have multiple test suites and associate them with a design derivative based on certain criteria. Additionally, test suites can also be associated with variant components thereby offering more testing flexibility. Test harness variant components comprising of model verification and validation blocks can also be created. Requirement linking is also possible at the component level and for each of the choices thus indirectly providing variable requirements.

From a behavioral modeling and simulation standpoint, the use of variants is particularly interesting in several Simulink design patterns [24] such as plant-controller, aircraft design, and various filter implementations. At a component level, variations in the behavior can be obtained by changing the parameters for a fixed structural implementation. However, variant semantics enable the structural variation of the algorithm. Variable interface definitions can also be created by using a maximal number of ports that can contain all variant choices' interface definitions. By judiciously using terminator blocks as input for unused ports, variability in the interface definition can be obtained. From a feature standpoint, variant components can be release-based giving the algorithm developer the option to carry out regression testing on new functionality or upgrades.

File-based componentization can be used as a template for integrating [25] component files which may be at different levels of fidelity with diverse ownership. If the interface definitions are decided early, the variant choices within a component can represent models of varying fidelity. To create a new release containing increasing fidelity, each component can be chosen using a control parameter, the entire model can then be built and tested to verify that it meets all the requirements. The time required to successfully complete this step is a function of the complexity of the components and their integration sequence. Engineers can explore all sequence permutations to identify a particular one that balances failure probabilities with build times.

A new concept design may require the testing of several alternatives at the component level. Traditional exploratory methods [26] have relied on parametric variations within a single model to identify optimal designs. The use of variants discretizes this space where each design derivative can be represented as a point. Since the framework uses control parameters accessible in the MATLAB Workspace, it is possible to solve each of these design problems in using data-parallel mechanisms [27].

From a C/C++ automatic code generation [28] perspective, variant components and their associated control parameters are mapped to separate functions and preprocessor definitions resulting in code variants. These definitions are used by the compiler to choose the valid function to be used for code generation as shown in Fig. 9. During the later stages of the development process, this enables the testing of variant components at the code level. Conversely, if there are hardware variants such as floating or fixed-point microprocessors, they will also require the use of variants upstream with different modeling implementations.

```
#if JUMBO_MODEL
        using_engine_model_1000();
#elif ENTRY_MODEL
        using_engine_model_100();
#elif MIDSIZE_MODEL
        using_engine_model_500();
#endif
```

Fig. 9 Code variants obtained through automatic code generation from a Simulink model with variant components

6 Conclusion

As product complexity manifests itself in diverse customer requirements, cost, and other considerations, organizations of the future will require support for variant technologies to meet those challenges. The competitive advantage that can be derived from designs that incorporate product platform ideas can result in substantial reduction in development time and costs.

Some variant-centric frameworks [29] advocate that the model be partitioned into common and the variable components. As attractive as the idea may seem, it may not be possible to realize such architectures in practice. For example, legacy reasons pose a huge hurdle for organizations to architect their designs to meet these criteria. Though modularization is a precondition to successfully employ design variant principles we have outlined, our method does not require that the model be modified to accommodate it. This is a positive consequence of decoupling the variant handling mechanism from the architecture which uses associations to map them. However, this still does not restrict fresh architectural designs to separate the common and variable features.

An elaborate variant management scheme with a user interface can be designed to meet the organization's custom needs. Care needs to be given to the variant representation and consistency. It is important to have a clear understanding of how the variant designs will ultimately map to the market needs. This may require agreement on market requirements involving cross-functional groups spread across marketing, sales, manufacturing, and development. We have outlined a set of best practices for the scripting methodologies that can serve as the backbone of such implementations.

For practitioners of Model-Based Design, variants offer new possibilities for managing their designs efficiently and will serve to complement other tool approaches. With automatic code generation, variant components in the software model are mapped to C function code variants that can be switched by simply modifying the preprocessor definitions. Design exploration where several alternatives exist for a component can now be managed efficiently to simulate every design alternative in a combinatorial fashion for a given test suite. For large-scale problems, these data-parallel problems could be distributed on a cluster of multi-core computers for overall speedup with our scripting methodology.

References

MATLAB and Simulink are registered trademarks of The MathWorks, Inc. See www.mathworks.com/trademarks for a list of additional trademarks. Other product or brand names may be trademarks or registered trademarks of their respective holders.

1. F-35 Variants, The F-35 Lightning II Main JSF Website, `http://www.jsf.mil/f35/f35_variants.htm`
2. Winchester, J.: Civil aircraft-passenger and utility aircraft: A century of innovation. Amber Books, London (2010)
3. Nicolescu, G., Mosterman, P.J.: Model-based design for embedded systems: computational analysis, synthesis, and design of dynamic systems. CRC Press, Boca Raton (2009)
4. Mosterman, P.J., Zander, J., Hamon, G., et al.: A computational model of time for stiff hybrid systems applied to control synthesis. Control Engineering Practice 19 (2011)
5. MathWorks, Simulink User Guide. MathWorks Natick, MA (2011)
6. Peak, R.S., Burkhart, R.M., Friedenthal, S.A., et al.: Simulation-based design using SysML—Part 2: celebrating diversity by example. In: INCOSE International Symposium, San Diego (2007)
7. Booch, G.: Object oriented analysis and design with applications. Addison-Wesley Professional (1993)
8. Kinnucan, P., Mosterman, P.J.: A graphical variant approach to object-oriented modeling of dynamic systems. In: Proceedings of 2007 Summer Computer Simulation Conference, San Diego, CA, pp. 513–521 (2007)
9. Mosterman, P.J., Vangheluwe, H.: Computer automated multi-paradigm modeling: an introduction. Simulation: Transactions of The Society for Modeling and Simulation International 80(9), 433–450 (2004)
10. Meyer, M.H., Lehnerd, A.P.: The power of product platforms: building value and cost leadership. Free Press, New York (1997)
11. Clark, K., Fujimoto, T.: New product development performance: strategy, organization, and management in the world auto industry. Harvard Business School Press, Boston (1991)
12. MathWorks, Stateflow user's guide. MathWorks, Natick, MA (2011)
13. Parnas, D.: On the criteria to be used in decomposing systems into modules. Communications of the ACM 15(12), 1053–1058 (1972)
14. Parnas, D.: On the design and development of program families. IEEE Transactions on Software Engineering (1976)
15. Zhang, H., Jarzabek, S.: XVCL: a mechanism for handling variants in software product lines. In: Science of Computer Programming, vol. 53, pp. 381–407. Elsevier (2004)
16. Minsky, M.: A framework for representing knowledge, the psychology of computer vision. McGraw-Hill (1975)

17. Bassett, P.: Framing software reuse—lessons from the real world. Yourdon Press, Prentice-Hall, NJ (1997)
18. Gomaa, H.: Designing software product lines with UML: from use cases to pattern-based software architectures. Addison-Wesley Professional (2004)
19. Possompès, T., Dony, C., Huchard, M., et al.: Design of a UML profile for feature diagrams and its tooling implementation. In: Proceedings of the 23rd International Conference on Software Engineering and Knowledge Engineering, Miami Beach, Florida, USA (2011)
20. Junior, E.A.O., Gimenes, I.M.S., Maldonado, J.C.: Systematic management of variability in UML-based software product lines. Journal of Computer Science 16(17), 2374–2393 (2010)
21. Taub, H., Schilling, D.: Digital integrated electronics. McGraw-Hill, New York (1977)
22. Barnard, P.: Graphical techniques for aircraft dynamic model development. In: AIAA Modeling and Simulation Technologies Conference and Exhibit, Providence, Rhode Island (2004)
23. Ghidella, J., Mosterman, P.: Requirements-based testing in aircraft control design. In: Proceedings of the AIAA Modeling and Simulation Technologies Conference and Exhibit, San Francisco (2005)
24. Chou, B., Mahapatra, S.: Techniques for generating and measuring production code constructs from controller models. SAE International Journal of Passenger Cars- Electronic and Electrical Systems 2(4), 127–133 (2009)
25. Walker, G., Friedman, J., Aberg, R.: Configuration management of the model-based design process. In: SAE World Congress, Detroit (2007)
26. Wakefield, A., Miller, S.: Improving System Models Using Monte Carlo Techniques on Plant Models. In: AIAA Modeling and Simulation Technologies Conference and Exhibit, Hawaii (2008)
27. Ghidella, J., Wakefield, A., Grad-Frielich, S., et al.: The use of computing clusters and automatic code generation to speed up simulation tasks. In: AIAA Modeling and Simulation Technologies Conference and Exhibit, South Carolina (2007)
28. MathWorks, Simulink Coder user guide. MathWorks, Natick, MA (2011)
29. Kang, K., Cohen, S.G., Hess, J.A., et al.: Feature-oriented domain analysis (FODA) feasibility study. Technical Report, CMU-SEI-90-TR-21, Software Engineering Institute, Carnegie Mellon University, Pittsburgh (1990)

Chapter 16
Safety and Security Interdependencies in Complex Systems and SoS: Challenges and Perspectives

Sara Sadvandi, Nicolas Chapon, and Ludovic Piètre-Cambacédès

Abstract. This paper has two objectives: raising awareness about the existence, nature and impacts of safety-security interdependencies in complex systems, and promoting the idea that System Engineering tools and methodologies may help to master them. Firstly, we illustrate and categorize the different types of safety-security interdependencies, before identifying their related stakes. Then, we highlight the links between safety and security ontologies, in theory and in practice. We also present some primary elements needed for a concrete application of System Engineering approaches on the safety-security issue. Finally, potential directions and future efforts needed to continue this research are discussed.

Keywords: safety, security, System Engineering, risk, hazard.

1 Introduction

Safety and security are two distinct but closely related disciplines. Nevertheless, they have been historically dealt with by different communities and considered separately ([10], [20]). Moreover, their respective specialists were themselves grouped outside the core development teams. Such a separation was not optimal, and didn't benefit from the many potential synergies between security and safety, but could be considered acceptable for simple systems. The growing complexity of modern industrial systems, their increasing reliance on information and communication technologies which promote interconnectivity and access, radically changes this situation. In various contexts, safety and security issues are now concerning the same

Sara Sadvandi
Sodius, System Engineering Department

Nicolas Chapon
Communication & Systèmes, Defence Systems

Ludovic Piètre-Cambacédès
Electricité de France, R&D Department

systems and have to be considered jointly. The adoption of such a joint treatment of safety and security is of paramount importance: the superposition of safety and security counter-measures is not mastered, and can lead to unexpected consequences. As discussed later in this article, it potentially has a strong impact on design and operational costs, as well as from a global risk management perspective. This paper aims to raise awareness about the existence, the nature and the potential impacts of such safety-security interdependencies.

Over the last decades, industrial experience in designing large and complex systems in various fields (military, aerospace, energy, etc.) has lead to the development of a whole new discipline, System Engineering (SE). SE may be defined as a structured and interdisciplinary approach which provides a framework to orchestrate the design of complex system. As a consequence, it has the potential to smoothly integrate security and safety-related requirements, and to coordinate the work of safety and security specialists as it has already done for other aspects in complex systems. The second objective of this paper is thus to stress the potential of System Engineering approaches, as a promising area of investigation to master the interdependencies between safety and security.

2 Safety and Security Interdependencies

2.1 Illustrating Safety and Security Interdependencis

Some interdependencies between safety and security may seem natural or obvious. For instance, it is commonly admitted that safe operations may need security [10, 14]: as a particular example, malicious modifications of measurements or configuration data in safety-instrumented systems may lead to unsafe conditions in an industrial infrastructure. Nevertheless, there are more diverse and subtler interdependencies to consider.

The example of an automated door has been used several times in the literature to illustrate this statement [3, 4, 22]: let's consider a system in charge of opening and closing an automated door, single entry to an access-restricted room. From a pure safety standpoint, such a system would have to be designed with a *fail-safe* behavior in case of electrical supply failure: the door would fail open in order to ease emergency operations, including evacuation and emergency team intervention, in case of fire for instance. From a security standpoint, such a system would have to be designed with a *fail-secure* behavior: the door would fail shut and remind shut, in order to prevent an intrusion, the electrical failure being potentially provoked by a malicious action. Thus, for the same system, depending on the adopted perspective, safety and security can lead to purely antagonistic requirements. Of course, once this antagonism is identified, several solutions are possible; this small example intends only to help the reader realize that less known types of safety-security interdependencies (here, an antagonism) exist.

2.2 *Types of Interdependencies*

An extended literature analysis made in [15] has lead to a categorization in four categories of interdependencies:

- **Conditional dependence:** fulfillment of safety requirements conditions security, or vice-versa. It is mentioned in scientific literature (e.g., ([10], [14], [24]) but also in industrial standards (e.g., cf. [7] for the nuclear industry).
- **Mutual reinforcement:** safety requirements or measures contribute to security, or vice-versa. Such situations enable resources optimization and cost reduction.
- **Antagonism:** safety and security requirements or measures lead, when considered together, to conflicting situations. The example given in Section 2.1 is a direct illustration.
- **Independence:** no interaction. The identification of a lack of independence between safety and security has intrinsic value and is needed for a complete view of risks: shortcuts are often made assuming that safety measures will automatically answer to security needs or vice-versa.

2.3 *Stakes*

Considering the different kinds of interdependencies, and concrete associated situations, two main types of stakes emerge:

- **Cost management**, depending on the identification and the leveraging of the possible synergies between safety and security.
- **Correctness of risk assessments**, by ensuring that the effects of safety-security interdependencies, such as antagonisms or reinforcements, are taken into account.

2.4 *State of the Art*

The issue of safety and security interdependencies is not new in itself: it has been identified for more than ten years, mainly in the defense and aeronautical industries (see for instance [4]). In these domains, several efforts have since been accomplished (e.g., the SafSec methodology [9] developed by Altran Praxis for the UK Ministry of Defense, the Federal Aviation Administration efforts in the US [23] or the work accomplished by EuroCAE WG72/RTCA SC-216 from an international perspective). Nevertheless, the existing results and initiatives are still either purely organizational, or more technical but not covering all types of interdependencies, and lacking industrial applicability. Moreover, they often assume an existing consistent ontology between safety and security, which is in fact lacking or inconsistent. They do not develop how engineering processes should be coordinated concretely. The next sections of this paper address both of these concerns.

3 Towards Consistent Security-Safety Ontology and Treatment

3.1 Unifying Safety and Security Ontologies

Recent security and safety standards, such as EBIOS [1] for security and ESARR [5] for safety, show that safety and security ontologies are mostly based on the same concepts and same phases. Others have made the same statement (e.g., [6], [21]). Based on EBIOS and ESARR, they can be summarized as follows:

- **Identification of the *hazards***, at the functional level, or capacity level, of the system. The hazards describe, in a generic way, failure modes that impair the safety or security of the system and its environment.
- **Identification of the *effects*** (or consequences) of the hazards and estimation of the severity of the effects.
- **Identification of the possible *causes*** (safety ontology), or ***threats scenarios*** (security ontology) that may induce the hazards, along with their ***probabilities/frequencies of occurrence*** (safety ontology), or ***likelihood*** (security ontology).
- Then **the concept of *risk*** comes naturally from the combination of identified severities and identified probabilities or likelihoods.

Thus, there is no reason, *in theory*, not to study safety along with security through the same ontology. However, in order to practically coordinate and enhance the treatment of safety and security, we have to distinguish between known and controlled risks, and unknown and uncontrolled risks. This distinction is inspired by the SEMA referential framework (System/Environment - Malicious/Accidental, cf. [17]) and can be seen as complimentary. It can provide significant help when managing risks with the complexity of today's industrial systems, as explained in the next subsections.

3.2 Distinguishing KC (Known and Controlled) from UKUC (Unknown or Uncontrolled) Risks

In security, known and controlled risks are related to a set of known threats and vulnerabilities, such as tools (software, hardware, fake documents, etc.), known attacks (man-in-the-middle, social engineering, etc.), and even known persons or known profiles, which may exploit known vulnerabilities inside the system. Conversely, unknown or uncontrolled risks can be considered as "zero-day risks". Zero-day risks correspond to the exploitation of vulnerabilities before the developer of the target knows about them. In most cases, zero-day exploits require a high level of technical skills and results from an indepth analysis of the target with a large investment in time. For example, Stuxnet, the famous malware that infected the Iranian nuclear facilities, was based on four so-called zero-day exploits [13].

In safety, what is usually known and controlled is the system itself, for example Mean Up Time (MUT), Mean Down Time (MDT) of each of its elements. Conversely, what is unknown and uncontrolled is the environment of the system, either the organizational, technical, or physical environment. The Fukushima disaster gives us a dramatic example of a partly-known and uncontrolled risk coming from the physical environment of the system.

3.3 Addressing UKUC Risks by Defense-in-Depth

Once security and safety risks are both classified into KC and UKUC categories, it is easier to treat them consistently. The UKUC part is the most difficult part to manage, for both security and safety issues; in many cases the decision maker, system architect and the manufacturer address these unknown risks by implementing a Defense in Depth (DiD) strategy[1].

The implementation of DiD is mainly decided at a strategic level, and therefore its implementation is malleable. For example, a decision maker is responsible for the storage of some critical materials. He has got an *a priori* choice: for the same cost, he can either build one special warehouse that can resist to earthquakes but not to kamikaze aircrafts; or build two warehouses that cannot resist to earthquakes, but one of the warehouses can be kept empty, thus strongly discouraging any kamikaze attacks. If he considers only the security, the second solution is the best; and if he considers only safety, the first is the best. A DiD strategy allows the deficiencies of such *a priori* choices to be overcome and enables a balanced coverage of both safety and security risks. The decision maker may decide to store the critical materials in the cheapest warehouse, in a military base a bit more expensive but for which the risk of earthquake is lower, and the money he spared will enable him to set up an additional ground-air weapon system. This simple example is meant to highlight these two simple ideas:

- **Safety and security are a decision-making concern**, and not only an engineering concern;
- When the decision maker wants **to address the UKUC part of the risk, the DiD strategy is key** for the consistent integration of security and safety.

3.4 Addressing the KC Risks with Formal Modeling

For the KC component of the risk, security and safety risks should be analyzed together through formal modeling to permit precise understanding, analysis and management of the safety and security of the system.

Mastering safety and security interdependencies implies formalization of both aspects. While for decades safety engineers have been using rigorous methods and formal modeling approaches (e.g., fault trees, Petri nets or Markov formalisms), it appears that on the contrary, risk formalization in the field of security is recent [12]. Even if several research initiatives and prototypes have already brought promising results (e.g., [12] or [16]), much effort is still required to bring the formalism of

[1] DiD strategy: since we cannot anticipate all the malicious or natural causes that may induce hazards, we therefore set up several independent barriers for each hazard, in order to bring enough confidence in the security and the safety of the system.

security modeling to the same level as safety modeling. Nevertheless, for the KC part of the security risks, this is an achievable objective. As a matter of fact, hackers try to exploit security flaws in the system by going through different well-known processes [18]. This includes sniffering communications, using precomputed tables attacks, social engineering, hijacking unauthenticated protocols, etc. Most of these hackers' habits can be formalized. If the unpredictable, uncontrolled threats can only be addressed by the DiD strategy, as seen in Section 3.3, the ordinary threats can be formalized and modeled rigourously, thus enabling a precise and complete risk analysis.

4 Harmonizing Safety and Security into a System Engineering Processes

4.1 Key Issues in Complex Systems

Some of the methodological problems of disconnected safety and security processes are:

- **Multiple types of design and conception:** independent design and conception of the same system are made by different disciplines such as safety and security.
- **Long term project life cycle:** security and safety being evaluated over time, the related activities apply to different phases of the system development, from conception and specification to verification and validation. Other difficulties arise due to possible updates or replacements of the methodologies during the system life cycle. Functional and dysfunctional architectures may also change.
- **Redundant analysis activities:** there are redundant activities due to the use of different vocabulary for the same task by the different teams.

An appropriate collaborative platform may address most of these issues. A collaborative platform aims at managing data coming from disconnected or semi-connected disciplines and at orchestrating various stakeholders (such as decision makers, architects, safety and security departments) in a specific corporate context. Some generic advantages of a collaborative platform include the possibility to collect common and valid data in different applications across the same project, to identify the core data relevant to different applications and to safety and security processes, to manage connected and disconnected data accessible between resources, the transparent access to a unified view, and a management of the project policies and procedures at the corporate level.

A collaborative platform aims to reduce the duplication, facilitate data transformation, improve the risk management, improve decision making, increase the information quality and simplify project development. As a consequence, it has the potential to coordinate safety and security efforts, and to deal efficiently with the interdependencies and the associated risks mentioned in Section 3. In the next sections, we propose an example of generic framework for a collaborative platform able to fulfil these objectives; then we present several possible architectures.

4.2 *Towards an Appropriate Framework to Deal with Safety and Security Interdependencies*

To accommodate all of the data formats, models and structures existing in a complex project, they need to be presented in a harmonized way. This implies that there must be a consolidated framework for exchanging and extracting data. This is specifically true for safety and security related elements, in order to be considered together in a global approach capturing their interdependencies, but also ensuring consistency with functional aspects of the system.

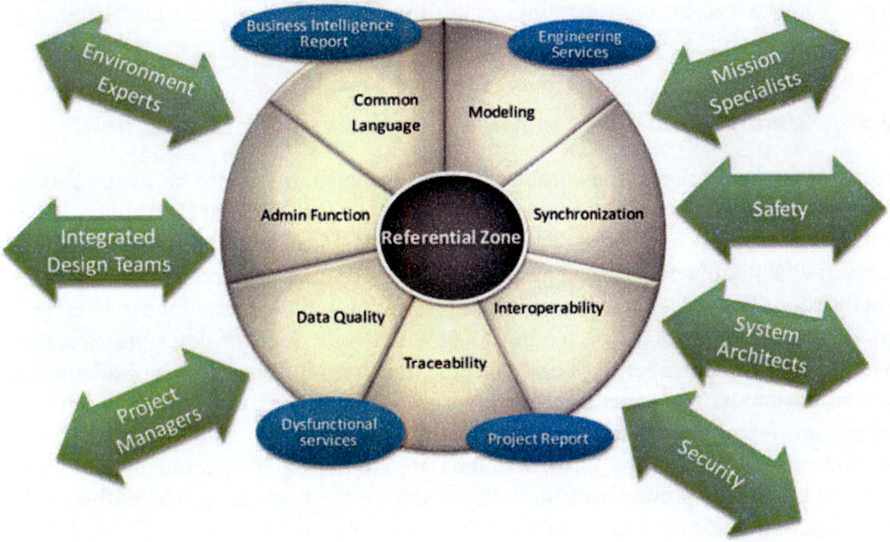

Fig. 1 An ideal collaborative framework

In Figure 1 we present a conceptual view of an appropriate collaborative framework. The framework should ensure and provide:

- *A common ontology*: common language or dictionary with precise descriptions, associated rules and attributes.
- *A referential zone*, which is based on a common ontology between safety and security (as discussed in Section 3.1). It enables mastering the data coming from different disciplines.
- *An administration function*: the framework administration is an important activity as it has to send the commands for saving, modifying, deleting or adding data.
- *Data quality management*: in a project each party can create, access, use, modify the data in different ways. Quality management ensures the coherence and consistency of data; the process of cleaning and normalizing the data plays an important role.

- *Traceability*: traceability links are defined between the system elements to identify the impact of changes on the system.
- *Integration and synchronization*: the framework should ensure consistency among the models and throughout the various data transformations.
- *Modeling and interoperability*: a set of consistent modeling languages that produces different points of view related to safety and security discipline. It can be based on formal languages, like Figaro or Altarica, with simulators.

This framework should also offer different auxiliary services such as a reporting service (e.g., to generate specification plans, traceability matrixes), business intelligence reports (e.g., benchmarking, performance metrics), verification and validation (V&V) services, interoperability and import/export services (model transformation).

4.3 Fundamental Steps to Make the Framework Operational

After determining the framework, a number of decisions have to be made; this includes the definition of a common ontology between the different disciplines, and in particular for safety and security which is applied to the referential zone. A common ontology facilitates data exchanges between the applications and within the different divisions and organizations. For example, in the framework defined in Section 3, some of the services need to exchange data with detailed sets of descriptions and attributes whereas other services use the same data with different levels of details and sets of attributes. The same data or model can be observed and used with different views; this fact increases the complexity of synchronization. A significant challenge in data mastering is to find out where the data are used and to define the strategies for standardizing, harmonizing, and consolidating their structure. The enforcement of a common ontology can be delineated in three phases (derived from [19], [11]):

Common Ontology Definition: identification of the common ontology, dictionary or language by analyzing the existing processes. This ontology is considered as an important part of the referential zone in the framework. In this phase, we first have to analyze the existing processes of each discipline (safety, security, functional architecture, etc.). This analysis includes key functionalities, input and output flows of the process, process life cycles, roles and responsibilities, definition of the interfaces between the disciplines, etc. From this analysis, a complete and coherent ontology, that will be used in the referential zone of the framework, is defined. The fundamental steps to define a usable ontology for the referential zone are as follows ([19], [11]):

- Establishing a general diagnosis from the existing safety and security processes,
- Analysis and diagnostic of the existing system.
- Defining the main data type associated with each discipline and their roles.
- Defining common characteristics (attributes, metadata, context, mode of governance, etc) for the datatype, (common dictionary).
- Identifying the roles, responsibilities and rules of governance associated with the content.

The outputs are (i) a dictionary, defining data types, functional and technical representations, (ii) the description of data life cycles, describing the transactions through all disciplines and applications.

Referential Urbanization and Process Management: setting up a management process for the referential zone. In this phase, we clarify and maintain the management processes of the ontology (e.g., creation, modification and elimination of data and information).

Data Governance: The governance of the referential contains the description of roles, security controls, implementation of the methods to detect defaults, assurance of the normalization of data.

4.4 Potential Architectures for Implementation

The architecture of the referential zone plays an important role as it is responsible for the validity and the quality of the information exchanged between the users [19, 2]. Architectural choices imply the definition and the description of the framework in terms of the relationships between its components, but also in terms of the strategy used to develop and maintain the referential zone. According to [19] and [11], we have identified three possible types of architecture to support the framework previously discussed:

Centralized: this architecture is based on a unique referential zone where the processes of data management are controlled by the same entities. A unique ontology and referential zone is shared by everyone. It is clear that creating such a referential is complex. Controlling the date input/output and the administration activities are difficult. Moreover, the evolution of one of the disciplines in terms of its meta model, makes updating the referential zone very complex. However, with this architecture we have a unique vocabulary and common model between all disciplines.

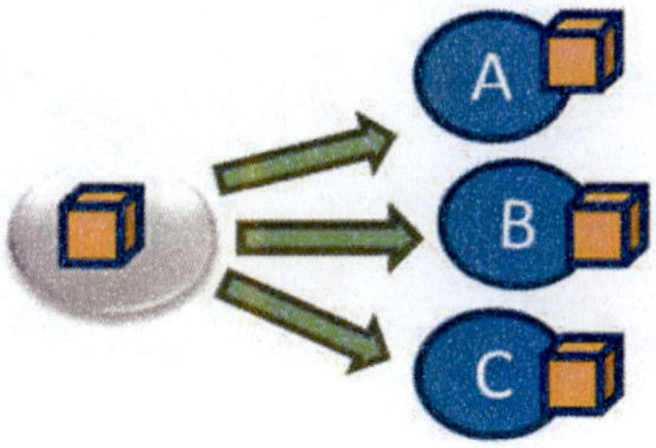

Fig. 2 Centralized architecture

Consolidated: this architecture is based on the normalization of the processes for each discipline, and a central Referential zone that covers all the data coming from the different disciplines (safety, security, architecture), with their own vocabulary and dictionary. In this architecture each discipline works with its own normalized data. Then their inputs are transferred to the standard referential format for saving. In order to save the data we need to launch a synchronization service to synchronize the referential with the normalized entry. In this architecture, defining the best strategy for quality management is an important factor. If a participant requests a data, the framework normalizes the valid data in the requested form and transfers it to that referred discipline.

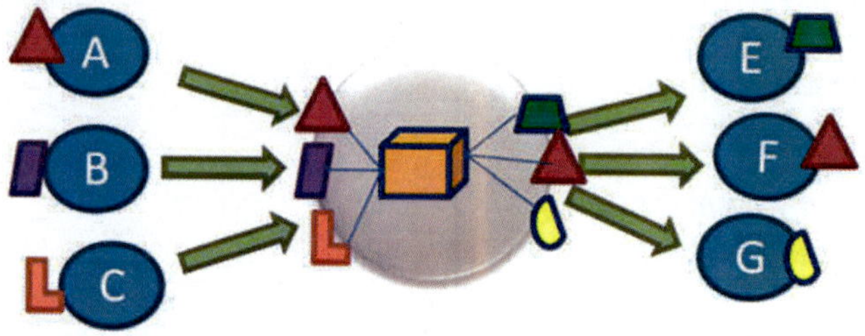

Fig. 3 Consolidated architecture

Cooperative: In this architecture, we have first to normalize the ontology and the vocabulary. So the Referential zone makes use of different normalized meta-models, where they are interconnected; it transfers and translates the information between the different disciplines. It means that the normalization and the synchronization of the data can be done by each discipline because at the beginning we have defined and applied a normalized strategy for all disciplines. On the other hand, they are dependent on the referential for the use of data coming from the other disciplines. Setting up such an architecture is complex in terms of control and synchronization, but it is highly scalable.

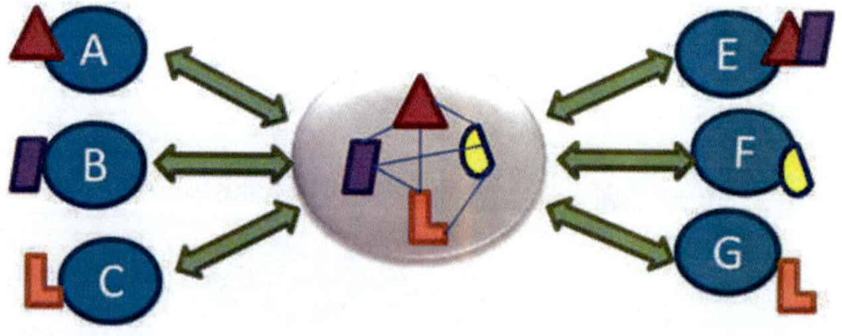

Fig. 4 Cooperative architecture

In summary, a centralized architecture could be considered as one of the best solutions for harmonizing safety and security disciplines as it has the best strategy for assuring the consistency of data transformation between teams by using a unique ontology. In a centralized architecture, the entire system is centralized and everyone uses the same vocabulary within the normalized process. While in a consolidated architecture each discipline uses its own normalized process and vocabulary, not having to change its habits. Nevertheless, the development and the control of the input/output and data transformation is more difficult. Cooperative architectures are interesting in the way that if one of the related processes evolves, adopting and updating the referential is easier compared to the other architectures, but setting up the cooperative architecture is difficult.

Finally, specific safety and security considerations may play a role when implementing the framework and choosing the architecture. Such aspects will be developed in a future paper.

4.5 Decompartmentalization of Normative and Collaborative Initiatives

There exist several initiatives aiming to develop a framework providing analysis of the interdependencies between safety and security. Several of them have been mentioned in Section 2.4 (e.g., see [8], [23] for the aeronautics in the past few years); others could be added here, like more recently the SEISES (Systèmes Embarqués Informatisés, Sécurisés et Sûrs) project in the aeronautics domain. Unfortunately, most of these initiatives are sector-oriented, although the issue of dealing with safety and security interdependencies is cross-sectorial.

Considering the stakes and the large room for progress in the area, transverse collaboration and a cross-sectorial approach is needed. In this context, a shared forum to foster and structure such transverse initiatives is currently lacking. Considering the role SE may play into tackling the safety-security interdependencies challenge, it would make sense if its related bodies like the AFIS (Association Française d Ingénierie Système) in France or the INCOSE (International Council on Systems Engineering) on an international level, took part in this process. Standardization bodies, like the ISO (International Organization for Standardization) or the IEC (International Electrotechnical Committee) could also play a useful role, as they both have working groups dealing with security, safety and system engineering, but up to now, on separated tracks.

5 Conclusion, Limits and Perspectives

This paper had two objectives: raising awareness about the existence, nature and impacts of safety-security interdependencies in complex systems, and promoting the idea that System Engineering may provide an appropriate set of tools and methodologies to master them.

On the first aspect, we have provided different examples and identified the underlying stakes: in a nutshell, unexploited synergies between safety and security lead to additional costs while undetected antagonisms and dependencies lead to inaccurate risk evaluation.

On the second aspect, in order to strengthen our message, we have presented some primary elements needed for a concrete application of S.E. approaches on the safety-security issue. The basis for a common ontology between safety and security has been proposed. Then, a high-level view of an appropriate framework, able to orchestrate safety and security concerns with the other engineering process, has also been described. Deeper analysis on how SE methods can be leveraged to the treat security-safety interactions is still needed and will be treated in future works.

Much more efforts are needed to give birth to operational solutions and tools. Cross-sectorial initiatives and applied research are in particular highly needed. In this objective, we strongly believe that setting higher level forums, multisectorial and multicultural, through the INCOSE or the ISO for instance, could help and could foster the process.

References

1. ANSSI France. EBIOS 2010: Expression des besoins et Identification des Objectifs de Sécurité (2010), `http://www.ssi.gouv.fr/`
2. Berson, A., Dubov, L.: Master data management and customer data integration for global enterprise. McGraw-Hill, Osborne (2007)
3. Derock, A., Hebrard, P., Vallée, F.: Convergence of the latest standards addressing safety and security for information technology. In: On-line proceedings of Embedded Real Time Software and Systems (ERTS2 2010), Toulouse, France (May 2010)
4. Eames, D.P., Moffett, J.: The Integration of Safety and Security Requirements. In: Felici, M., Kanoun, K., Pasquini, A. (eds.) SAFECOMP 1999. LNCS, vol. 1698, pp. 468–480. Springer, Heidelberg (1999)
5. Eurocontrol. Eurocontrol SAfety Regulatory Requirement. Eurocontrol Safety Regulation Commission (2001)
6. Deleuze, G.: Un cadre conceptuel pour la comparaison sûreté et sécurité de filières industrielles. In: Proceedings of the 2nd Interdisciplinary Workshop on Global Security (WISG 2008), Troyes, France (2008)
7. International Electrotechnical Commission (IEC). Nuclear power plants – instrumentation and control important to safety – requirements for computer security programmes. IEC Committee Draft 62645 (April 2010)
8. Jalouneix, J., Cousinou, P., Couturier, J., Winter, D.: Approche comparative entre sûreté et sécurité nucléaires. Technical Report 2009/117, Institut de Radioprotection et de Sûreté Nucléaire (IRSN) (April 2009)
9. Lautieri, S., Dobbing, B.: SafSec: Integration of Safety & Security Certification, SafSec Methodology: Standard (3.1) (November 2006)
10. Line, M.B., Nordland, O., Røstad, L., Tøndel, I.A.: Safety vs. security? In: Proceedings of the 8th International Conference on Probabilistic Safety Assessment and Management (PSAM 2006), Nouvelle-Orléans, États-Unis (May 2006)
11. Loshin, D.: Master data management. The MK/OMG Press (2009)
12. Nicol, D.M., Sanders, W.H., Trivedi, K.S.: Model-based evaluation: From dependability to security. IEEE Transactions on Dependable and Secure Computing 1(1), 48–65 (2004)
13. Falliere, N., Murchu, L.O., Chien, E.: W32. Stuxnet Dossier, version 1.4. Symantec reports (2011)

14. Nordland, O.: Making safe software secure. In: Proceedings of the 16th Safety-Critical Systems Symposium, Improvements in System Safety, SSS 2008, Bristol, UK, pp. 15–23 (February 2008)
15. Piètre-Cambacédès, L.: Des relations entre sûreté et sécurité. PhD thesis, Télécom Paris-Tech (2010) (in French)
16. Piètre-Cambacédès, L., Bouissou, M.: Attack and Defense Modeling with BDMP. In: Kotenko, I., Skormin, V. (eds.) MMM-ACNS 2010. LNCS, vol. 6258, pp. 86–101. Springer, Heidelberg (2010)
17. Piètre-Cambacédès, L., Chaudet, C.: The SEMA referential framework: avoiding ambiguities in the terms "security" and "safety". International Journal of Critical Infrastructure Protection 3(2), 55–66 (2010)
18. Provadys: Top 10 Corporate Networks Security flaws (2009), `http://www.checkmates.eu/`
19. Régnier-Pécastaing, F., Gabassi, M., Finet, J.: MDM, enjeux et méthodes la gestion des données, Dunod (2008)
20. Schoitsch, E.: Design for safety and security of complex embedded systems: a unified approach. In: Proceedings of the NATO Advanced Research Workshop on Cyberspace Security and Defense: Research Issues, Gdansk, Poland, pp. 161–174 (September 2004)
21. Stoneburner, G.: Toward a unified security-safety model. IEEE Computer 39(8), 96–97 (2006)
22. Sun, M., Mohan, S., Sha, L., Gunter, C.: Addressing safety and security contradictions in cyber-physical systems. In: Proceedings of the 1st Workshop on Future Directions in Cyber-Physical Systems Security (CPSSW 2009), Newark, USA (July 2009)
23. U.S. Federal Aviation Administration (FAA). Safety and security extensions for Integrated Capability Maturity Models (September 2004)
24. Winther, R., Johnsen, O.-A., Gran, B.A.: Security Assessments of Safety Critical Systems Using HAZOPs. In: Voges, U. (ed.) SAFECOMP 2001. LNCS, vol. 2187, pp. 14–24. Springer, Heidelberg (2001)

Chapter 17
Simulation from System Design to System Operations and Maintenance: Lessons Learned in European Space Programmes

Cristiano Leorato

Abstract. Simulation is a key activity that supports the specification, design, verification and operations of space systems. Traditionally, for the same European Space Programme, different simulation facilities are developed for different phases of the development life-cycle of space systems. Those facilities are often developed by different, independent, and geographically distributed parties, utilizing heterogeneous technologies and standards. In the last few years, a more comprehensive reuse strategy across the space mission life-cycle has been adopted in specific European Space Agency missions, namely in the Automated Transfer Vehicle and in the Lisa Pathfinder mission. This paper reports on those experiences. Guidelines are identified, allowing understanding, at early phases, the technical implications of reuse. The applied methodology also gives indications to the management, helping assess the actual feasibility and cost of the reuse of the existing simulators.

Keywords: Simulation, life-cycle, space, reuse, Model Based Design, Automated Transfer Vehicle, Lisa Pathfinder.

1 Introduction

Simulation is a key activity that supports the specification, design, verification and operations of space systems [4]. System modelling and simulation can potentially support a number of use cases across the spacecraft development life-cycle, including activities as system design validation, software verification & validation, spacecraft unit and sub-system test activities, etc.

Cristiano Leorato
ESA/ESTEC Keplerlaan, 1 - 2201 AZ Noordwijk, The Netherlands
Cristiano.Leorato@esa.int

As shown in Fig. 1, different simulation facilities are required during the different phases of the development life-cycle of space systems. Design simulators are developed at the beginning of the project life-cycle, in order to support the design of the space mission, e.g. to evaluate performances and the budget for critical resources (mass, power, etc.). They can be developed at different levels, and their actual scope is very diverse, as it ranges from:

- Definition of the system concept.
- Simulation of one or more hardware subsystems.
- Simulation of the algorithms to be implemented in the on-board software.
- End-to-end simulation, including both the ground segment and the flight segment.

Later in the life-cycle, Software Validation Facilities (SVFs) are required to support the validation of the on-board software. Finally, training, operations and maintenance simulators are used to ensure that the ground segment and operations team are ready to support the operations activities post-launch. It is nowadays largely recognized that maintaining the coherency between the different simulation facilities is essential to properly support system engineering [4].

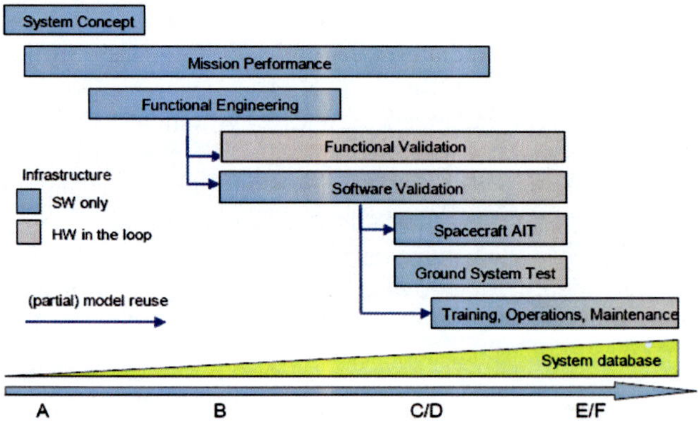

Fig. 1 Simulation facilities across the system development life-cycle (ECSS-E-TM-10-21A)

The development of the space simulation facilities is a challenging enterprise, sharing a number of commonalities with Systems of Systems [7] development. Technologies between the different simulation facilities are highly heterogeneous. In European Space Agency (ESA) missions, design simulators often use Model Based technology (typically Matlab/Simulink), whereas operational simulators are built upon component oriented technology, and based on the Simulation Model Portability (SMP) standard [5].

Furthermore, each facility is targeted at specific classes of end-users (e.g. system engineers, On Board Software engineers, operations engineers), but dependencies between facilities exist.

Finally, geographically distributed manufacturers concurrently develop the simulation facilities for the same ESA mission. The governance itself of each development may also be distributed across different entities or different ESA establishments.

In the last few years, specific ESA missions have achieved significant improvements, paving the way for a more comprehensive reuse strategy across the space mission life-cycle:

- In the Automated Transfer Vehicle (ATV) mission, the design simulators of major subsystems have been reused as part of the real-time ATV Test Facilities. ATV Test Facilities include, among others, the Functional Simulation Facility (FSF), the Software Validation Facility (SVF), and the ATV Ground Control Simulator (AGCS). The ATV Test Facilities have been developed by ASTRIUM ST. The first ATV (Jules Verne) was launched in March 2008. The second ATV (Johannes Kepler) was launched in February 2011. Three more ATVs are scheduled for launch until 2015.
- In the Lisa Pathfinder (LPF) mission, the DFACS (Drag Free Attitude and Control System) design simulator has been reused and is now one of the core components of the LISA Pathfinder simulator for the Science Technology & Operations Centre (STOC), located at ESA/ESAC. LISA Pathfinder is scheduled for launch in 2012. The DFACS Simulator was developed by ASTRIUM Gmbh. It was originally conceived for the analysis of the Drag Free and Attitude Control System performance, and served as a prototype for the on-board control algorithms.

In the near future, the simulator for the PROBA3 mission is foreseen to support a wide range of phases in the project life-cycle. It will use the Formation Flying Test Bed (FFTB), which provides the means to develop simulators addressing the specific needs of formation flying missions. FFTB also supports the reuse of Matlab/Simulink models, usually developed as part of the initial system design, towards later phases, including software validation and operations.

This paper reports on recent ESA experiences in the reuse of simulators across the project life-cycle. It especially focuses on the reuse of design simulators in an operational context. Guidelines are identified, allowing understanding, at early phases, the technical implications of reuse. The applied methodology can also give indications to the management, helping assess the actual feasibility and cost of the reuse of the existing simulators.

2 The Lisa Pathfinder STOC Simulator

During the operational phase of the LPF mission, the STOC will be in charge of the operations of the LPF experiments. For the STOC to be able to perform its planning activities, an experiment simulator including the spacecraft and its environment is required. The STOC simulator will support the validation of the LPF Technology Package (LTP) run procedures and the data analysis activities. It will execute the run procedures generated by means of the MOIS tool [3] and provide measurements to the LPF Data Analysis tools. The STOC simulator will also be used for the training of STOC personnel and to incrementally update the modelling of the LPF spacecraft and in particular of its LTP experiment using flight data.

2.1 The Reused Simulators

The STOC simulator is based on two other simulators that were already available in the LPF project. The Drag Free and Attitude Control System (DFACS) is a Windows Matlab/Simulink based simulator for the analysis of the DFACS performance. It has served as a prototype for the onboard control algorithms.

The Software Validation Facility (SVF) is part of the Astrium Model-based Development and Verification Environment (MDVE), where it is used for the integration and debugging of the onboard software.

The reuse of the DFACS and SVF facilities is the main constraint in the development of the STOC simulator. The DFACS and SVF provide two alternative simulation strategies, which are highly complementary in terms of efficiency and accuracy of the results. The main intent of the STOC simulator is to provide, in a unique framework, the seamless integration of the underlying simulation strategies, together with common simulation services. Those services include: recording, monitoring, injection of telecommand (TC) sequences, injection of external stimuli, and definition of parameters of the simulation models.

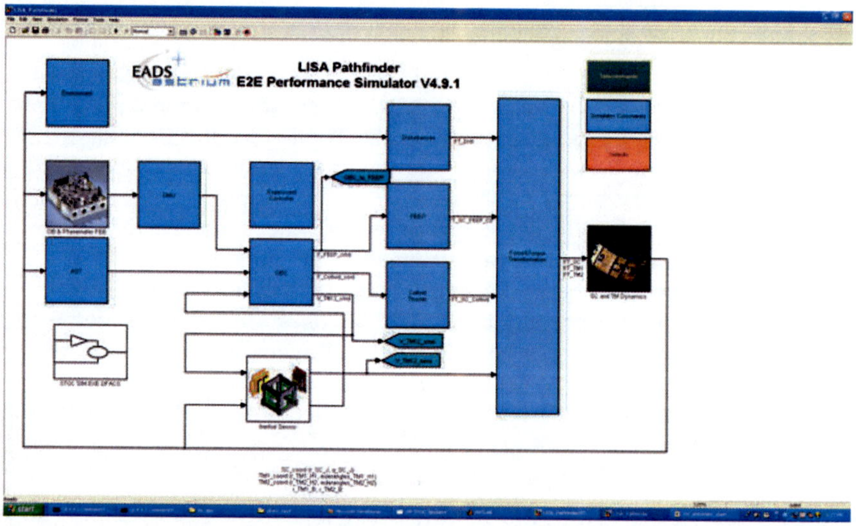

Fig. 2 The DFACS Simulator.

These simulation services necessarily rely on different mechanisms provided by the DFACS and the SVF, but it is the objective of the STOC simulator to hide as much as possible to the user the different implementation details provided by the two systems. The inputs to each service (e.g. variables to be recorded, initialization of parameters in the simulation models, etc.) are provided by the user defining a set of artefacts by means of a common user-friendly Man Machine Interface (MMI).

A detailed description of the LPF STOC Simulator can be found in [2]. The remainder of this section highlights the main issues found in reusing the DFACS simulator.

2.2 The Coupling Problem

Already before the Preliminary Design Review of the LPF STOC Simulator (held in April 2008), it was clear that, although the DFACS status was rather stable, modifications from ASTRIUM could be expected. In order to limit the impact of configuration control issues, the DFACS extensions needed for the STOC purposes (e.g. monitoring, recording, etc.) are decoupled, as much as possible, from the ASTRIUM line of development. Those extensions are implemented as separate libraries, directly depending on the Matlab/Simulink layer. The Software Reuse File document is carefully maintained, so that new versions of the ASTRIUM DFACS can be safely and efficiently incorporated.

Furthermore, the interface between the STOC Simulator MMI and the DFACS environment is mainly based on intermediate text files. This increases observability, and minimizes the coupling between the two systems.

2.3 The Scope Problem: Commanding and Initializing the System

Given the more limited scope of the DFACS simulator with respect to the operational context, an important factor to the success of the project was the development of an appropriate DFACS adapter, allowing mapping, when needed, the spacecraft TCs to their counterparts in the design simulator. The mapping itself is defined in an XML file. The maintenance of this XML file is a critical task. Specific tools (see Fig. 3) have been designed and embedded into the STOC Simulator, allowing:

- Viewing the XML mapping by means of a MMI.
- Showing a report of the dependencies between different TCs, as they may be implicitly specified in the XML mapping.
- Validating the XML mapping w.r.t. a given version of the TM/TC database.

Initialization of the simulators is also a concern. Given the different nature of the two simulators, the mechanism to initialize the simulation variables is very different in the DFACS and in the SVF environment. In the SVF, data integrity is ensured by a dedicated simulator database, defining the relevant properties of the simulation variables. For example, relevant properties are: default value, dimension, range, etc. The documentation provided by ASTRIUM for the numerical models is consistent with the format and naming conventions specified in the simulator database.

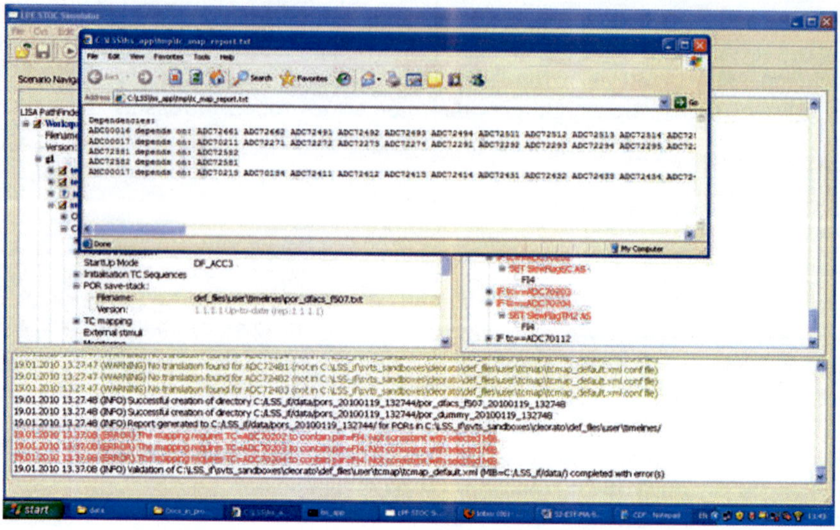

Fig. 3 The LPF tool validating the DFACS TC mapping.

On the other hand, the initialization on the DFACS is based on Matlab scripts. The association with SVF variables is not obvious, because:

- The mapping is not defined in the simulator database: the simulator database only "knows" the SVF simulator; it "does not know" the DFACS simulator.
- Specific conventions and rules have not been enforced in the LPF project since the beginning of the DFACS development.

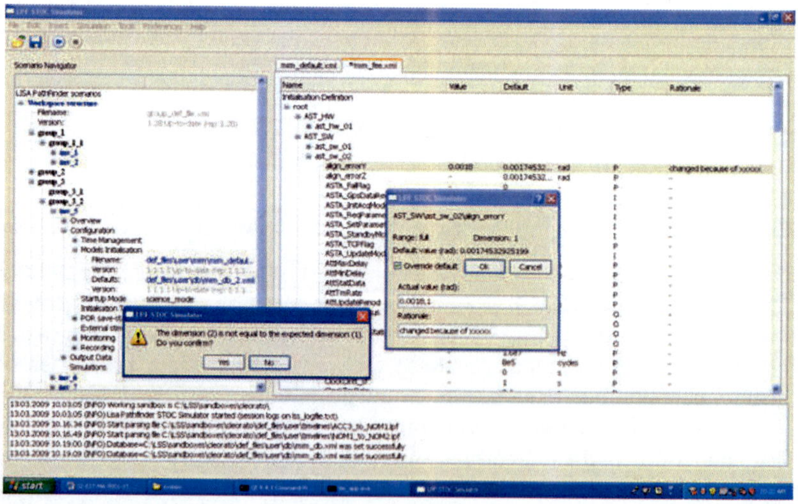

Fig. 4 Editing the initialization of simulation variables.

Nonetheless, in the LPF project, the mapping is made available by ASTRIUM, in the form of additional text files. These text files are exploited by an appropriate DFACS adapter, which can derive the actual initializations to be performed in the Matlab environment. This process is transparent to the user. The user is only requested to specify the initialization in a common MMI (see Fig. 4), using essentially the same format and naming conventions specified in the simulator database, and in the documentation provided by ASTRIUM for the numerical models.

2.4 The Restore Problem

The LPF STOC simulator user requirements included the capability of defining "breakpoints", intended as the capability of saving/restoring the simulation status. From a technical point of view, the complete implementation, verification, and validation of this requirement introduce significant complexity on the development.

However, after more detailed discussions with the final users, it was concluded that only a limited number of breakpoints are needed by the STOC. Furthermore, given the high speed factor of the DFACS simulator (25x real-time), the time to reach each any of the identified required breakpoints from the default initial DFACS status is actually negligible.

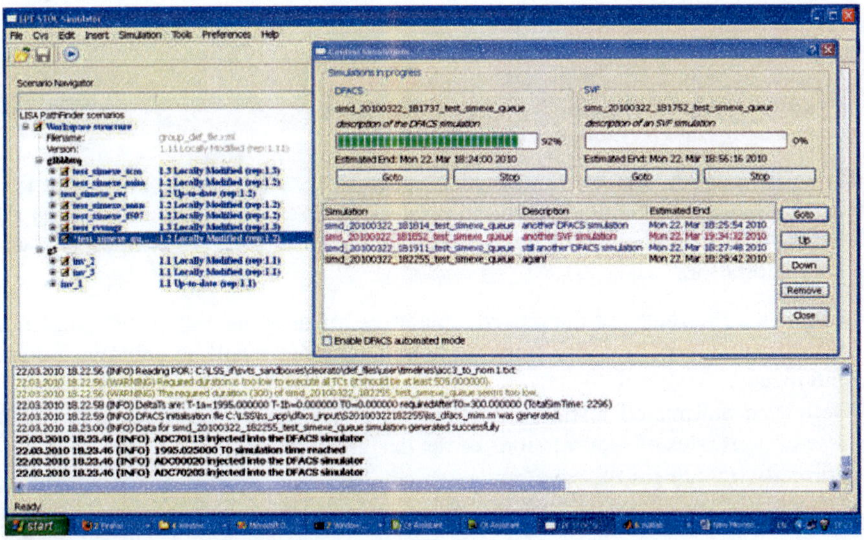

Fig. 5 Executing simulations in batch mode.

Therefore, instead than implementing the full save/restore capability, a light approach has been chosen:

- Special initialization sequences have been identified. The user can easily select them, and the sequence is fully simulated, any time it is required to reach the conditions corresponding to the required "breakpoint".

- Because it is fully acceptable for the STOC to analyze the simulation results off-line, the STOC simulator allows queuing simulations, executing them in batch mode (see Fig. 5).
- If required, the STOC simulator may be easily enhanced, so that different DFACS simulations can be executed in parallel, exploiting the different cores of the STOC Simulator machine.

3 The ATV Ground Control Simulator

The development of the ATV Ground Control Simulator (AGCS) has leveraged from the other Test Facilities [6] which had been developed for the ATV, e.g.:

- The Functional Simulation Facility (FSF).
- The Software Validation Facility (SVF).

The usage of appropriate Simulink programming standards has allowed meeting the required real-time performances. Numerical models could be extended, e.g. implementing the specific failures required to train the ATV Control Center operators.

However, it shall be noted that the ATV real-time Test Facilities include the design simulators of major hardware subsystems, including:

- The Solar Array subsystem.
- The Propulsion subsystem.
- The Docking and Refueling subsystem.

Because the design simulators were developed in textual languages (C and Fortran), an additional effort has been required in order to incorporate them into the Matlab/Simulink development chain. For this purpose, a rigorous approach was needed, in order to:

- Handle the interfaces of the reused simulators towards the rest of the simulation infrastructure, e.g. to define how the reused simulator shall be commanded and initialized.
- Define an automated testing environment, to reduce the testing overhead, in case of deliveries of new versions of the design simulators.
- Correctly implement the save/restore capability.

The analysis of all these three aspects has contributed to the definition of the overall design. It has also contributed to the definition of appropriate and consistent procedures and tools, as shown in Fig. 6. Further details are provided in [1].

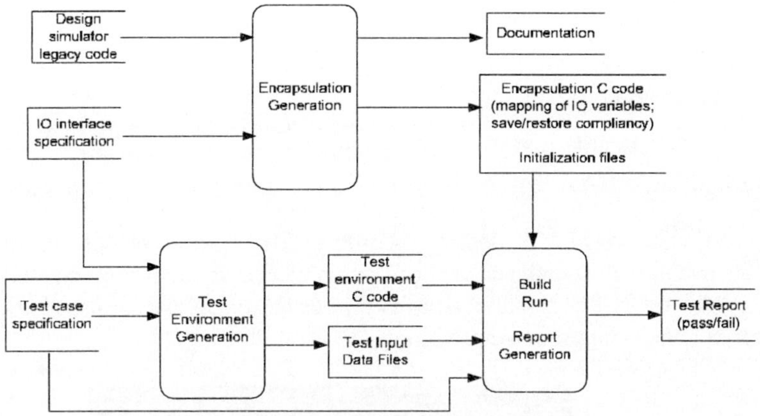

Fig. 6 The established workflow to reuse ATV design simulators.

It is important to notice that the three aspects mentioned above closely mirror the three key issues identified in the previous section, regarding the reuse of the LPF DFACS simulator:

- Identify the scope of the simulator, in terms of its interfaces (e.g. commanding and initialization of the reused system) with respect to the overall operational context.
- Identify and minimize the coupling of the design simulator with the overall operational infrastructure, i.e.: define and maintain the (possibly automated) procedure to incorporate the changes in the design simulator into the operational system.
- Analyze the rationale for the save/restore requirement, and identify a suitable and long-term viable strategy, in order to satisfy the actual user needs.

4 The Methodology

This section summarizes the natural generalization of the experiences reported above. In order to assess, for a given project, the actual feasibility and cost of the reuse of design simulators in an operational context, it is essential to formalize and analyze at least three key views of the design simulators. These views respectively focus on:

- The scope of the reused simulators.
- The coupling of the reused simulators with respect to the overall operational simulator.
- The implementation of the save/restore capability, and of other capabilities specific to operational simulators.

The analysis of each one of these three aspects shall rely on the project context, including:

- The project background.
- The items already developed for the project, their relationships, and their governance.
- User requirements.

The analysis of each view will identify specific needs in terms of design, processes, and tools. Once this analysis has been completed, it is therefore essential to review the results under the global perspective, in order to achieve a consistent design, as well as consistent processes and tools.

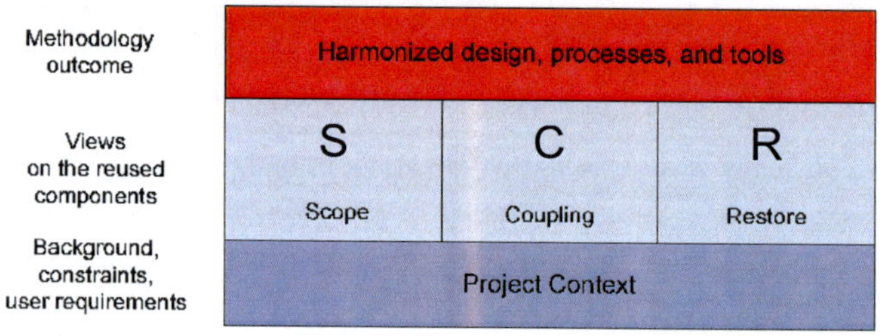

Fig. 7 Reuse of design simulators in operational contexts: the Global view.

4.1 The Scope View

Operational simulators have a different scope than design simulators. An effective reuse requires the development of adapter layers, e.g.:

- To map spacecraft TeleCommands (TCs) to their counterparts in the design simulator.
- To initialize the status of the design simulator.

It is essential that the required adapter layers are maintained up-to launch and beyond. For this purpose, the development of appropriate auxiliary tools may also be needed. Auxiliary tools typically include Man Machine Interface (MMI) editors, tools validating the adapter layers, comparison tools with data in reference databases, etc.

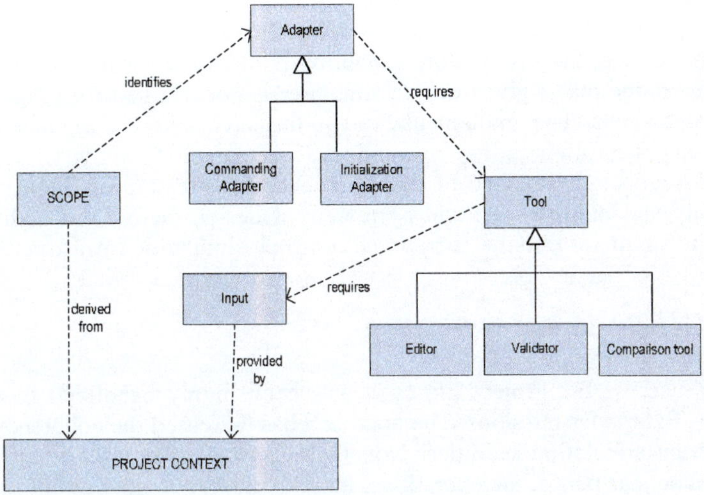

Fig. 8 The Scope view.

It is recommended to identify, at an early stage, the boundary between the legacy systems and the new operational infrastructure, taking particular care to also ensure that any inputs required by the auxiliary tools are available and will be maintained in the long-term, according to the project context.

4.2 The Coupling View

Coupling with reused simulators shall be minimized. Taking into account the project context, the risks deriving from concurrent updates of the reused simulators shall be carefully assessed. Appropriate procedures shall be defined and maintained, in order to safely incorporate the changes into the overall operational system.

4.3 The Restore View

Specific requirements need to be considered for the development of an operational simulator:

- Modelling of specific equipment failures.
- High speed-up factor and/or the capability to save the simulation status, and restore it when requested by the users.

Modelling of specific equipment failures is not usually an issue. In case the associated development shall only be part of the operational simulator, this shall be taken into account in the Coupling view.

On the other hand, a critical aspect in operational simulators is its throughput in terms of significant simulations which can be completed in a given time interval. Throughput is a monotone function of the speed-up factor and of the level of save/restore compliancy (e.g.: 0= save/restore is not supported, 1= save/restore is fully supported). The objective is therefore to optimize the throughput.

In practice, the save/restore capability is often the aspect which is more likely to determine that a given design simulator is not retained for usage in the operational context. The implementation of the save/restore compliancy on existing simulators is often a time consuming and error prone activity. Procedures and tools shall be established to ensure that the save/restore implementation is thoroughly validated, by test and by review of design, before embedding the reused design simulator into the operational run-time simulation infrastructure.

5 Conclusions

Reuse across the project life-cycle has been highly beneficial in the ATV and LISA Pathfinder missions. The approach has increased the coherency between the different simulation facilities. Nonetheless, when it is intended to reuse design simulators as part of the operational infrastructure, potential pitfalls exist. In future missions, risks can be mitigated by applying a set of proven guidelines.

References

[1] Leorato, C., Guidolotti, E., Segneri, D.: Reusing Legacy Code in a Simulink Environment for Real-time Industrial Simulations: Techniques and Prototype Support Tools. In: Proceedings of the DASIA Conference (2005)

[2] van der Plas, P., Leorato, C.: The LISA Pathfinder Simulator for the Science and Technology Operations Center: Simulator reuse across the Project Life-cycle: practical experiences and Lessons Learned. In: Proceedings of the DASIA Conference (2010)

[3] Camino, O., Blake, R., Heinen, W.: Smart 1 Scheduler – A Cost Effective Mission Scheduler compatible with SCOS 2000. In: Proceedings of the 4th International Workshop on Planning and Scheduling for Space, ESOC, Darmstadt (2004)

[4] Walsh, A., Wijnands, Q., Lindman, N., Ellsiepen, P., Segneri, D., Eisenmann, H., Steinle, T.: The Spacecraft Simulator Reference Architecture. In: Proceedings of the 11th International Workshop on Simulation & EGSE facilities for Space Programmes, ESTEC (2010)

[5] Fritzen, P., Segneri, D., Pignede, M.: SWARMSIM - The first fully SMP2 based Simulator for ESOC. In: Proceedings of the 11th International Workshop on Simulation & EGSE facilities for Space Programmes, ESTEC (2010)

[6] Leorato, C., Picard, C.: Reusable cross-platform design for the ATV Test Facilities: Incremental Development and Model Based Design Patterns. In: Proceedings of the DASIA Conference (2006)

[7] Gianni, D., Lindman, N., Fuchs, J., Suzic, R.: Introducing the European Space Agency Architectural Framework for Space-based Systems of Systems Engineering. Submitted to the Complex Systems Design & Management Conference, pp. 335–346, Paris (December 2011)

Chapter 18

ROSATOM's NPP Development System Architecting: Systems Engineering to Improve Plant Development

Mikhail Belov, Alexander Kroshilin, and Vjacheslav Repin

"It has never been more important to have coherent systems engineering processes that can be applied across multi-party teams."
INCOSE Systems Engineering Vision 2020, September 2007 page 14

"Conway's law suggests that effective systems design emerges from system-oriented organizations and the processes employed therein."
INCOSE Systems Engineering Handbook v. 3.2, January 2010 page 43

Abstract. This article is devoted to the architecting of the NPP unit development system (NPPDS) for ROSATOM corporation in a new NPP design initiative, VVER TOI^{TM}. The NPPDS is considered to be a set of processes, organization structures, data, databases, software and interfaces, and documentation.

The rationale for selecting the architecture framework approach for NPPDS architecting is given and the specific original Capital Project Architecture Framework (CPAF) previously developed is described.

The stakeholders and their concerns are defined, as are the views developed based on CPAF viewpoints. Views cover functions and processes, organization structure, data and aspects of software systems of the NPPDS.

Mikhail Belov
Deputy CEO, IBS
mbelov@ibs.ru

Alexander Kroshilin
Deputy CEO, VNIIAES
kroshilin@vniiaes.ru

Vjacheslav Repin
Head of PLM Department, IBS
vrepin@ibs.ru

The importance of processes modeling and processes development for NPPDS architecting is discussed. The key issue is that it is exactly engineering processes that ensure the interoperability of the organization entities, data and software in the development and utilization of NPPs and similar capital projects.

Developing and tailoring both the CPAF – for the purpose of architecting the plant utilization system – and the reference model are considered to be additional research avenues.

1 Introduction

The efficiency of a capital project considerably depends on computer systems which support engineering and construction processes; therefore, the NPP unit development system (NPPDS) is considered to be the key tool of the NPP life cycle management of ROSATOM'S VVER-TOI™ initiative.

ROSATOM is a Russian state-owned corporation which carries out all aspects of nuclear energy generation – from uranium mining and enrichment all the way to NPP decommissioning and site remediation. ROSATOM operates ten NPPs in Russia, which constitute around 14% of Russian energy consumption.

Many ROSATOM affiliates are involved in different activities at all NPP life cycle stages. Some of them design NPPs and nuclear reactors, while others construct or operate NPP units.

At the moment, ROSATOM management is implementing the so-called VVER-TOI™ initiative, whose goal is a new NPP unit design (VVER technology-based); the design must be:

- replicable based on a developed "off-the-shelf" design;
- optimized from the standpoint of economic criteria;
- based on the information model principle and systems engineering approaches.

The VVER-TOI™ initiative is intended to reduce NPP unit development costs & time through:

- a replicable design of the NPP;
- cheaper and easier changes;
- knowledge-based engineering;
- concurrent engineering.

The NPPDS is treated as an integrated set of processes, organization structures, data, databases, software and interfaces, and documentation. It is the basic tool to develop the plant and the "reservoir" to save the results: data and documents. It is also one of the key tools to implement the plant construction and plant commissioning processes.

The NPPDS is a "system of systems", comprising several design and development organization entities and different software systems used by those entities, each of which is a system in and of itself.

An important requirement for contemporary plant development software systems is that they support not only the plant design process, but also planning, technology development and simulation for complicated construction work executed by various construction entities that require absolutely different resources, logistics planning, etc. And what is important is that all these plans and schedules be synchronized and serve as the sole integrated plan of the capital project or for establishing a plant.

Actually, an NPPDS development project is the "assembling" of several software platforms, the consolidation of different design and construction organization entities, the integration of engineering data and the framing of unified processes. Each of these tasks alone is very difficult, and fully executing them demands a concurrent analysis of a great number of divergent needs, very different data about plant design and the plant construction process, the development of engineering solutions for different plant subsystems and the consideration of diverse organizational entities and subsystems at different life cycle stages.

Special approaches are needed to carry out such complicated NPPDS development and implementation tasks, and this article is devoted to one such approach: the use of the architecture framework for system architecting.

2 VVER-TOITM Initiative and NPP Development

To meet the main goals of the VVER TOI™ initiative, ROSATOM management has launched the NPPDS with the following tasks:

1. To implement contemporary design and development approaches, software tools, etc;
2. To create a common integrated IT environment for all VVER TOI™ parties;
3. To create an NPP unit information model and to make it usable during the whole NPP life cycle (around 60 years).

The first task has been applicable to several ROSATOM entities for a long time, and the efforts within the VVER TOI™ initiative should be focused on consolidating such activities with different parties and on creating common solutions for all VVER TOI™ parties.

Implementing the second task should exclude paper exchanges, which is presently common practice in ROSATOM, and implementing a fully electronic engineering data exchange.

The third task should involve a data-centric approach, as opposed to the current document-centric practice.

Replication of the NPP design for a concrete unit is the core idea of the VVER TOI™ initiative, so the VVER TOI™ NPP design considerable differs from any concrete NPP.

The life cycle of a concrete NPP includes several stages, such as a feasibility study and site selection (see Fig. 1).

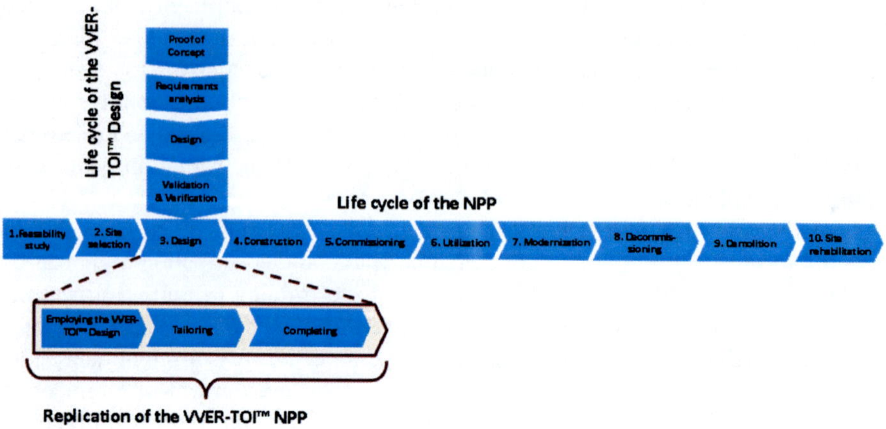

Fig. 1 VVER TOI™ lifecycles stages

The VVER TOI™ NPP life cycle does not coincide with the concrete NPP life cycle, nor is it even a subset, because the target system of VVER TOI™ is the design data and documents, not the NPP itself.

The VVER TOI™ NPP unit design is developed as a "template", so the design stage of a concrete NPP in reality constitutes employing and tailoring a previously developed VVER TOI™ design.

The VVER TOI™ NPP document set includes technical documents and the preliminary licensing documents (nuclear safety analysis, etc).

The construction set of documents should be prepared at the later stages of the life cycle of a concrete nuclear power plant tailored to the site.

According to the above, the NPPDS should support all the tasks arising during the NPP unit replication and during the usage of its information model across all life cycle stages of an NPP created using VVER TOI™.

The NPP unit information model as the set of knowledge and data about the unit is the main content of the NPPDS in the VVER TOI™ initiative, and is kept in electronic form.

Based on life cycle management principles, the NPP unit information model should be fixed and saved after completing the life cycle stages with the appropriate NPP unit configuration data.

The concept stage is the first stage of NPPDS creation, and the architecture is the core element of the NPPDS concept, where the technical and organizational requirements for the NPPDS are also defined. Once the first stage has been completed, the activities of the second stage can be executed. The core work of the second stage is the integration of existing software platforms and the implementation of new essential platforms. In addition, the NPP unit information model should be created during this stage. At the third stage, the ISO 15926 standard should be used as the foundation for the main software platform interfaces and to store the information model. The software-independent ISO

15926 format will ensure availability of the NPP information model data for the full life cycle period.

3 Our Approach to NPPDS Development

The process approach and the architecture framework idea of ISO/IEC FDIS 42010 standards are at the heart of the development of the NPPDS VVER-TOI™.

The architecture framework idea was first introduced by John Zachman at IBM in 1987 [9]. Since then, more than ten different architecture frameworks have been developed: government frameworks (Federal Enterprise Architecture Framework – FEAF, produced in 1999 by the US Federal CIO Council, Federal Enterprise Architecture [1], Treasury Enterprise Architecture Framework – TEAF [6]); defense industry frameworks (DoDAF, US Department of Defense Architecture Framework [7], MODAF, UK Ministry of Defence Architecture Framework, etc.); open source and consortia-developed frameworks (TOGAF – the Open Group Architecture Framework [4], Generalized Enterprise Reference Architecture and Methodology – GERAM, TRAK, under the control of the TRAK Steering Group, chaired by the UK Department of Transport [5], etc.); firm-developed frameworks (Zachman Framework [9], Integrated Architecture Framework from the Capgemini company [8]); and others. The ISO 42010 standard (the current release was issued in 2007) also provides recommendations for system architecting and architecture framework usage.

The NPPDS architecting process consists of the following stages:

- Determining the stakeholders;
- Defining the stakeholders' concerns;
- Development views and models to cover all concerns;
- Architecting and architecture description development.

Based on our own experience, we can conclude that the list of stakeholders and their concerns do not vary significantly for plants in different industries, and so a standard stakeholder list and set of concerns might be defined. Such standardized lists are fundamental for the plant development system architecture framework – the Capital Project Architecture Framework. Based on an analysis of such typical stakeholders' concerns, we have defined the following four viewpoints as the core element of the CPAF:

- Processes and Functions viewpoint;
- Organizational Structure viewpoint;
- Data viewpoint;
- Information Systems viewpoint.

All this viewpoints have been applied to the NPPDS architecting as the appropriate views, and we describe all these views and the models below.

The general approach to architecture views development used in the NPPDS architecting is described below.

This approach includes analyzing the existing engineering processes ("as is" analysis), developing new ones ("to be" development) and using these "to be" processes as the foundation for architecture development from the perspective of software and data.

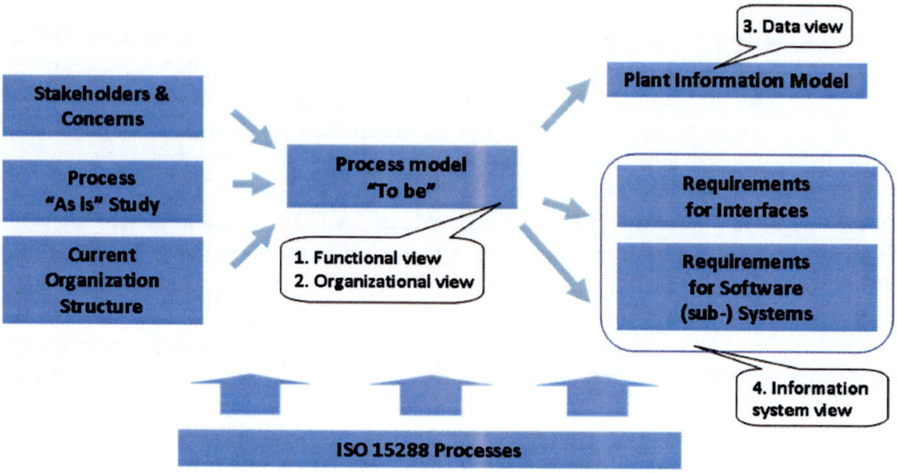

Fig. 2 Our architecting approach

The "as is" analysis of the capital project participants' processes is the first stage of the NPPDS architecting; these participants are also future NPPDS users. Such an analysis is carried out through interviews and studying enterprise quality control documents. The "as is" process model is the result of this step and includes a description of the functions, roles, deliverables and software used for each process studied. All this data is the input for developing the "to be" process model.

The ISO 15288 systems engineering standard process model defines not only the list of reference system life cycle processes, but also their functions, deliverables and interconnections. Thus, this process model is used as the reference model for "to be" processes. The detailed process design is based on the best systems engineering practices, PLM philosophy and systems engineering standards. The results of this stage include process descriptions (functions, roles, software, etc.) and instructions, guidelines and other documents which establish the execution of the engineering process encompassing all plant development participants.

The ARIS Toolset software by IDS Sheer (http://www.softwareag.com/) is used as the core process development tool, providing the process model development as an integrated database consisting of common catalogs of functions, roles, events, organization entities, documents, etc. This feature is very important for process modeling, inasmuch as it allows checking overlap and other process conflicts and generating NPPDS architecture documents as reports from the process model database. For example, model database queries and reports are used to produce the requirements for NPPDS software systems and their interfaces, as well as for the NPPDS data model.

4 NPPDS Stakeholders and Concerns

The list of NPPDS stakeholders in the VVER TOI™ initiative comprises:

- plant owner/operator – ROSENERGOATOM concern;
- plant designer(s) – Atomenergoproekt engineering company;
- nuclear steam supply system designers – OKB Gidropress engineering company;
- scientific R&D entities – Kurchatov scientific center, VNIIAES engineering company;
- construction entity (or entities) – Atomstroyexport and others;
- PDS users – all entities mentioned above;
- PDS integrator/developer – VNIIAES engineering company, IBS system integrator company;
- software vendors – Intergraph, Siemens PLM, Dassaul Systeme, etc.

Stakeholder interviews were conducted during the NPP life cycle study and IT audit, which were completed at the initial stage of the NPPDS creation. The list of stakeholders' concerns compiled as a result includes:

1. Fully automated design in 2D, 3D paradigms;
2. Support for the "building information model" and similar notions;
3. Support for systems engineering and other modern engineering practices, such as parallel engineering and knowledge-based engineering;
4. Support for a data-centric idea and a life cycle management concept, which have become standards in current engineering practice;
5. The idea of "creating information once";
6. The integration of various IT platforms. CAx and PDM systems from different vendors, the requirements management system, the project planning system and other IT platforms are used to design the plant and its subsystems. Widely used engineering approaches require a thorough integration of those IT platforms;
7. Collaboration of geographically scattered groups. Employees from several enterprises participate in the plant development, so the NPPDS should provide for collaboration, not only inside one group in one room, but also among geographically dispersed working groups;
8. Paperless information exchange. Effective work can not be carried out on the basis of paper exchanges; therefore, the NPPDS should support completely paperless technologies;
9. Support for standardized processes (requirements management, configuration management, change management, documents and data management);
10. Design data availability for decades;
11. Simplification of procedures for making changes and making them more cost effective;
12. Industry standards support.

5 NPPDS Architecture Views and CPAF Viewpoints

We describe in this section NPPDS architecture views and models. The views are developed according to CPAF viewpoints, and the models are developed according to CPAF model kinds. Thus, this section describes, in fact, both the CPAF itself and its usage in a concrete example of architecture: the VVER TOI™ NPPDS architecting. As mentioned above, the following views are used for the description of the architecture of the VVER TOI™ NPPDS:

- Processes and Functions view;
- Organizational Structure view;
- Information Systems view;
- Data view.

5.1 Processes and Functions View

The ISO/IEC 15288 standard is the methodical basis for this view development, and the ARIS process modeling software by the IDS Scheer company is used as the development tool. The models of this view define the NPP life cycle stages and the processes that are carried out during these stages.

The following models are included in the view:

- Life Cycle Model (LCM);
- Life Cycle Function Model (LCFM);
- Process Model (PM).

The LCM contains the high-level function description: NPP life cycle stages and end-to-end support processes that are carried out during the whole life cycle. Process definition is executed by gathering and analyzing the deliverables of the life cycle stages (data, documents and products). The Value-Added Chain diagram (of the ARIS tool) is used as a graphical representation for this model.

The LCFM represents a functional description with more detail than the LCM by the enumeration of the basic functions carried out by parties during the NPP life cycle stages. The LCFM is the intermediate functional model between the LCM and the PM, and first and foremost serves as the compact integrated representation of the whole functionality within a one-page diagram. The notation of this model is one or more freestyle diagrams.

The most detailed functional description is represented by the PM, which consists of the processes diagrams. In contrast to the LCFM, the PM describes the life cycle processes in more detail and contains not only the name, but also the characteristics and function attributes (executors, input and output data, software systems, etc.), as well as the interrelations between them. The EPC (Event-driven Process Chain) and EPC-Row diagrams are used in this model.

The LCFM and the PM are developed as an integrated ARIS database, where each document, software system or executor is defined once and is used in several places. Such an approach considerably facilitates the analysis of processes by using sophisticated database reporting options.

The development of the LCFM and the PM is executed iteratively in close co-operation with the NPP development initiative parties.

Engineering processes are the key element of the NPP development system, because it is exactly these processes that determine how each project party operates and that define the system's overall functioning, goals and efficiency. The key issue is that exactly engineering the processes ensures the interoperability of the organization entities, data and software, along with the development and utilization of NPPs and similar capital projects.

We define core engineering development processes which are fundamental to the development of an NPP based on the principles of systems engineering. Due to the presence of a considerable number of different parties on the NPP development team, these processes should be common to all these parties and should incorporate all the parties.

Let us consider the following common or standardized NPP design processes:

- Requirements management (including verification and requirements analysis);
- Configuration management;
- Change management (a sub-process of the configuration management process);
- Documents and data management (information management).

These processes cover the entire NPP design and development stages and are substantially interlinked with each other.

We define those processes that are conventional and mandatory for development and modeling at the NPPDS architecting stage.

All of these processes in the NPP development are defined according to ISO standards (ISO 15288, ISO 10007 and ISO/IEC 29148.2 draft), and thus do not need to be described here in detail. Below, we will give general information about each of the processes and will describe the specific features of the NPPDS system where applicable.

Requirements management envisages the creation and usage of the requirements breakdown structure in the requirements management software system, which consists of internal and external requirements (for VVER-TOI™), the requirements are transferred to appropriate PDM systems for tracing and further verification and validation. In addition, corresponding documents are issued from the requirements management system. The key issue for VVER-TOI™ is support for NPP design replication. To implement this, the requirements structure is divided into fixed and changeable parts corresponding to respective parts of the VVER-TOI™ NPP design. This separation allows analyzing the effect of specific requirements of a concrete NPP on the pre-engineered VVER-TOI™ design.

The configuration management envisages the creation and usage of the baseline configuration of the NPP unit and baseline configuration of the Nuclear Steam Supply System in the PDM systems. The baseline configuration is defined based on the hierarchical structure of the NPP unit and includes all data and documents approved. The hierarchical breakdown structure consists of a permanent part – several upper levels, which are common to all VVER-TOI™ parties. Hierarchical structure facilities and installations (like nuclear reactors or turbines) are assigned to each development team for engineering. Several NPP configurations should be fixed after some NPP life cycle stages. The main configurations are: "as required/ordered", "as designed", "as built" and "as modified". Baseline configurations and appropriate data and documents should be saved electronically for use in operation and maintenance.

The main objective of change management in VVER-TOI™ is the formalization of the development procedures of the NPP baseline configuration, which includes only the approved documents and data.

The change management process is based on the following principles:

- Timely checking and defining of possible consequences due to changes to the NPP configuration, data and documents throughout the life cycle;
- Conducting necessary changes to the NPP configuration, data and documents according to valid standard documents;
- Identifying and verifying the traceability of all configuration changes, with associated data and documentation throughout the life cycle.

The main goal of the documents and data management is to provide all project parties with the required current, complete and authentic documents and data throughout the whole life cycle – from data or document creation to destruction.

To achieve this objective, the following tasks must be executed:

- Definition of the documents and data that should be managed;
- Classification of the documents and data;
- Document life cycle management;
- Management of documents and saving of data;
- Requirements definition of the formats of the documents and saving of data;
- Access management;
- Registration and tracing of documents and data;
- Documents and data quality assurance;
- Collaboration work with documents and data (including parallel engineering).

5.2 Organizational Structure View

The presence of many organizations as members of the VVER-TOI™ team demands a special description of the organizational structure, and the Organizational Structure view (which consists of a similar model) is used for this purpose. It represents the administrative and role structure of NPP development team entities. It is important to note that this model is closely linked with the Functions and Processes view, and this link reflects the assignment of processes and processes activities to executing entities. The organizational structure model elements are used for the LCFM and PM development to define the entity responsible for the function execution.

The organization structure diagrams of ARIS should be used for this model representation

5.3 *Information Systems View*

This view describes the NPPDS architecture as the set of software systems and their requirements and integration. The following models are included in the view:

- Software Systems Model (SSM);
- Requirements for the software platforms implementation and requirements for the platforms integration model (RM).

The Process Model is the initial information for SSM development, and contains data on the software systems used for the support and execution of various process functions. Also, a description of the information flow constitutes the initial information for the development of the software application integration concept.

The SSM is drawn up in the form of one or more freestyle diagrams.

The set of software systems used for NPP development is shown in Figure 3.

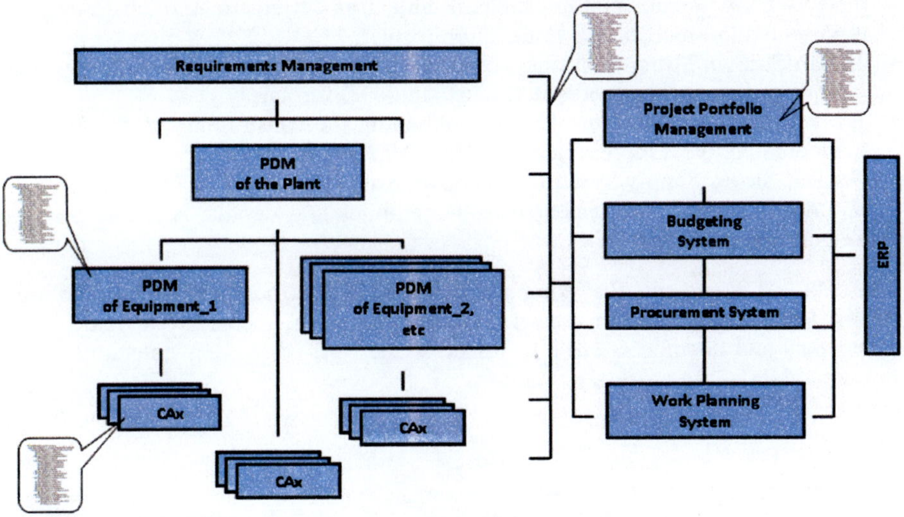

Fig. 3 Software systems used for NPP development

The following software systems are used in the VVER TOI™ NPPDS:

- Requirements Management System – Rational DOORS by IBM;
- Plant Data Management – SmartPlant® Foundation by Intergraph;
- Product Data Management – TeamCenter by Siemens PLM Software;
- Project Portfolio Management – Primavera by Oracle;
- Budgeting System – Atomsmeta, custom developed software;
- Procurement System, Work Planning System, ERP – different modules of SAP Business Suite by SAP AG.

The requirements model contains detailed requirements for the software and the interfaces (the structure and the direction of transferred documents, data, etc.), which might be developed based on reports from the ARIS database (PM).

The requirements model is drawn up as a textual description and is an initial stage in project planning to implement NPPDS software subsystems.

5.4 Data View

Various kinds of data are created and used in NPP development, and practically all of them are represented in several hierarchical structures:

- Requirements Breakdown Structure Model, containing the hierarchy of the requirements for the NPP, from high level general requirements to very detailed ones, is represented in Rational DOORS format;
- Plant Breakdown Structure Model, containing the hierarchy of the NPP unit functional systems and subsystems, is represented in SmartPlant® Foundation format;
- Documents Breakdown Structure Model, containing the structure of the complete set of documentation created and transferred to the customer, is represented in SmartPlant® Foundation format;
- 3D Breakdown Structure Model, containing the hierarchy of the buildings and their elements, is represented in SmartPlant® 3D format;
- Work Breakdown Structure Model, containing the construction work structure to erect the NPP, is represented in PRIMAVERA format;
- Nuclear Steam Supply System Breakdown Structure Model, containing the hierarchy of the main equipment in subassemblies and details, is represented in TeamCenter format.

The data and documents that completely describe the structure elements (requirements, technological systems, etc.) are linked to the elements of each structure. These data and documents are also linked to the PM, where they constitute input or output data of the processes functions.

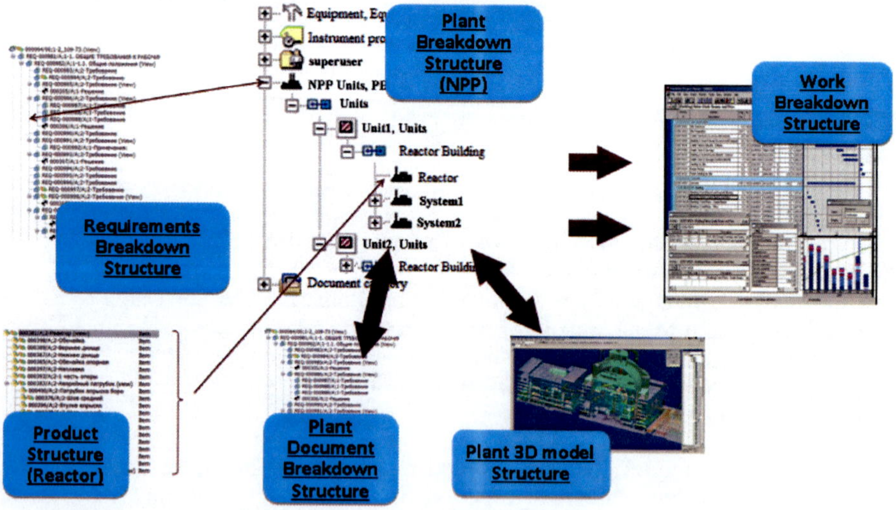

Fig. 4 NPP development data

This set of data models contains the full set of NPP data, and the hierarchical data model is easily interpreted by NPP developers and users. The simple hierarchical diagrams used for all these models are easily understood by designers and other professionals from the NPP development team, and so are quite suitable for communication purposes.

The models of all the views described above are interlinked, and the information contained in each of them is used to develop other ones.

6 Conclusion

In this paper, the approach taken and experiences gained in introducing the CPAF as an architecting tool for the VVER TOI^{TM} NPP development system are presented.

We regard the following as the means to work further on capital project information system development: developing and tailoring the CPAF for the purpose of architecting the plant utilization system – an integrated IT system which supports plant utilization and which usually includes enterprise asset management (EAM) systems, manufacturing execution systems (MES), enterprise resource planning systems (ERP), etc.

CPAF and the approach described in the paper might be successfully used for architecting plant development systems in other industries, like oil & gas or petrochemicals. This prospect is based on the fact that the list of stakeholders and their concerns does not vary significantly for plants in different industries, and so the viewpoints and views are just the same.

In addition, due to those similarities, the views and models developed for the NPPDS might be used as reference models in other architecting projects. Generalized engineering development processes which are common to the development of any plant of any industry, data models and software systems models are common to the development system of any plant of any industry. Reference models and reference architectures are quite widely-used approaches in systems engineering and information system development. The concept of the reference model of Open Distributed Processing, the Core Architecture Data Model of DoDAF [7] and the Federal Enterprise Architecture Framework reference model of the FEA [1] might be mentioned as examples. A reference model or architecture might be used in order to standardize and simplify Plant Development/Plant Utilization System development projects, while its use improves communications among all parties of capital projects, and it might also be used for educational purposes in software projects which accompany the capital project itself.

References

1. FEA PMO, FEA Consolidated Reference Model Document Version 2.3 (October 2007), `http://www.whitehouse.gov/sites/default/files/omb/assets/fea_docs/FEA_CRM_v23_Final_Oct_2007_Revised.pdf` (accessed June 20, 2011)

2. INCOSE, Systems Engineering Handbook v. 3.2 INCOSE (January 2010)
3. Sessions, R.: A Comparison of the Top Four Enterprise-Architecture Methodologies. MSDN Library (May 2007), http://msdn.microsoft.com/en-us/library/bb466232.aspx
4. The Open Group, The Open Group Architecture Framework (TOGAF) - Version 9, Enterprise Edition (2009), http://pubs.opengroup.org/architecture/togaf9-doc/arch/ (accessed June 20, 2011)
5. UK Department of Transport TRAK Steering Group, TRAK Enterprise Architecture Framework (2010), http://trak.sourceforge.net/ (accessed June 20, 2011)
6. US Department of the Treasury Chief Information Officer Council. Treasury Enterprise Architecture Framework. Version 1 (July 2000)
7. US Department of Defense, DoD Architecture Framework Version 2.0 (2009), http://cio-nii.defense.gov/docs/DoDAF%20V2%20-%20Volume%201.pdf (accessed June 20-27, 2011)
8. van't Wout, J., Waage, M., Hartman, H., Stahlecker, M., Hofman : The Integrated Architecture Framework Explained. Springer, Heidelberg (2010)
9. Zachman, J.: A Framework for Information Systems Architecture. IBM Systems Journal 26, 276–292 (1987)

Chapter 19
Systems Engineering in Modern Power Plant Projects: 'Stakeholder Engineer' Roles

Roger Farnham and Erik W. Aslaksen

Abstract. Systems Engineering is a methodology for handling complexity, developed initially to handle the complexity found in the development and operation of large defence and telecommunications systems. This paper addresses a somewhat different domain in which industry is faced with increasing complexity: the domain of large infrastructure projects The paper focuses on the complexity of both the functional and contractual interactions of the numerous stakeholders involved in such projects, and demonstrates that the principles and processes of systems engineering can be also be applied here. The examples that are used are from the power and energy sector. Over the last few years these projects have become more complex from a contractual point of view, with the key parties engaging their own engineering advisors. In this paper, these roles are described as 'Stakeholder Engineer' roles.

1 Introduction

Systems Engineering is a methodology for handling complexity, developed initially to handle the complexity found in the development and operation of large defence and telecommunications systems. This paper discusses a different domain, an industry which is faced with increasing complexity: the domain of large infrastructure projects. There is complexity in both the functional and contractual interactions of the numerous stakeholders involved in such projects, and demonstrate to you that principles and processes of systems engineering can be beneficially applied here also.

Roger Farnham
Sinclair Knight Merz http://www.globalskm.com
Cadell House, 27 Waterloo Street, Glasgow, G2 6BZ, UK
e-mail: RFarnham@globalskm.com

Erik W. Aslaksen
Sinclair Knight Merz
100 Christie Street, PO Box 164, St Leonards, Sydney, NSW 2065, Australia
e-mail: EAslaksen@globalskm.com

This paper uses examples from the power and energy sector, and while many of the conclusions are just as valid for other sectors of the infrastructure industry as well, such as transportation, health care, and education. The authors work for a large international multi-discipline engineering consultancy with major involvement in the power sector and are involved in various engineering support roles in large projects. In [1] they discussed that commercial infrastructure projects did apply Systems Engineering but that the number of project phases and contractors involved in the design process caused disjointedness as each contractor often saw its phase as 'the project'. That paper also explored the whole range of stakeholders which were present in projects. In [2] Aslaksen argues that stakeholder management is a key activity in Systems Engineering, stating that stakeholder requirements should be defined prior to the start of a project, and that there should be a defined process for handling the inevitable changes to the requirements. But a clear and detailed definition of the stakeholder group is less likely to have been agreed and the management of the interfaces as they develop throughout the project is a complex task.

This paper uses power projects as examples to explore how Systems Engineering can come to the support of the disparate group of organizations which make up the stakeholders in a large power project.

2 Independent Power Projects

2.1 Overview

The paper will focus on projects which are generally described as Independent Power Projects (IPPs) and now quite commonplace in the power industry. There are a number of stakeholders are in these projects, and we shall focus on those with interfaces requiring most management, and the key contractual agreements that need to be in place.

A typical IPP will have a consortium building a plant, often to satisfy the need of a country which cannot afford to finance such a project itself. The consortium / developer is rewarded by a long term contract to supply electricity. IPP arrangements are very common in both post-privatisation countries and in those with fast-developing economies.

So what Systems Engineering processes are already applied, and where sometimes these get sidelined because hierarchical contractual relationships come into play. How much Systems Engineering good practice is currently applied to this sort of project? We look at both the general Systems Engineering standards and a recent US Department of Energy standard which is actually very appropriate to this industry. Finally we consider if the processes we apply to recent power plant projects can be applied to new build nuclear projects, which is progressing more quickly in the UK than in other countries. These projects are not actually IPPs, as major extant utilities in the UK are developing the projects, however, the authors consider much of the stakeholder model which has developed for IPP projects appears to be applicable.

2.2 Engineering Roles

There are a number of engineering roles within large projects, the most labour intensive being that of Owners Engineer, the 'OE' acting as the technical customer; however, there are other engineering technical advisor roles. Over the last few years these projects have become more complex from a contractual point of view, and the various key parties in the project engage their own engineering advisors [1]. Likely roles include a Lender's Engineer and an Independent Engineer between the owner and the power taker, the latter being the electricity grid company or, in some cases, another party that has contracted to take the power generated so that they can trade that power. I'm going to use the term off-taker as a generic description of this stakeholder - the party which will dispatch the plant, i.e. define how much power the plant will generate to match its customers' needs. There are other stakeholders which we shall discuss later.

The Architect-Engineer role is becoming more commonplace; for this paper we shall consider that it is a role within the Owner's Engineer role.

2.3 Contract Models

The most popular form of contract is the EPC model, which combines the engineering, procurement and construction (or integration) of the equipment and systems required to build a plant. The specific EPC *engineering* activity relates to detail design, and is often part of the EPC contractor's organisation. A large number of sub-contractors will be involved, and many of the subcontracts will supply systems which are 'packages' with their own control systems. There is a growing trend for modular sub-systems, skid-mounted or containerised, if they can be made small enough. This speeds up the construction process and moves much of the quality control back to the more controlled environment of the factory.

An EPC contract will generally deliver a plant ready to operate, normally a turnkey solution, complete with all the support documentation, the EPC contractor also having provided the training for the owner's operations & maintenance staff. If the developer is a smallish organization then it will appoint an owner's engineer to manage the EPC contractor.

In the past a Build, Own & Operate (BOO) project would normally have be delivered by a large organization such as Siemens, but more recently consortia have been encouraged to deliver such plants. With a single large organization, all of the above developer/owner and EPC roles will be taken up by a single organization, and there are generally no major roles here for independent advisors. However, in the case of a consortium approach, there is both a need to establish the consortium to allow a bid to be made to the off-taker, and then there is a need to tune the working arrangements with the successful consortium to ensure the project proceeds successfully. Both these consortium development stages benefit from technical advisor roles.

An EPCM is a variation on the EPC model. Here a supporting engineering company delivers the complete detail of the design as part of the EPC team, and often also provides the project management. The engineering company remains involved in the contract until commissioning is completed, or at least until one generation unit is fully commissioned.

3 Relationships within IPP Projects

3.1 An Example as Introduction

The figure below maps the typical contractual relationships within a modern independent power project.

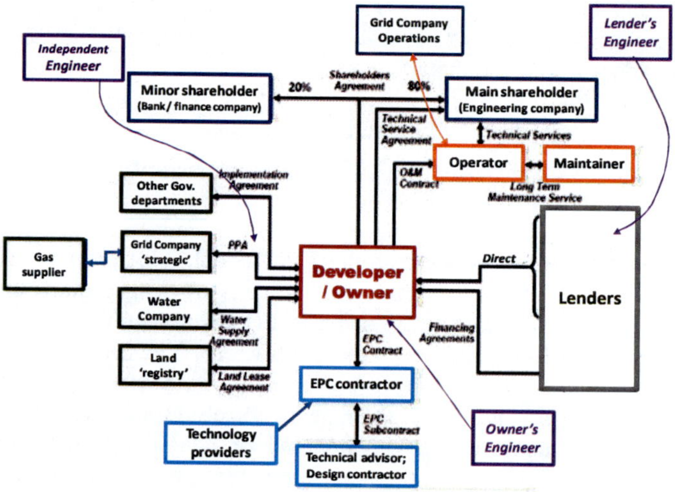

Fig. 1 An example of contractual relationships within a modern IPP

At the centre of the figure we see the developer, who will later be referred to as the owner of the plant. The developer is invariably a consortium, or a joint venture between a large international organisation and a smaller local company which knows the local market and the local regulatory issues and, dare we say it, gets some sympathy from the government of the country which must approve the project. However, invariably the consortium will require expert advice both on the technology to be installed and the environmental issues. A nuclear plant will also introduce a raft of safety issues. GDF Suez and Iberdrola have established a joint venture company to build new nuclear plants internationally. In the UK they have teamed with a British owned UK generator, Scottish & Southern Electricity, to form NuGeneration (NuGen). In other countries they would partner with local companies.

The off-taker here is the National Electric Power Company which operates the grid. Other stakeholders shown in the figure include: the parent companies which have formed the consortium, which we generally refer to as the developer, and eventually as the owner/operator of the plant; the banks who are funding the project; the fuel supplier; the EPC contractor, the the key equipment suppliers.

There are two key technical advisor roles shown here: the developer's Owner's Engineer, and an Independent Engineer (SKM in this case) between the off-taker and the developer, mainly representing the interests of the off-taker, but often providing some support to the developer, in the overall interests of the off-taker.

3.2 Competing Quality Management and Systems Engineering Systems

If all major projects were built by major plant suppliers, acting as the EPC contractors, then everything should fall into place easily; however, as Fig. 1 shows, that is not the real world. There are likely to be separate contractor systems as follows: a large EPC contractor will have developed its own quality and design processes, and at least one major supplier with equally mature processes. And of course, there are the demands of the developer, and the off-taker. All is not lost: some Systems Engineering is applied by the various parties, albeit not always announced in Systems Engineering terminology. The processes executed as the project develops are generally those required by ISO 15288 [3], and the overall development is usually structured according to the Vee-model.

3.3 Systems Engineering for Contractual Interfaces

The process of interface management is just as applicable to contractual interfaces as to technical interfaces, and the principles of system architecting, with its aim of simplifying the interfaces, apply just as well to the architecting of the contracting structure as to the architecting of the physical structure.

In the United States, the Department of Energy (DoE) has recently published a directive [4] addressing large capital projects. Here the DoE represents the position of the off-taker. This initiative has developed an *order*, a 'firm' standard, on programme management for the acquisition of large capital assets such as a power station. More interesting to the Systems Engineer is that there is also an associated guide [5] on the role of Systems Engineering within projects working to this order. Stakeholders are rarely individuals, usually organisations, but some will argue that they can be systems or standards. In general our stakeholders will be organisations.

The table below shows the well established technical processes as defined in ISO/IEC 15288 [3], its associated software engineering guide [7] and the INCOSE Handbook [8].

a)	Stakeholder Requirements Definition	X
b)	Requirements Analysis Process	X
c)	Architectural Design	X
d)	Implementation	x
e)	Integration	x
f)	Verification	X
g)	Transition	x
h)	Validation	X
i)	Operation	
j)	Maintenance	
k)	Disposal	

For IPP projects, the large '**X**'s indicate where there is a major role for the 'Stakeholder Engineer', with clearly defined deliverables; the smaller 'x's a reviewer's role.

It is very important to identify all the stakeholders early in a project, and to identify what technical support is required for those stakeholders. Not only for the work that they have to undertake themselves, but also for activities that need to be undertaken in conjunction with other stakeholders, and which may even sometimes be funded by others.

4 DOE O 413.3-B and DOE G 413.3-1

DOE O 413.3-B [4] and DOE G 413.3-1 [5] are the US Department of Energy's new order, and the associated guide for Program and Project management for the acquisition of capital assets. A key theme in the order is what most of us describe as gated processes, but which the standard describes as *critical decision points*. Critical decision point zero (CD-0) requires a needs statement to be finalized. The guide goes on to list the deliverables required for the CD-1 project definition:

- Conceptual design report;
- Developed acquisition strategy;
- Preliminary project execution plan;
- Project data sheet;
- Preliminary security assessment;
- Preliminary hazard analysis;
- Environmental documents;
- PED (Project Engineering & Design) request.

The standard is a fairly wide ranging commentary on Systems Engineering best practice. Of interest is its approach to managing a complex set of work packages. DOE G 413.3-1 champions the use of the Dependence Structure Matrix (DSM). This is obviously a proven SE tool for analysing interfaces. For many years power sector project managers have been analysing programmes/schedules from a PERT (program evaluation & review technique) perspective to check the logic of project programs. PERT has similarities to DSM, so the approach is not a radically new concept to the non-Systems Engineer. However, DSM is more methodical and easier to check for completeness. The guide provides good advice on topics such as managing dependencies showing a backward flow of information.

Of course, one has to first *define* the interfaces; this is an important Systems Engineering task. Both the work packages and their interfaces are developed in a *top-down fashion*; this is how Systems Engineering handles the complexity, and a key contribution that Systems Engineering methodology can bring to project management.

The guide also provides good checklists relating to actual power project design deliverables, and will provide a good reference for checking completeness of plans. It has also been developed with nuclear projects in mind and lists the various stages in the development the safety case documentation. For non-nuclear projects, some of this can be ignored, however, non-nuclear plants are becoming more and more scrutinised from a *Safety in Design* perspective with such as that promoted in the UK's recent standard on the Optimal Management of Physical Assets [6]; the project manager would be advised to ignore this advice at his peril.

5 Key Relationships in a Recent IPP Project

This figure below provides a different view of the relationship between a developer and the EPC contractor, and the EPC contractor's sub-contractors who will deliver most of the actual equipment and systems.

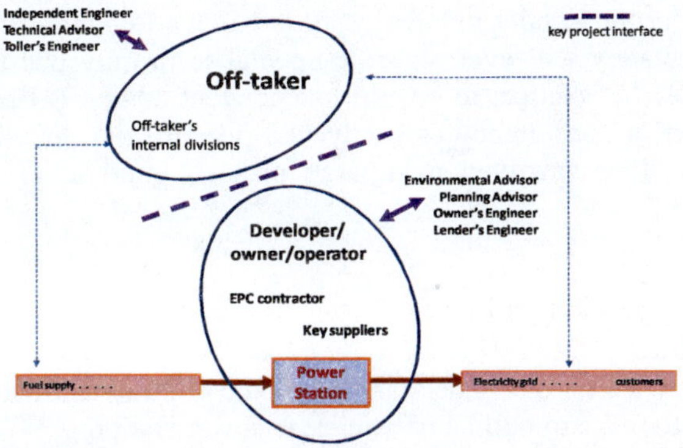

Fig. 2 A key project interface.

In this case, the key interface is considered to be that between the developer and the off-taker. Typically this is defined by a Power Purchase Agreement (PPA), which has been negotiated earlier, as part of the total project bid, and usually on an open tender basis; neither the developer nor the off-taker will have won their absolute preferred contract terms.

Technical advice will be required both by the off-taker, which may be the national grid company or a company which is leasing the plant on long term basis, and by the developer. Often the off-taker accepts the responsibility for procuring the fuel for the plant, but they will want proof that the plant developer has got *all* other necessary agreements in place, such as long term service agreements, and long term funding agreements. There will also be interaction between the various bodies advising the developer and the lenders; these activities are often described as due diligence exercises.

We have already discussed some of the requirements of the developer's camp. Early requirements are the assessment of local planning requirements, the development of an Environmental Impact Assessment, and the development of the conceptual design of the plant.

The off-taker will have produced a high level functional specification (user's needs) as part of its basic requirements. The developer needs to assess how to take this forward into a technical specification to be tendered by an EPC contractor. The developer needs either an in-house engineering team or an Owner's Engineer

(OE) to manage the contract with the EPC contractor. This role is a long term relationship, so much care is taken selecting such a contractor. Normally, the OE will:

- develop the technical specification for issue to the tenderers;
- assess the tenders;
- help negotiate the contract;
- review the design;
- inspect equipment during manufacture;
- act as the developer's representative during factory and site acceptance tests;
- provide a main project management role.

Lenders will be approached for the majority of the funding, and unsurprisingly, there is a need for the Lender to have expert advice: a Lender's Engineer. No engineering consultancy will ever claim to specialise in only one the roles shown here, or in Fig.1. In practice, to ensure independent advice is being delivered to each stakeholder, a consultancy can only win one of the roles. So, a number of consultancies will be involved in a large project; good news for engineering consultancies.

6 New Nuclear Projects

The figure below shows our understanding of the contractual relationships which are being put into place to build a new nuclear power station.

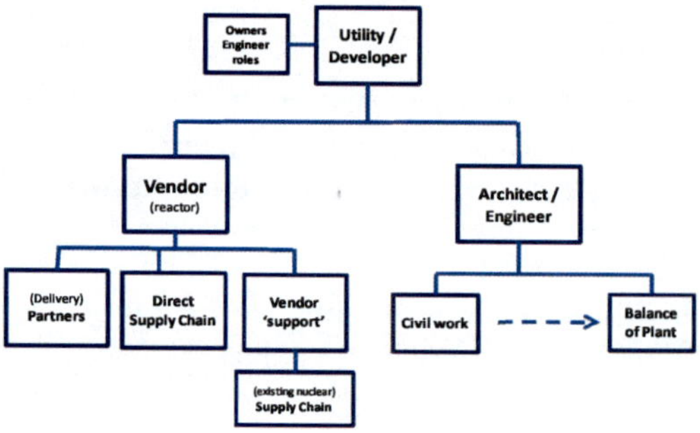

Fig. 3 Typical contractual relationships within a nuclear power station project

From a UK perspective, all the developers to date are utilities which are major players in the UK market place. The major difference between the nuclear sector and the non nuclear sector is that the main equipment supplier, the reactor vendor, needs to have pre-qualified its design with the nuclear regulator. There are a limited number of reactor vendors. With the current plans to build new plant in a number of countries, none of these vendors are interested in being total plant EPC contractors, but wish to concentrate on their extent of supply - the nuclear island.

Consortia are required to deliver specific projects. The other half of plant supply, the non-nuclear island requires a 'separate EPC contractor', and it is the integration of the reactor vendor's role with the non-nuclear other plant which is generally being referred to as the *architect/engineer* role. This covers not just the non-nuclear island, but also the interfaces with the outside world, interfaces that will be key to delivering a successful project. However, there is an added complication in that the total plant design details are required to meet the safety case.

In the UK there has been a safety case pre-qualification process in the Generic Design Assessment (GDA) [9] which has effectively agreed a baseline acceptable reactor designs. However, specific site details will have to be addressed via site-specific safety cases, and these have to be agreed with the safety regulator via the developer, because it is the operator which will hold the site licence to operate the plant.

The figure below shows the developer's side of our fence populated with main players as per the last slide.

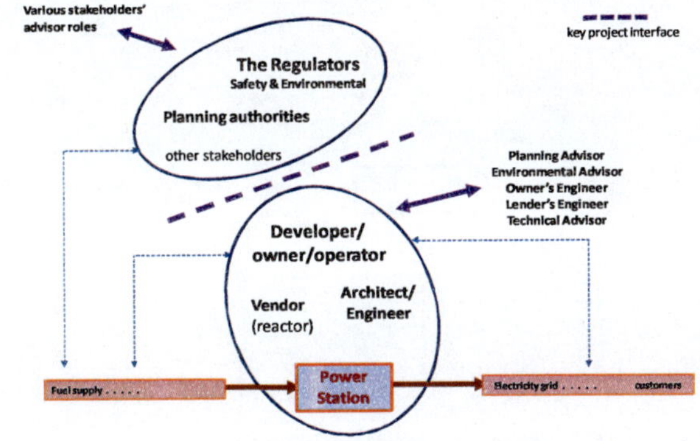

Fig. 4 The key interface between Regulators and the developer

On the other side we see the regulators and the various planning authorities which require information to allow the project to proceed.

In IPP projects the off-taker will invite a number of consortia to bid; the off-taker may have already identified a preferred supplier for the main plant. This has a number of analogies to what is happening in nuclear in the UK. So, as discussed earlier, the consortium is a three way arrangement rather than the two parties in the non-nuclear project. In the IPP consortia projects we described a Technical Advisor role, which was something of a marriage guidance role between the various members of the consortium. In the nuclear project, all the consortium members will be well aware of the need to satisfy the regulator's requirements, however, each party will be a large organisation with its own systems. Each of the parent companies will require to be satisfied that it has sufficient information of its joint venture activities.

On the regulator's side of fence, there is generally only a need for specialist contract support in either safety case or environmental. The bodies which do require more traditional stakeholder support are the local planning authorities and some local development agencies who require support to advise their own stakeholders.

7 Managing the Stakeholder Interfaces

We now look at how we might manage these various interfaces by using a basic fuel in, electricity out model, for a recent European IPP power station project; the facility had an international developer / now owner. The figure below shows the key stakeholders.

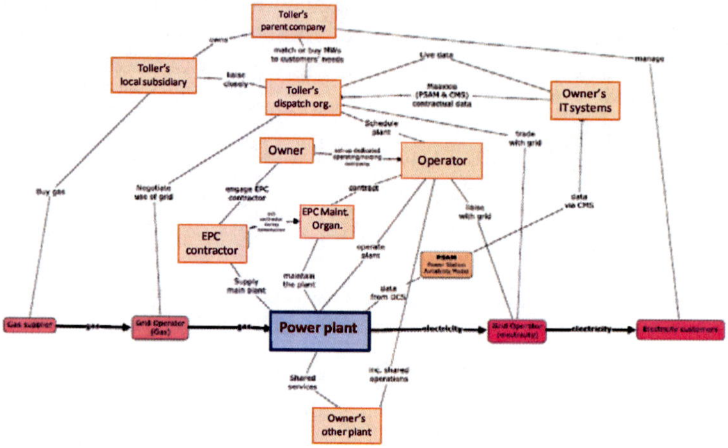

Fig. 5 Key stakeholders in a recent IPP

The plant is dispatched from an off-taker, or Toller, in another country; the plant is leased on a long term (15 years) basis. The owner, via an operating subsidiary company, operates the plant, but does not schedule the plant's output, nor purchases the fuel to generate that electricity. The Toller instructs the Operator (dispatches) what load to generate, and purchases the fuel in a combination of long term and spot market purchases. The challenges here were to ensure that the off-taker (our client):

- was assured that the plant would be available for the duration of the lease period;
- received information (to the Toller's dispatch team) that would allow it to generate instructions to the plant to meet the demands of its customers and the requirements of the electricity grid, and ensure that fuel is purchased to match electricity generation required;
- received adequate performance information, both planned, historical, and real time, to allow the Toller to manage the generation capacity, whether due to plant problems arising, or due to ambient weather conditions, or the need to clean compressors, etc.

This figure also helps illustrate that, overlaid on the whole technical/contractual structure, there is an *economic* model with interfaces in the form of material and information flows; we should not forget that the purpose of these projects is to make a profit, and that the various Stakeholder Engineers must satisfy that need of the developer/owner and other stakeholders.

All of these models are built in a top-down, step-wise fashion as the development progresses through the life cycle stages, starting with very rough, high-level models at the pre-feasibility stage. This top-down approach, with backward traceability at each step, is a core feature of Systems Engineering. This "not losing sight of the main purpose" is what is needed to keep these large, long duration projects on track. Traditional Systems Engineering tools can be used: the interfaces can be managed via various schedules and risk register techniques, with more sophisticated approaches employing relational databases. For example, the interfaces are defined in terms of deliverables generated by the work packages or organisational entities. The schedules or databases will help manage the interfaces by documenting their scope, completeness and issues which need to be resolved.

8 Conclusions

We've looked at some typical recent projects to explore if we can create a generic Systems Engineering model. We've identified a number of technical advisor roles or 'Stakeholder Engineer' roles.

Large power plant projects are well managed, with various aspects of Systems Engineering already in place, however, there can be clashes between the systems of the various stakeholders delivering a project. Often a stakeholder will require its technical advisor to review the suitability of the project-specific systems being delivered, particularly their compatibility with its own systems.

The US Department of Energy standard may have been developed specifically for US public contracts, however, it does provide a useful generic benchmark and should provide a useful contractual and discussion baseline. It has also been written with nuclear in mind, and integrates safety case deliverables with other deliverables. And of course, it champions Systems Engineering.

Of course there's an opportunity for a high level 'guiding principles' monitoring of our *system of systems*, however, the needs of the various stakeholders are always going to be different, so the advisors are still likely to be a number of parties. If the system of systems approach was is be encouraged, the project initiating stakeholder - generally the off-taker - must identify that requirement, and if it is to be delivered in practice, it will require to be a contractual deliverable. The US DOE order and guide is effectively an example of a key stakeholder setting the benchmark.

If the various (Stakeholder Engineer) project reviews are delivered using the same techniques and terminology, then hopefully, we can look forward to improved project management and the ability to address the complex project arrangements that are probably here to stay.

References

1. Farnham, R., Aslaksen, E.W.: Applying Systems Engineering to Infrastructure Projects. In: INCOSE Spring Conference, Nottingham, UK (2009)
2. Aslaksen, E.W.: Designing Complex Systems, Ch. 2. CRC Press (2009) See ch. 2

3. ISO/IEC 15288 Systems Engineering – System life cycle processes. See section 5.5
4. US Department of Energy Order DOE O 413.3-B. The 400 series directives include policies, notices, manuals and guides relating to Work Processes
5. US Department of Energy Guide DOE G 413.3-1. A guide to applying Systems Engineering for use with DOE O 413.3B
6. UK standard PAS 55, Optimal management of physical assets
7. ISO/IEC 19760, Software Engineering Guide to ISO/IEC 15288
8. INCOSE Systems Engineering Handbook (2007)
9. The UK's Generic Design Assessment of reactor designs: `http://www.hse.gov.uk/newreactors/background.htm`

Chapter 20
Self-Organizing Map Based on City-Block Distance for Interval-Valued Data

Chantal Hajjar and Hani Hamdan

Abstract. The Self-Organizing Maps have been widely used as multidimensional unsupervised classifiers. The aim of this paper is to develop a self-organizing map for interval data. Due to the increasing use of such data in Data Mining, many clustering methods for interval data have been proposed this last decade. In this paper, we propose an algorithm to train the self-organizing map for interval data. We use the city-block distance to compare two vectors of intervals. In order to show the usefulness of our approach, we apply the self-organizing map on real interval data issued from meteorological stations in France.

1 Introduction

In real world applications, data may not be formatted as single values, but are represented by lists, intervals, distributions, etc. This type of data is called symbolic data. Interval data are a kind of symbolic data that typically reflect the variability and uncertainty in the observed measurements. Many data analysis tools have been already extended to handle in a natural way interval data: principal component analysis (see for example [1]), factor analysis [2], regression analysis [3], multidimensional scaling [4], multilayer perceptron [5], etc. Within the clustering framework, several authors presented clustering algorithms for interval data. Chavent and Lechevallier

Chantal Hajjar
SUPELEC Systems Sciences (E3S) - Signal Processing and Electronic Systems Department, Plateau du Moulon, 3 rue Joliot Curie, 91192 Gif-sur-Yvette cedex, France
Université Libanaise, Beirut, Lebanon
e-mail: Chantal.Hajjar@supelec.fr

Hani Hamdan
SUPELEC Systems Sciences (E3S) - Signal Processing and Electronic Systems Department, Plateau du Moulon, 3 rue Joliot Curie, 91192 Gif-sur-Yvette cedex, France
e-mail: Hani.Hamdan@supelec.fr

[6] proposed a dynamic clustering algorithm for interval data where the prototypes are elements of the representation space of objects to classify, that is to say vectors whose components are intervals. In this approach, prototypes are defined by the optimization of an adequacy criterion based on Hausdorff distance [7, 8]. Bock [9] constructed a self-organizing map (SOM) based on the vertex-type distance for visualizing interval data. Hamdan and Govaert developed a theory on mixture model-based clustering for interval data. In this context, they proposed two interval data-based maximum likelihood approaches: the mixture approach [10, 11, 12] and the classification approach [13, 14]. In the mixture approach, a partition of interval data can be directly derived from the interval data-based maximum likelihood estimates of the mixture model parameters by assigning each individual (*i.e.* a vector of intervals) to the component which provides the greatest conditional probability that this individual arises from it. In the classification approach, a partition is derived by maximizing the interval data-based likelihood over the mixture model parameters and over the identifying labels for mixture component origin of each vector of intervals. Chavent [15] presented an algorithm similar to that presented in [6] by providing the L_∞ Hausdorff distance between two vectors of intervals. De Souza and De Carvalho [16] proposed two dynamic clustering methods for interval data. The first method uses an extension for interval data of the city-block distance. The second method is adaptive and has been proposed in two variants. In the first variant, the adaptive distance has a single component, whereas it has two components in the second variant. De Souza *et al.* [17] proposed two dynamic clustering methods, based on Mahalanobis distance, for interval data. In both methods, the prototypes are defined by optimizing an adequacy criterion based on an extension for interval data of the Mahalanobis distance. In the first method, the used distance is adaptive and common to all classes, and prototypes are vectors of intervals. In the second method, each class has its own adaptive distance, and the prototype of each class is then composed of an interval vector and an adaptive distance. El Golli *et al.* [18] proposed an adaptation of the self-organizing map to interval-valued dissimilarity data by implementing the SOM algorithm on interval-valued dissimilarity measures rather than on *individuals-variables* interval data.

In this paper, we propose a self-organizing map based on city-block distance for *individuals-variables* interval data. Working with interval variables may prevent the loss of information caused by the utilization of single valued variables. In fact, by using minimal and maximal values to record the temperature, we get a more realistic view on the variation of the weather conditions instead of using average temperature values.

The paper is organized as follows. In Section 2, we give a definition of the self-organizing maps and their training algorithms. In Section 3, we present the SOM algorithm for interval data. In Section 4, we show the results of implementation of our approach on real interval data observed in meteorological French stations. Finally, in Section 5, we give our conclusion.

2 Self-Organizing Maps

The Self-Organizing Maps (SOM) are invented by Kohonen [19, 20]. A SOM is a kind of artificial neural network that uses unsupervised learning to map high dimensional data consisting of n vectors $x_i \in I\!R^p$ $(i = 1, \ldots, n)$, into a low dimensional space, usually of two dimensions, called a map grid. The map consists of K neurons that can be arranged either on a rectangular or on a hexagonal lattice. Figure 1 represents a square map grid of 7 rows and 7 columns. To each neuron k $(k = 1, \ldots, K)$ is associated a prototype vector w_k in the input space $I\!R^p$, and a location r_k in the output space formed by the line number and the column number of the neuron on the map grid. A SOM is often used for data clustering where each neuron constitutes a cluster. By giving the map dimensions (number of neurons) and the data set as input to the network, data vectors will be allocated to the neurons of the map in order to produce organized data clusters as output (see Figure 2). Similar input vectors will be allocated to the same neuron or to a neighbor neuron on the map. In the following, we present two variants of the SOM training algorithm: the incremental training algorithm and the batch training algorithm.

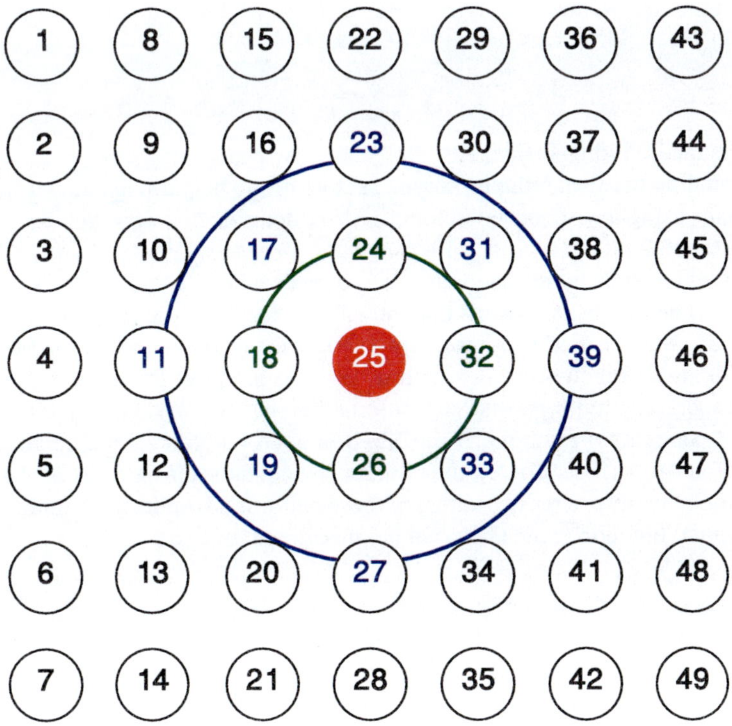

Fig. 1 Square map grid of 49 neurons.

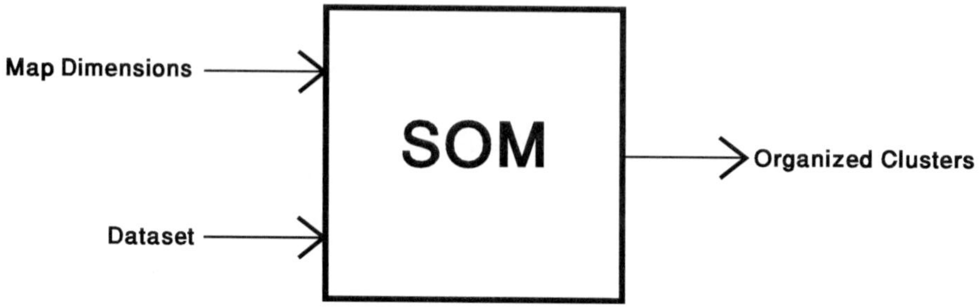

Fig. 2 SOM for clustering.

2.1 *Incremental Training Algorithm*

In the map training, each training step consists of presenting randomly an input vector x, chosen from the data set, to the network and then finding the Best Matching Unit (BMU) of this input vector. The BMU is the neuron whose prototype vector is closest to the input vector in term of Euclidian distance. If we denote c the BMU of input vector x and w_c the prototype vector of this BMU, c is defined as the neuron for which:

$$d(x,w_c) = \underbrace{min}_{k=1,\ldots,K} d(x,w_k) \tag{1}$$

where d is the Euclidian distance.

After finding the BMU, the prototype vectors of the map are updated toward the input vector using a neighborhood function that determines how much the neurons are close to the BMU.

Let $h_{ck}(t)$ be the neighborhood function between neuron c and neuron k at time t, and let $N_c(t)$ be the set of neurons that lie within a certain radius around neuron c at time t. This radius, called neighborhood radius, takes large values at the beginning of the training and then decreases with time. In its simplest form, the value of the neighborhood function $h_{ck}(t)$ is 1 if neuron k belongs to the set $N_c(t)$ and 0 otherwise. For example, in Figure 1, if the BMU is neuron 25 and if the neighborhood radius is equal to 1 (inner circle), then the set N_{25} contains the neurons $25, 18, 24, 26$ and 32, and only the prototype vectors of these neurons are updated. A more flexible neighborhood function is the Gaussian neighborhood function:

$$\begin{aligned} h_{ck}(\sigma(t)) &= \exp\left(-\frac{d^2(r_c,r_k)}{2\sigma^2(t)}\right) \\ &= \exp\left(-\frac{\|r_c - r_k\|^2}{2\sigma^2(t)}\right) \end{aligned} \tag{2}$$

where r_c and r_k are respectively the location of neuron c and neuron k on the grid and $\sigma(t)$ is the neighborhood radius at time t. The width of the gaussian function is defined by $\sigma(t)$.

Equation (3) shows the updating of the prototype vectors w_k $(k = 1,\ldots,K)$ at iteration $(t+1)$.

$$w_k(t+1) = w_k(t) + \alpha(t)h_{ck}(\sigma(t))\,[x(t) - w_k(t)] \qquad (k = 1,\ldots,K) \tag{3}$$

where x is the input vector, h_{ck} is the neighborhood function, c is the BMU of x, σ is the neighborhood radius and α is the learning rate. The learning rate takes large values at the beginning of the training and decreases with time.

2.2 Batch Training Algorithm

In the batch algorithm, the prototype vectors are updated after presenting the whole data set to the map according to Equation (4). Each prototype vector is replaced by a weighted mean over the data vectors. The weights are the neighborhood function values.

$$w_k(t+1) = \frac{\sum_{i=1}^{n} h_{kc(i)}(\sigma(t))\,x_i}{\sum_{i=1}^{n} h_{kc(i)}(\sigma(t))} \qquad (k = 1,\ldots,K) \tag{4}$$

where $h_{kc(i)}(\sigma(t))$ is the neighborhood function between the neuron k and the BMU $c(i)$ of the input vector x_i, and $\sigma(t)$ is the neighborhood radius. All the input vectors associated to the same BMU have the same value of the neighborhood function.

In the last few iterations of the algorithm when the neighborhood radius tends to zero, the neighborhood function $h_{kc(i)}(\sigma(t))$ will be equal to 1 only if $k = c(i)$ (k is the BMU of input vector x_i) and 0 otherwise. The input data set is then clustered into K classes. The center of each class C_k is the neuron k whose prototype vector w_k is a mean of the data vectors belonging to that class. This implies that the updating formula of Equation (4) will minimize, at convergence of the algorithm, the L_2 distance clustering criterion:

$$G = \sum_{k=1}^{K} \sum_{x_i \in C_k} d(x_i, w_k) \tag{5}$$

In addition, using the values of the neighborhood function as weights in the weighted mean defined in Equation (4) will preserve the topology of the map. We notice that this method is similar to the dynamical clustering method (in this case K-means) with the advantage that clusters that are close to each other are mapped to neighboring neurons on the map grid.

3 Self-Organizing Maps for Interval Data

Let $\mathscr{R} = \{\mathscr{R}_1, \dots, \mathscr{R}_n\}$ be a set of n symbolic data objects described by p interval variables. Each object $\mathscr{R}_i$ is represented by a vector of intervals $\mathscr{R}_i = ([a_i^1, b_i^1], \dots, [a_i^p, b_i^p])^T$ where $[a_i^j, b_i^j] \in \mathbb{I} = \{[a,b];\ a \in \mathbb{R},\ b \in \mathbb{R},\ \textit{and}\ a \leqslant b\}$.

3.1 City-Block Distance between Two Vectors of Intervals

The city-block distance between two vectors of intervals $\mathscr{R}_i$ and $\mathscr{R}_{i'}$ is defined by:

$$d(\mathscr{R}_i, \mathscr{R}_{i'}) = \sum_{j=1}^{p} \left[\left| a_i^j - a_{i'}^j \right| + \left| b_i^j - b_{i'}^j \right| \right] \tag{6}$$

3.2 Optimizing the Clustering Criterion

Let P a partition of K clusters, $P = (C_1, \dots, C_K)$. Each cluster C_k is represented by its class prototype w_k:

$$w_k = \left([u_k^1, v_k^1], \dots, [u_k^p, v_k^p]\right)^T$$

The prototypes w_k $(k = 1, \dots, K)$ that minimize the clustering criterion:

$$G = \sum_{k=1}^{K} \sum_{i \in C_k} d(\mathscr{R}_i, w_k) \tag{7}$$

is given by [16]:

- u_k^j $(j = 1, \dots, p)$ is the median of the set $\{a_i^j, i \in C_k\}$.
- v_k^j $(j = 1, \dots, p)$ is the median of the set $\{b_i^j, i \in C_k\}$.

It's an L_1 distance optimization problem.

3.3 The Algorithm

The set $\mathscr{R}$ is used to train a self-organizing rectangular map of K neurons.
Each neuron k of the map has an associated p-dimensional prototype consisting of a vector of intervals $w_k = ([u_k^1, v_k^1], \dots, [u_k^p, v_k^p])^T$. The proposed algorithm is based on the batch training algorithm (Section 2.2). The city-block distance defined in Equation (6) is used to measure the proximity between two vectors of intervals. The gaussian neighborhood function defined in Equation (2) is used to determine the neighborship between neurons. The prototype vector w_k of neuron k is updated as follows:

- $u_k^j(t+1)$ $(j = 1, \dots, p)$ is the weighted median of the set $\{a_i^j,\ (i = 1, \dots, n)\}$. The weights are the values of the neighborhood function $\{h_{kc(i)}(t),\ (i = 1, \dots, n)\}$.

- $v_k^j(t+1)$ $(j=1,\dots,p)$ is the weighted median of the set $\{b_i^j, (i=1,\dots,n)\}$. The weights are the values of the neighborhood function $\{h_{kc(i)}(t), (i=1,\dots,n)\}$.

The weighted median of a sorted vector of weighted elements is the element of this vector for which the cumulative sum of weights of all preceding elements is less than or equal to the half sum of all weights, and the cumulative sum of weights of all following elements is less than or equal to the half sum of all weights.

The algorithm is listed as follows:

1. Initialization: $t=0$
   - Choose the map dimensions $(lines, cols)$ and lattice (rectangular or hexagonal). The number of neurons is $K = lines \cdot cols$
   - Choose the initial value (σ_i) and final value (σ_f) of the neighborhood radius.
   - Choose the total number of iterations $(totalIter)$.
   - Choose the first K input vectors as the initial prototype vectors.
2. Allocation:
   - For $i=1$ to n, compute the Best Matching Unit $c(i)$ of the input vector $\mathscr{R}_i$ $=([a_i^1,b_i^1],\dots,[a_i^p,b_i^p])^T$. $c(i)$ is the neuron whose prototype vector is closest to the data vector $\mathscr{R}_i$ in term of city-block distance:

$$d(\mathscr{R}_i, w_{c(i)}) = \underbrace{min}_{k=1,\dots,K} d(\mathscr{R}_i, w_k) \quad (8)$$

   - For $k=1$ to K, compute the values of the neighborhood function $h_{kc(i)}(\sigma(t))$; $(i=1,\dots,n)$
3. Training: For $k=1$ to K, update the prototype vectors of the map.
4. Increment t and reduce the neighborhood radius $\sigma(t)$ according to Equation (9). Repeat step 2 until t reaches the maximum number of iterations $(totalIter)$.

$$\sigma(t) = \sigma_i + (\frac{t}{totalIter}) \cdot (\sigma_f - \sigma_i) \quad (9)$$

4 Experimental Results

The data set used as experiment concerns the monthly average of daily minimal temperatures and the monthly average of daily maximal temperatures observed in 106 meteorological stations spread all over France. These data are provided by Meteo France.

Table 1 shows an example of the interval data set. The lower bound and the upper bound of each interval are respectively the average minimal and average maximal temperatures recorded by a station over a month for the year 2010.

The data set consists of $n=106$ vectors of intervals, each one of dimension $p=12$. The map is a square grid composed of $K=16$ neurons. The training of the map is performed according to the algorithm described in Section 3.3. The initial

Table 1 Stations minimal and maximal monthly temperatures.

Number	Description	January	February	...	December
1	Abbeville	[-1.5,3]	[0.9,5.9]	...	[-2.4,2.2]
.	.	.	.	.	.
.	.	.	.	.	.
.	.	.	.	.	.
53	Langres	[-3.7,0.5]	[-1,4]	...	[-3.4,1.2]
.	.	.	.	.	.
.	.	.	.	.	.
.	.	.	.	.	.
106	Vichy	[-2.4,3.8]	[-0.2,7.6]	...	[-3.1,5.1]

neighborhood radius is $\sigma_i = 3$ and the final neighborhood radius is $\sigma_f = 0.1$. The total number of iterations is $totalIter = 200$. The initial prototype vectors are equal to the first K input vectors.

4.1 Visualization of the Map and the Data in Two-Dimensional Subspace

In order to be able to visualize the map and the data, we used interval principal component analysis (Method of Centers) [1] to project the prototype vectors and the data vectors on a subspace spanned by two eigenvectors of the data with greatest eigenvalues. Figure 3 shows the results of PCA projection of the data vectors and the prototype vectors connected with their centers. We notice a good degree of map deployment over the data and a good degree of map topology preserving.

4.2 Clustering Results and Interpretation

The trained network leads us to a partition of 16 clusters. Figure 4 shows the repartition of the data vectors over the 16 neurons. Figure 5 represents the map of France containing the 106 stations. All stations of the same cluster are drawn with the same color. We can conclude that the stations located near each other geographically or having approximately the same climate, tend to be assigned to the same neuron or to a neighbor neuron on the SOM grid. Stations of coastal regions are allocated to neurons on the first column of the SOM grid. Stations of colder regions are found as we move up and to the right. The neuron on the fourth row, first column contains stations installed in the warmest regions of France. The neuron on the first row, fourth column contains stations installed in the coldest regions of France. The appendix of this chapter contains the list of the 106 stations.

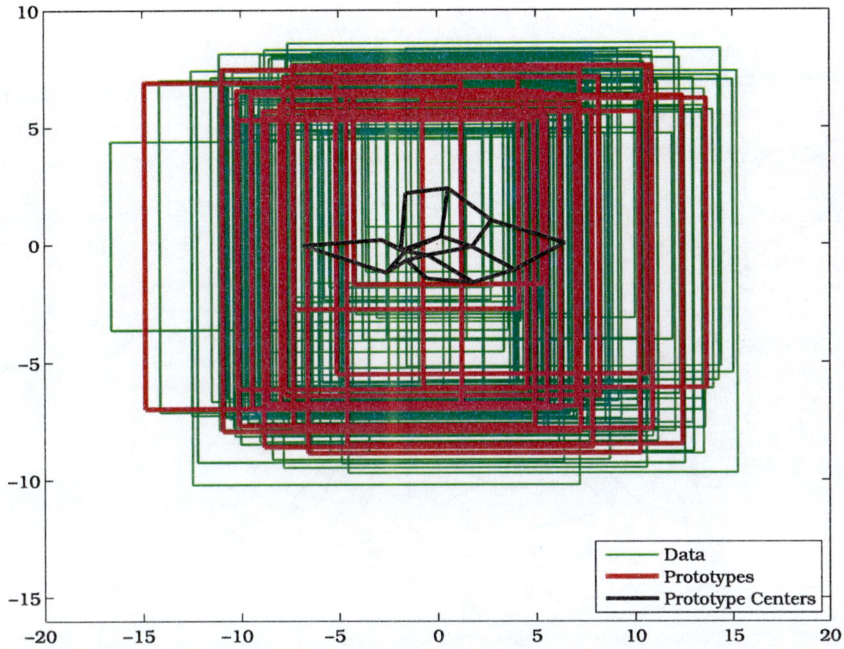

Fig. 3 PCA Projection of the prototype vectors and the data.

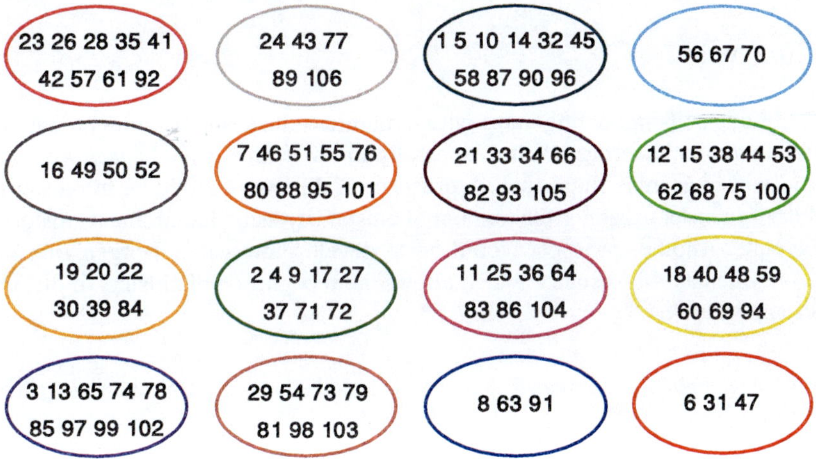

Fig. 4 SOM grid and stations clustering results.

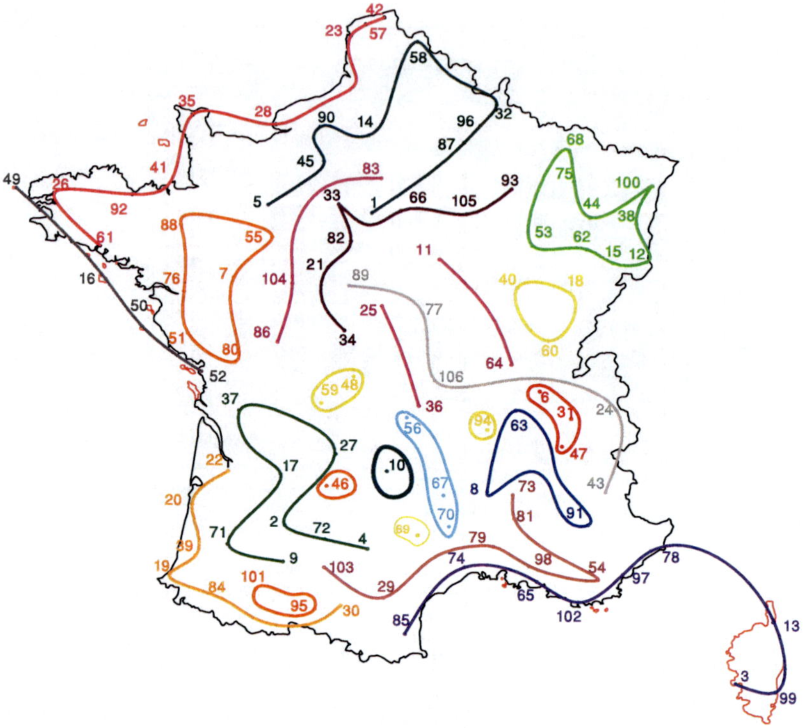

Fig. 5 Clusters distribution on the geographical map of France.

5 Conclusion

By combining self-organizing maps with dynamical clustering for interval data, we can obtain good clustering results, especially in multidimensional cases where the proximity of clusters is important. A prospect of this work might be to cluster the SOM itself in order to reduce the number of clusters by using hierarchical clustering, for example. Another prospect would be to develop interval self-organizing map based on Mahalanobis distance which allows the recognition of clusters of different shapes and sizes.

Acknowledgements. The authors would like to thank the Lebanese University for partially financing this work.

Appendix - List of Stations

Table 2 lists the 106 French meteorological stations.

Table 2 Description of the French meteorological stations.

Number	Description	Number	Description	Number	Description
1	Abbeville	37	Cognac	73	Montélimar
2	Agen	38	Colmar	74	Montpellier
3	Ajaccio	39	Dax	75	Nancy-Essey
4	Albi	40	Dijon	76	Nantes
5	Alenon	41	Dinard	77	Nevers
6	Ambérieu	42	Dunkerque	78	Nice
7	Angers	43	Embrun	79	Nimes-Courbessac
8	Aubenas	44	Epinal	80	Niort
9	Auch	45	Evreux	81	Orange
10	Aurillac	46	Gourdon	82	Orléans-Bricy
11	Auxerre	47	Grenoble	83	Paris-Montsouris
12	Bâle-Mulhouse	48	Guéret	84	Pau
13	Bastia	49	Ile d'Ouessant	85	Perpignan
14	Beauvais	50	Ile d'Yeu	86	Poitiers
15	Belfort	51	La Roche-sur-Yon	87	Reims
16	Belle-île	52	La Rochelle	88	Rennes
17	Bergerac	53	Langres	89	Romorantin
18	Besanon	54	Le Luc	90	Rouen
19	Biarritz	55	Le Mans	91	Saint-Auban
20	Biscarrosse	56	Le Puy	92	Saint-Brieuc
21	Blois	57	Le Touquet	93	Saint-Dizier
22	Bordeaux	58	Lille	94	Saint-Etienne
23	Boulogne-sur-Mer	59	Limoges	95	Saint-Girons
24	Bourg-Saint-Maurice	60	Lons-le-Saunier	96	Saint-Quentin
25	Bourges	61	Lorient	97	Saint-Raphal
26	Brest-Guipavas	62	Luxeuil	98	Salon-de-Provence
27	Brive-la-Gaillarde	63	Lyon-Bron	99	Solenzara
28	Cap-de-la-Hève	64	Mâcon	100	Strasbourg
29	Carcassonne	65	Marignane	101	Tarbes
30	Cazaux	66	Melun	102	Toulon
31	Chambéry	67	Mende	103	Toulouse-Blagnac
32	Charleville-Mézières	68	Metz	104	Tours
33	Chartres	69	Millau	105	Troyes
34	Châteauroux	70	Mont-Aigoual	106	Vichy
35	Cherbourg-Valognes	71	Mont-de-Marsan		
36	Clermont-Ferrand	72	Montauban		

References

1. Cazes, P., Chouakria, A., Diday, E., Schektman, Y.: Revue de Statistique Appliquée XIV(3), 5 (1997)
2. Chouakria, A.: Extension des méthodes d'analyse factorielle à des données de type intervalle. Ph.D. thesis, Université Paris 9 Dauphine (1998)
3. Billard, L., Diday, E.: In: Kiers, H., Groenen, P., Rasson, J.P., Schader, M. (eds.) Data Analysis, Classification, and Related Methods, Proc. IFCS 2000, Namur, Belgium. Springer, Heidelberg (2000)
4. Denœux, T., Masson, M.: Pattern Recognition Letters 21(1), 83 (2000), doi:http://dx.doi.org/10.1016/S0167-8655(99)00135-X
5. Rossi, F., Conan-Guez, B.: Classification, Clustering, and Data Analysis. In: Jajuga, K., Sokolowski, A., Bock, H.H. (eds.) pp. 427–436. Springer, Berlin (2002)
6. Chavent, M., Lechevallier, Y.: Classification, Clustering and Data Analysis. In: Jajuga, K., Sokolowski, A., Bock, H.H. (eds.) pp. 53–60. Springer, Berlin (2002); Also in the Proceedings of IFCS 2002, Poland
7. Barnsley, M.: Fractals Everywhere, 2nd edn. Academic Press (1993)
8. Chavent, M.: Analyse des données symboliques. une méthode divisive de classification. Ph.D. thesis, Université de PARIS-IX Dauphine (1997)
9. Bock, H.H.: Journal of the Japanese Society of Computational Statistics 15(2), 217 (2003)
10. Hamdan, H., Govaert, G.: XXXVème Journées de Statistique, SFdS, Lyon, France, pp. 549–552 (2003)
11. Hamdan, H., Govaert, G.: CIMNA 2003, premier congrès international sur les modélisations numériques appliquées, Beyrouth, Liban, pp. 16–19 (2003)
12. Hamdan, H., Govaert, G.: IEEE International Conference on Fuzzy Systems, Reno, Nevada, USA, pp. 879–884 (2005)
13. Hamdan, H., Govaert, G.: IEEE International Conference on Systems, Man and Cybernetics, The Hague, The Netherlands, pp. 4774–4779 (2004)
14. Hamdan, H., Govaert, G.: IEEE International Conference on Cybernetics and Intelligent Systems, Singapore, pp. 410–415 (2004)
15. Chavent, M.: Classification, Clustering and Data Mining Applications. In: Banks, D., House, L., McMorris, F.R., Arabie, P., Gaul, W. (eds.) pp. 333–340. Springer, Heidelberg (2004)
16. De Souza, R.M.C.R., De Carvalho, F.A.T.: Pattern Recognition Letters 25(3), 353 (2004)
17. De Souza, R.M.C.R., De Carvalho, F.A.T., Tenório, C.P., Lechevallier, Y.: Proceedings of the 9th Conference of the International Federation of Classification Societies, pp. 351–360. Springer, Chicago (2004)
18. El Golli, A., Conan-Guez, B., Rossi, F.: JSDA Electronic Journal of Symbolic Data Analysis 2(1) (2004)
19. Kohonen, T.: Self Organization and Associative Memory, 2nd edn. Springer, Heidelberg (1984)
20. Kohonen, T.: Self-Organizing Maps, 3rd edn. Springer, Heidelberg (2001)

Chapter 21
Negotiation Process from a Systems Perspective

Sara Sadvandi, Hycham Aboutaleb, and Cosmin Dumitrescu

Abstract. Systems engineering covers a wide range of activities, which require know-how transmission between people and disciplines working in different domains. Negotiation is mentioned as one of these activities, in (IS/IEC 15288) and is essential to the overall system engineering process. It can also occur as a result of specific project organization issues, between different departments or organizations, negotiation between stakeholders for contradictory requirements, allocation of internal tasks in large organizations, system design trade-offs etc.. The lack of a thorough description of the negotiation process, as well as its complexity, that results from the wide range of possible scenarios, richness and uncertainty of the contextual information, diverse human behavior, leaves its organization and implementation to the preference of the systems engineer or program manager. Conflict situations, decisions shared between multiple actors, acquisitions in systems engineering, are all resolved through negotiation and the use of laws, rules and process formalization. We believe that a system approach can greatly improve understanding of the potential scenarios and the relation to constraints coming from within and outside the group, to the objectives of each actor, as well as to decisions taken inside the group in the course of the negotiation process. It also facilitates achieving an optimal outcome of the negotiation process in relation to the purpose of each participant. In this paper we analyze the negotiation process from a system's perspective by defining a framework for analysis and optimal resolution of a given negotiation case. We propose a general interpretation that is compatible with formalized processes meant to close deals between different actors.

1 Introduction

The negotiation can be defined as a form of communication where participants make arrangements on each others' requirements, responsibilities and limits to find the

---

Sara Sadvandi
sara.sadvandi@gmail.com

Hycham Aboutaleb
hycham.abou-taleb@polytechnique.edu

Cosmin Dumitrescu
cadumitrescu@gmail.com

best possible agreement that leads to reaching all required objectives. Whether between individuals, corporate or nation states, it is nothing more than a process in which all stakeholders reach an agreement that serves everyone's best interest.

In system engineering standard ISO/IEC 15288, negotiation is presented as an activity of the "Agreement Processes" which include "Acquisition Process". The standard does not enforce any method, but it specifies in the form of a note that the agreement establishes:

- System product and service requirements,
- Development and delivery milestones,
- Verification, validation acceptance conditions,
- Exception handling procedures,
- Change control procedures,
- Payment schedules,
- Rights and restrictions concerning intellectual property (see [1, 2]).

The system engineer or the organization is free to specify the process details. The system integrator often faces complex situations that require negotiation between multiple stakeholders and the engineer is required to do tradeoff analysis for the system design and careful choice of the suppliers in relation to multiple criteria (system maintenance, component performance, price, availability etc.).

The increased complexity of the process is given by several factors like: number of participants (parties), the influence of context elements, reciprocal visibility over the participants' objectives, interests or constraints, sustainable relation between participants, which requires attaining an optimal outcome.

This article proposes formalization for the negotiation process, which analyzes the process with a systemic approach by taking into account the negotiation group or participants in a given context and evaluating different scenarios.

With careful observation and formalization of all aspects of negotiation, we propose a systemic approach that allows handling the complexity of the negotiation in an optimal agreement that follows the negotiation process and that eases reaching a desired outcome [5]. This systemic approach is based on a simplification of negotiation process to be well analyzed, in order to ensure a global and structured understanding of the whole process.

The article explains the subject by emphasizing on the complexity of the process. We begin with the current introduction and continue by describing our approach to understanding the complexity layers of a complex system which applies to negotiation. Then we describe the concepts used in the analysis of the negotiation process and establish the relationship between these different concepts. This is followed by the description of negotiation with a systemic approach, by pointing out that increased complexity is induced by the increased number of possible scenarios, uncertainty and rich context information. We show how the proposed approach helps to understand the process and to describe a set of steps that can steer the negotiation towards an optimal outcome.

2 Characterization of Complex Systems Aspects

As different aspects become too complex for the mind to easily understand or operate with, different approaches are possible in order to better understand a complex system (see for instance [3, 4]). Three concepts have been taken into account in this systemic method: abstraction level, decomposition level, and view [8].

While abstraction level allows the observer to have a holistic view of a system but in respect to different aspects that we shall describe further, the level of decomposition partitions the problem space and allows a localized understanding of the different dimensions of a system. As each person understands a given problem in his/her particular manner, it is of common sense that we can analyze a system from different points of view that are perfectly coherent with each other see [13, 14].

- *Abstraction:* holistic view of the system that is relative to both the level of detail through decomposition and the type of information captured., but one does not need to consider these layers to understand a general phenomenon or one that is possible only in certain conditions.
- *Decomposition:* isolate system components for a detailed analysis, given that all information of the context of the analyzed element is regarded.
- *Perception:* the point of view of each actor that limits or filters the available information, it allows building different models or representations of the problem. In the case of negotiation each actor, or participant, has a different perspective of the situation and visibility on the interests of others.

Layers of abstraction:

- *Structural layer:* characterize the form of the physical elements of the system, including actors. Actors, in turn, may be represented by persons, organizations, nation states etc. The same applies for identifying the structures that are in interaction with the system.
- *Dynamical layer:* characterize changes over time, as well as time-based properties such as milestones in project or the timescale in which the system is operating.
- *Behavioral layer:* relate to the emergent behaviors of actors resulting from the evolution and dynamism of the process. Behavior is influenced, led by, steered by the elements in the decisional layer, which may be seen as "forks" in the path of the existing scenario.
- *Decisional layer:* any decision that has an impact on the overall system, and in consequence future evolution of the process or the following scenario. In technical systems where control physical process is required, decisions may come in the form of a control system, but in this case decisions are taken by the actors involved in the process of system. This renders the situation more complex since none has global knowledge of the other's intentions, so the following steps may be characterized by a certain level of uncertainty.
- *Conceptual layer:* represents the major lines of the process. It reduces the space of the possible scenarios or actions by defining basic rules or constraints. The concept refers to the core elements supporting the actions within the system. For example social systems require communication, but some means of communication can require a certain level of formalism, and others can allow freedom of expression and make the communication skills of the participants a real asset.

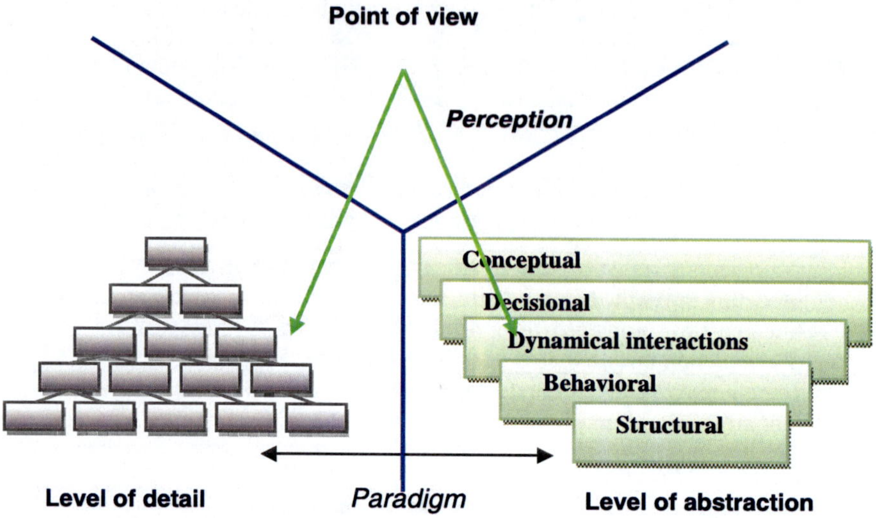

Fig. 1 Characterization of complex systems aspects

The figure 1 presents the three different concepts that we take into account in a system's analysis. Global decisions can steer the systems interactions and induce a different behavior. At the same time the system is designed in respect to certain concepts. For example negotiation is viewed as a communication process where constraints and objectives are adjusted in order to reach overall satisfaction; in consequence the "communication" is one of the core concepts [8].

3 Formalization of Negotiation

There are multiple sources of complexity in the negotiation process, induced by the human factor, or by the need to evaluate different outcomes. Thus, a systemic approach must help in several areas, such as: identifying context factors, tailoring the process model, contributing to obtaining the best outcome and making implicit assumptions explicit. Thus, it is needed to clearly define elements, phases and steps of a negotiation (see [7, 9, 10, and 11] for an exhaustive survey on the subject).

Elements:

Context: circumstances under which the negotiation takes place. A context is unique for each negotiation and consists of:

- *Stake holder*: a party with a right to attend an interest in the negotiation.
- *Actor:* is a stakeholder that is directly involved in the negotiation on his behalf or on the behalf of other stakeholders. Actors interact with each other according to their interests and their stakes.
    - *Point of view:* what the actor knows about the other agents (which of the above elements are known to him)

- *Constraint*: is defined as an element or a factor that restricts an actor in the negotiation process from achieving his potential with reference to his goal. It can restrict any aspect in the problem spaces (abstraction levels) presented before: restriction of decisional authority, restriction on exercised behavior or "structural" constraints in the sense of number of participants at the negotiation representing a single entity, organization etc.
- *Objective*: an element, factor, benefit that represents the final purpose of an actor in respect to the negotiation. The objective can also be seen as a compulsory, necessary interest, with the highest priority among other interests belonging to the same actor, or as shown later on the highest priority stake.
- *Interest*: in the context of each negotiation the actors define their own goals and desired outcomes. These goals and desired outcomes are equivalent to interests of lower priority, or their achievement is not compulsory.. Interests might represent: self-interest (desired by only one actor) or shared-interest (desired by all the actors), visible-interest (visible) or hidden-interest.
  - *Interest boundaries:* each agent has a certain space of decision, and a margin in respect to his interest and objective.
  - *Priority:* the degree to which an interest is desired as an outcome.
- *Stake*: it is the interest with the highest priority. A negotiation cannot be successful unless the stake of each actor is achieved.

As mentioned before, all these elements can be public or private, not visible for the other participants. Meanwhile, each actor has a different perception about the other participants and on the general set of possible scenarios.

- *Private/Shared information:* private information is not known to all participants, but may be perceived by or shared with some of the participants. An actor may have a special position towards a given party defined in the case of a partnership, however a second actor that participates to the negotiation may not benefit from the same advantages.
- *Public information*: the public information exposed by each actor represents data (constraints and interests) that is known to all participants and there is no difference on the perceived information in respect to this data. One common practice for obtaining the best offer is to send to actors, on an anonymous basis, in order to obtain the "Best and Final Offer" (BFO). This renders the other participants' offers visible in the course of the negotiation.
- *Option (for cooperation):* during negotiation the interest might not be achievable. Hence an option consists in combination of achievable interests. This combination can vary due to the dynamic aspect of negotiation.
- *Criteria of legitimacy:* the use of objective, fair standards or criteria is an effective method of dealing with claims or proposals based upon emotions or desires during negotiation.

Phases:

- *Preparing:* to participate successfully in negotiation the actors should have a clear idea about their own stakes, interests, priorities and those of other actors.
- *Sharing:* to have a better understanding of the context the actors reveal their interests and objectives to each other.
- *Trading:* to reach a compromise, an actor gives up a low priority interest in order to get a high priority interest.
- *Implementing:* to enforce what actors have agreed on once the deal is closed.

4 Systemic Approach and Negotiation Complexity

4.1 Scenario Space Complexity Analysis

A systemic approach is needed to simplify and layer the negotiation process to be well analyzed, in order to ensure a global and structured understanding of the whole process, including stakeholders, constraints, interests and associated priorities. These factors and their dynamic characteristics induce a huge number of possible scenarios and associated states [5] [13].

To analyze and model a cooperative negotiation system, a hierarchical method is proposed to control and handle the space of scenarios in a dynamic environment. The key step in this methodology is the identification of system dimensions as a way to organize and control the variety of the existing scenarios. Such a systemic approach needs an analysis phase during which the system is decomposed and reviewed. This approach simplifies the understanding of the dynamic system with a large range of scenarios by applying transformations that handle its complexity.

4.1.1 Identification of Scenarios and Induced Complexity

The negotiation evolution is impacted according to each actor perception of other actors' constraints and interests. However, constraints and interests being identified beforehand, all possible states and scenarios can be identified in relation to these identified elements. Visibility on these elements is different for each actor, so the scenarios that each could imagine may be different and also change as the negotiation advances and new constraints are added.

A *state* is characterized by the set of interests and their associated status, whether it has been met or not.

A *scenario* is characterized by the states and the decisions that led the current state.

All states and scenarios can be identified and represented as follows:

A transition occurs when a new interest is met.

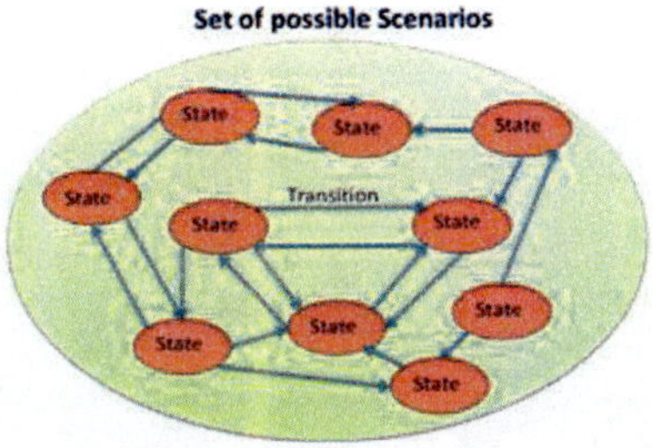

Fig. 2 All possible states and scenarios

4.2 Handling the Negotiation Complexity

We use the following approach to tackle with the complexity, by interpreting the negotiation process through the framework presented in part 2. The three dimensions: abstraction, decomposition, perception, allow understanding of the different aspects of a system.

4.2.1 Negotiation Group Structure and Holistic View

By analyzing the discussed process through the different layers of abstraction, the analysis can be more comprehensible, by having a global view to the whole system [6]. Holistic views leads to visualize, understand and assimilate a complex and critical situation of negotiation by global point of view. The holistic view contain: the actors, their interactions and interfaces, exchanges information and rules and criteria for governing the negotiation.

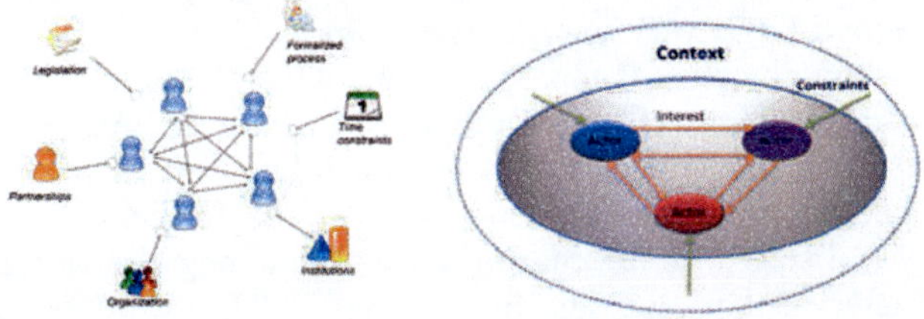

Fig. 3 Holistic view

4.2.2 Negotiation Group Structure and Actor Perception

Group composition shows that there is no overall supervisor. Each actor has a limited view angle: apart from publicly announced interests, constraints and objectives, each person is able to observe and anticipate only a part of the information and behavior of the other participants to the game.

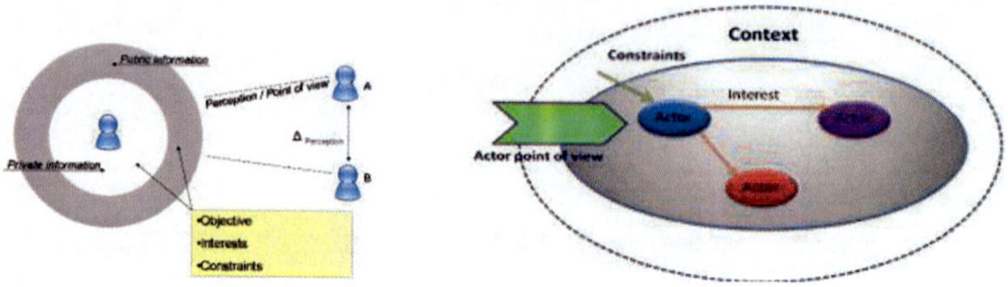

Fig. 4 Actors point of views

4.2.3 Level of Details

4.2.3.1 Structural Layer: (Physical Dimension)

Objective: the main objective is to characterize the form of the physical and social elements of the negotiation process, including stakeholders, actors, infrastructures, organization, etc. By organizing these elements in a well-defined structure, we have clear overviews of the actors and their interface with their environment and with each other are established.

Interface with other layers: The structural layer has an impact on the other layers and is impacted by them:

- ✓ Defining the organizations involved in negotiation, which have their own culture (behavioral layer)
- ✓ The organization might change in time, if the timescale of negotiation is large enough (dynamical layer)
- ✓ The way negotiation is organized impacts how the decisions are made and the final outcome (decisional layer)

Output: Negotiation context structural analysis

4.2.3.2 Dynamical Layer: (Temporal Dimension)

Objective: the main objective is managing time, which leads to schedule the tasks and work to meet deadlines. It characterizes changes over time, as well as time-based properties such as milestones and scheduling. In the case of a formal process like tendering and post tendering negotiation for example, the activities are done in several stages. In the case of a less formal negotiation we can still define temporal intervals between significant changes. Thus we define a time frame in the case of a negotiation process as the interval between two milestones or actions of the participants: modifications of constraints, interest limits, interest priorities or decisions. A decision is the action that takes the process further to the next step or time frame.

Interface with other layers: The dynamical layer has a direct impact on the other layers.

The Temporal layer has impact on:

- ✓ Preparation of planning
- ✓ Periodic inspection of the negotiation
- ✓ Estimation of the percentage of each phase
- ✓ Control of the passage of milestones
- ✓ Adjustment

- ✓ Definition of the necessary steps or activities (structural layer)
- ✓ Definition of delays
- ✓ Strengthening of human resources (structural layer)
- ✓ Behavioral and cultural restrictions (behavioral layer)
- ✓ Future estimation (decisional layer)
- ✓ Definition Milestones, baselines(structure and decisional layer)

Output: negotiation scheduling over time.

4.2.3.3 Behavioral Layer: (Social / Human Dimension)

Objective: The main objective is to identify the emergent behaviors of actors resulting from the evolution and dynamism of the negotiation process. The goal is to take into account all social and human aspects that might impact the course of a project, yielding to misunderstandings, and to integrate all these aspects in the management of the project.

Interface with other layers: The behavioral layer has an impact on the other layers and is impacted by them:

- ✓ Reaction to information changes in time (dynamical layer)
- ✓ Decisions impact and are impacted by behaviors (decisional layer)
- ✓ Organizational and social context impacts the behaviors (structural layer)
- ✓ If timescale is large enough, culture might change over time (dynamical layer)

Output: Set of parameters to take into account, since they impact decisions and negotiation evolution

4.2.3.4 Decisional Layer

This aspect of the systems represents decisions that are taken along, as the scenarios unwrap, it characterizes any decision that impacts the whole evolution of the negotiation process.

Objective: The purpose of the analysis at this level is to document critical points in the negotiation process that can lead to different outcomes and further facilitate the decision making activities.

By decision we refer to a change brought upon the objectives, stakes and constraints, change that is initiated by one of the actors that take part in the negotiation.

Interface with other layers: Each change in the system state, in our case in the actor objectives, stakes, constraints, shall influence further steps in the negotiation and the outcome, but may have a potential influence o external elements as well. As stated before the information captured at this level has an impact on lower abstraction layers, both in internal evolution of the system but also on external elements through the interactions with the environment.

Output: It brings support in understanding and documenting the decisions taken by each actor and collectively visible decisions that can further enrich knowledge for future actions. Each decision also has to be regarded in respect to the individual and global objectives as detailed below.

4.2.3.5 Conceptual Layer

The conceptual layer supports all elements that enable and characterize the normal functioning of the system. The concepts emerge from the context, trends in the environment, existing knowledge and tools and influence the behavior of the system in profound ways. For example the means of communication between actors is to be represented on this level.

Objective: The purpose is on the one hand to reduce the number of scenarios as much as possible in respect to elements that are invariant from the environment and within the system itself, and on the second hand to understand the intimate relationship between these elements and the way the system behaves in order to anticipate potential profound changes. It should take into account both human and technology aspects. For example passing from older means of communication to social networks today, can greatly improve the knowledge we dispose of but also the information we expose, with potential influence in the negotiation process.

Interface with other layers: Since the information captured here enables the functioning of the system, we can also consider it impacts all abstraction layers below, that capture different information about the system.

Output: Information about elements that are invariant on a short time span (usually much longer that the negotiation process itself) and that define how the system will behave and enable its normal functioning.

4.3 Negotiation and Systemic Approach

Each actor that participates to the process is motivated by an objective, which might be known or not to all participants. The objectives can be in depending on the case in opposition, as in an auction, where buyers are competing and offering increasingly higher prices, or complementary objectives as in the relationship between buyer-seller for example.

It is more difficult to achieve a global win-win situation with opposed objectives, since the competition concept itself involves a winner. In this case a balance should be reached between complementary objectives on one side and opposed objectives on the other, by distribution of the gain from the first side to the second.

The *maximal outcome space* represents the set of all possible outcomes do not violate any constraint, *lowest* acceptable limit of interests or objectives. Each actor stake shall at least have been met.

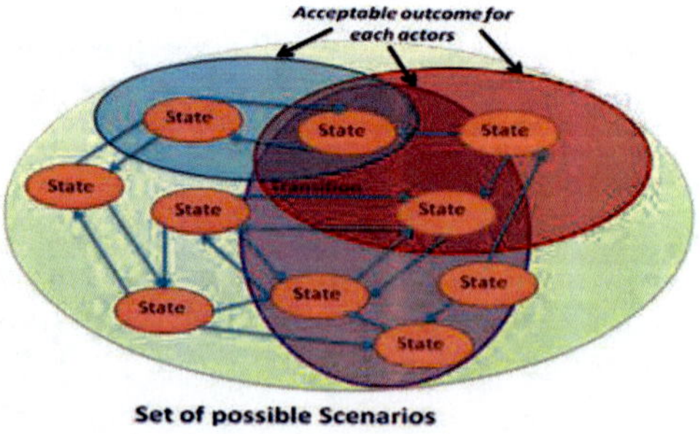

Fig. 5 Acceptable outcome for each actor

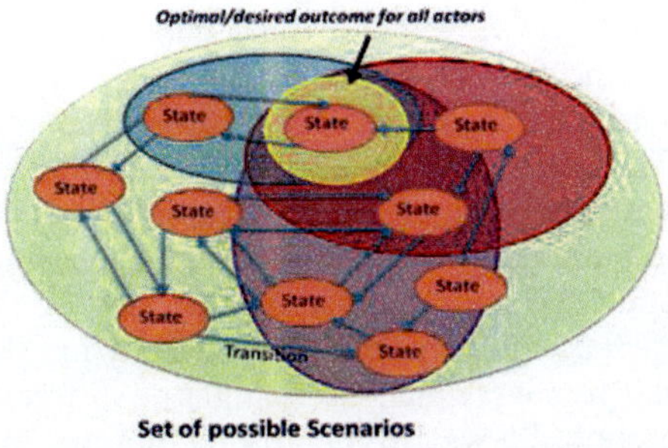

Fig. 6 Desired outcome for the negotiation

The *acceptable outcome space* represents the set of all possible outcomes that is included in the maximal space, but is larger than and includes the desired space. It is a situation that is considered acceptable, but not optimal. Each actor stake and few high priority interests have been met.

The *desired outcome space* represents the set of all possible outcomes that do not violate any constraint, and has *highest* achievable value for interests and objectives. The stake and the greatest number of interests of the actors have been met.

A first purpose is to reach this space, so a minimal level of satisfaction is ensured for all participants. The difficulty resides in the fact that not all information is public, so there is a deviation between what every actor perceives and the global objective data.

5 Advantages of the Proposed Systemic Approach

In this proposed systemic approach with a main goal to organize scenarios space and handle their complexity. Usually, when using traditional methods to analyze negotiation, it is impossible to simplify the understanding of the negotiation context and its dynamic behavior. System approach for negotiation has the following advantages:

- Allow freedom of behavior while being able to understand and master the negotiation scenario. While we propose a general interpretation that is compatible with formalized processes meant to close deals between different actors, the formalization can inhibit intuition, creativity, and innovation. Very formal and constrained procedures can go so far that they inhibit negotiation.
- Understanding the process in general terms and with a holistic view is more important than constraining in order to improve predictability of the process.

6 Conclusion

System complexity is usually due to the recursive intricacy and the interactions between the subsystems. However, human behavior makes a system far more complex and complicated due to its uncertainty. Since negotiation is clearly and mainly based on the human behavior, the outcome is often unpredictable: actors usually don't have the holistic view that enables understanding of the situation and taking into account all the factors and elements that can impact the negotiation process. Moreover, negotiation objective might be irrational or emotional due to human psychology. To handle the complexity of a negotiation process, we proposed a better understanding through a system perspective. We identified the several levels (layers) a system can have, and analyzed according to each dimension, taking into consideration the perception (point of view) of each actor. It is expected that this approach could be used for any system, but the analysis order (the order in which system layers were analyzed) might be different for other systems.

In this paper, we proposed a way to identify the optimal expected outcome to a negotiation. We identified the way to reach the optimal outcome, by analyzing the different scenarios leading to it. The global optimal outcome contributes to a sustainable relationship between actors, the three system dimensions define a way to understand negotiation as a system, while the process formalization defines a way to collaborate for achieving the final win-win situation.

A next step would be to simulate a negotiation using this approach and try to find a generalization or identify properties a social system needs to have to be able to follow this approach. It represents at the same time a first attempt that can be continued through a detailed description of each step, and supported by specific tools that capture information and help decision making at each step of the process.

References

[1] ISO/IEC 15288. Systems and software engineering

[2] Guideline Systems Engineering for Public Works and Water Management, Ministry of Water Management, 2nd edn. The Nederlands (May 2008)

[3] Simon, H.A.: The Architecture of Complexity (1962)

[4] Rittmann, S.: A methodology for modeling usage behavior of multi-functional systems

[5] Bériot, D.: Manager par l'approche systémique. Editions d'Organisation (2006)

[6] Aboutaleb, H., Boutin, S., Monsuez, B.: A hierarchical approach to design a V2V intersection assistance syste. In: CSDM 2010 (2010)

[7] Fisher, R., Ury, W.: Getting to Yes: Negotiating Agreement Without Giving In (1981)

[8] Rhodes, D.H., Ross, A.M.: Five aspects of engineering complexes systems. In: 4th Annual IEEE Systems Conference in San Diego, California, April 5-8 (2010)

[9] Cai, X., McKinney, D.C.: A Multiobjective Analysis model for Negotiations in Regional Water Resources Allocation, AM. In: ASCE 1997 (1997)

[10] Huillon, J.-P.: La logique systémique au service de la négociation (2010)

[11] de Weck, O.L.: Multivariable Isoperformance Methodology for Precision Opto-Mechanical Systems (September 2001)

[12] Mostashari, A.: Stakeholder-Assisted Modeling and Policy Design Process for Engineering Systems (2005)

[13] Krob, D.: Architecture of complex systems: why, what and how? Stanford University (2007)

[14] Krob, D.: Eléments d'architecture des systèmes complexes. CNRS Editions, pp. 179–207 (2009)

Chapter 22
Increasing Product Quality by Implementation of a Complex Automation System for Industrial Processes

Gulnara Abitova and Vladimir Nikulin

Abstract. Modern control methods are suitable for increasing production capabilities of industrial enterprises by employing information technologies, which are often used for modernization of technological processes without the need of acquiring new equipment. Therefore, automation of the industrial processes is one of the most important directions of the technological progress, as it leads to improvement and modernization. A technological process of tellurium production includes a number of stages, which require automatic regulation of many parameters with a high degree of precision. In this work, we analyze and propose a unified concept of application of modern means of automation that can be used to build a three-level hierarchical control system intended for assuring the product quality.

Keywords: automation of industrial processes, information technology, control systems, complex systems, sensors, devices.

1 Introduction

One of the main objectives in developing an industrial enterprise is optimal control and management of the main assets. Modern control methodologies are capable of increasing productivity of industrial cycles by means of automatic control, monitoring, and management systems. As a result, modernization can be achieved

Gulnara Abitova
Eurasian National University, Astana, Kazakhstan
Gulya.abitova@gmail.com

Vladimir Nikulin
Binghamton University, Binghamton, NY, USA
vnikulin@binghamton.edu

without a need for purchasing new and expensive equipment. The components used for such control and monitoring tasks, including different sensors and devices, are usually tightly integrated with automatic systems that control the entire technological process. Hence, this approach facilitates the reduction of expenses used to service the main assets, allows increasing output productivity, reducing downtime, and achieving better management of personnel and the entire enterprise. Therefore, automation of technological processes is one of the most important directions for enhancement, modernization and an intensification of the production facilities.

2 State-of-the-Art in Process Control

2.1 Technological Parameters of Tellurium Production

When designing automated technological cycles for metallurgical processes or for other branches of industry, many parameters may have to be taken into consideration and controlled to achieve optimal operation [1]. At the same time, it is necessary to control every step of the entire production cycle to assure correct flow of every process, to observe its state, and to facilitate monitoring and efficient use of resources.

Currently available physical and chemical processes of industrial tellurium extraction are very well know and commonly used in practice. The production of tellurium from plumbic materials is classified as a very sophisticated physical and chemical process, which requires regulation of technological regimes within certain bounds. The task of control requires collecting considerable amounts of data from a large number of devices, sensors and actuators.

A technologic process of tellurium extraction includes operations that require automatic regulation of the following parameters with high degree of precision: the flow rates of technological liquids, the level of liquids in a reactor, pulp color during sulfation stage, weight of dry ingredients on various technological stages, etc.

2.2 System of Control and Management of the Tellurium Production

In the recent practice, a widely used regulation approach relied on controllable valves and pumps working at full capacity for supplying fluid components at the required rates, dry products were typically measured on scales, and control of the color of pulp was based on visual estimation by the operator involved in the industrial process.

In metallurgical production, the most popular approach to supplying technological fluids is based on application of throttles. The idea of this approach is to adjust the flow based on a signal from a special level sensor (LS), which measures the level of technological fluids in a tank and sends its signal to a level control regulator. This regulator controls the flow rate of the technological fluid according to a pre-defined procedure. This is achieved by sending a signal to an actuation mechanism (AM), which in turn controls a throttle-type device. At the same time, a pump (P) is used to create continuous pressure to push technological fluids through the system, as shown in Fig. 1.

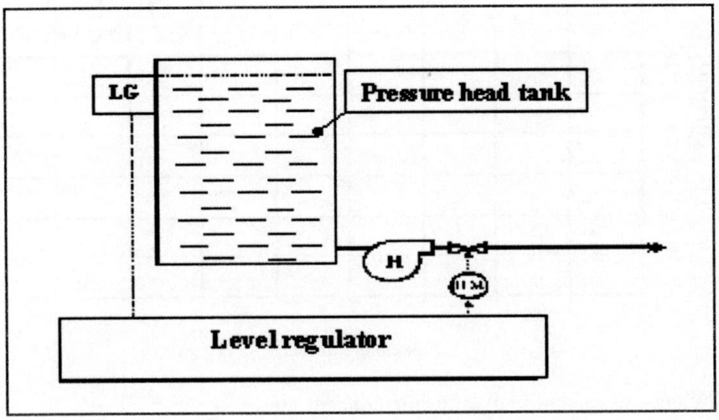

Fig. 1 Throttle control approach.

It should be mentioned that modern technologies are capable of very precise regulation of the main parameters of an industrial process [3]. However, it is common industrial practice to use quantitative approaches to regulation, such as the throttle control system discussed above. A very popular device used in metallurgical processes is a flow rate controller with an adjustable diaphragm. A control method based on this device is very simple, but it leads to the loss of almost half of the power of the pump which is spent for overcoming resistance of the throttle, which is extremely inefficient.

3 Optimization of Production by Complex Automation of Technological Processes

Most of the measurement approaches used to obtain the values of parameters and physical signals are based on different physical phenomena and fundamental laws and implemented using state-of-the-art microelectronics, microprocessors, solid-state semiconductors, electro-chemical elements, and so on.

Technological lines for extracting rare earth metals, e.g. tellurium, are characterized by a large number of transients; therefore, optimal regimes of operation require an accurate system for monitoring and regulation of technological parameters. In practice, tellurium oxidation is monitored by laboratory testing or visually to detect deviation of parameters from the desired values. An example of laboratory results is presented in Fig. 2.

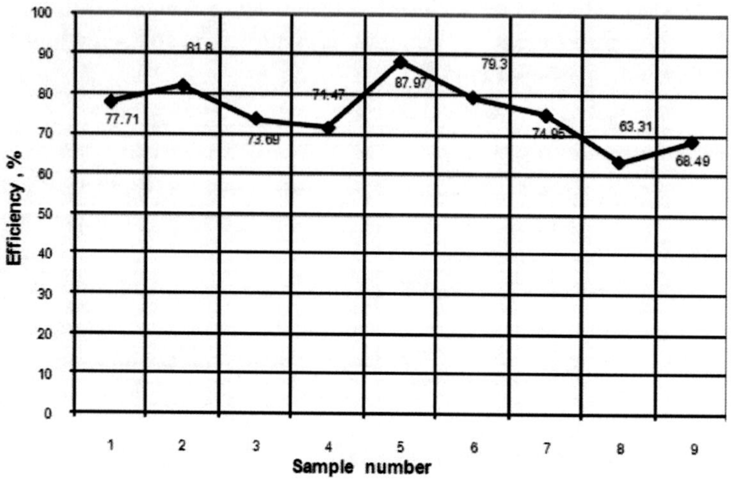

Fig. 2 Efficiency of oxidation of tellurium concentrate

Efficient production can only be guaranteed by automated monitoring and control of technological processes aimed at maintaining all parameters at their optimal levels. Therefore, in this work, we offer a uniform concept of implementation of modern means of automation based on three-level hierarchy for control and monitoring of product quality [2]. By doing so, we create complex systems for automation of technological processes (CSATP) that include a multi-layer structure.

3.1 First Level of CSATP

Sensors and devices that monitor technological parameters and the state of equipment, as well as other mechanisms and actuators, comprise the first, or lowest, level of the proposed CSATP.

On this level, we developed techniques and employed a spectrum of devices for precise management of operation of the pumps used for transporting liquefied products obtained during main and auxiliary stages of the physical and chemical processes used for extracting rare earth metals (for example tellurium). Such techniques and devices include:

- Directly controlled asynchronous electric drive actuators used for pumping technological fluids
- Qualitative control of the pumps used in the extraction stage
- Vector speed control of the electric drives for technological fluid pumps
- Vibrodosimeters of dry substances
- Dosimeters of dry chemical reagents with vibro-drive
- Level, temperature, flow rate, color, and pressure sensors.

3.2 Microprocessor-Based Approach to Controlling Asynchronous Electric Drive

While performing control of any technological cycle, as well as in the process of management or monitoring of the main production tasks, it is often necessary to adjust the operational characteristics of motor-driven pumps depending on the following parameters:

- values of the required and actual consumption of chemical reagents according to a specific technological regime;
- values of the required and actual levels of technological fluids in the reactor while pumping chemical reagents.

While controlling or monitoring technological processes, we need to collect information from sensors installed in different parts of the system, such as pipes, and feed this information to a logic controller to obtain such characteristics as the pressure change across a narrowing tube, temperature and pressure of liquids or gases inside the tube. As a result, we can determine the flow of reagents and the associated energy consumption.

One advantage of this method is its simplicity; however, the systems that are based on this approach also have certain disadvantages. In particular, we can expect reduced efficiency, especially when the flow rates are small. This is explained by the use of throttling mechanism. In this case, angular speed of the motor that runs the pump remains constant, and is essentially at the maximum value. At the same time, most of the power is used to overcome resistance of a partially closed valve (throttle).

It can also be concluded that an alternative apparatus that would implement direct scalar control of the pump, which performs pumping of technological fluids, may have disadvantages of its own. Those include low static accuracy, low dynamic accuracy, and complexity of implementation.

Since one of the typical tasks of complex automation of industrial processes requires precise control of the amount of gases and liquids used in the production, it is necessary to create an apparatus that would regulate flow rates of technological liquids and gases by adjusting the speed of actuating pumps, which are typically driven by asynchronous motors [3].

The proposed approach allows improving economic efficiency of the technologic equipment by using adjustable-speed asynchronous electric drives for the pump. The control is facilitated as shown in Figure 3.

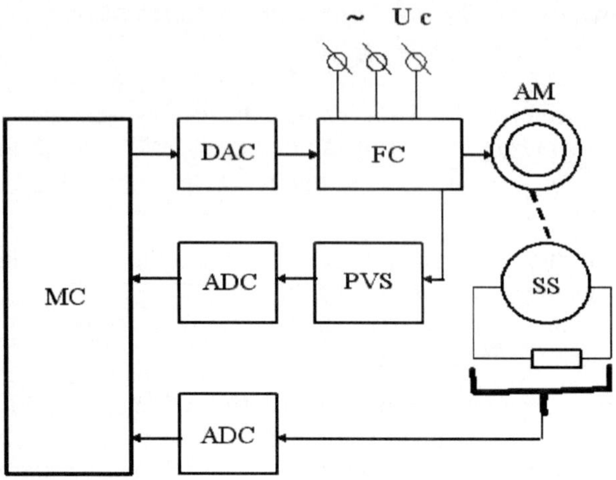

Fig. 3 Control system for the asynchronous electric drive
MC – microcontroller, DAC – digital-to-analog converter, ADC – analog-to-digital converter, FC – frequency converter, PVS – phase voltage sensor, AM – asynchronous motor, SS – speed sensor

This approach is based on microprocessor-aided control of asynchronous motors. Each individual electric actuator has two important interrelated functions: electro-mechanical energy conversion and control of a technological process of a particular piece of equipment.

Hence, in this paper, we present an approach to increasing the efficiency and the quality of control characteristics, including dynamic response and accuracy, of a system that maintains flow rates of technological fluids without a need for complex structures. This is achieved by implementing a device capable of regulating flow rates by adjusting the speed of rotation of an actuating mechanism [5]. The proposed system is based on a vector speed control approach that uses a commutation table to assure proper operation.

A commutation table is a core of this automatic control system. Both static and dynamic characteristics of the proposed electric drive are highly dependent on how accurately the commutation table is compiled. Proper design of the table assures fast response of the electric drive to both a change in the reference input signals and a sudden change in external perturbations. Details of the commutation table are presented in Fig. 4.

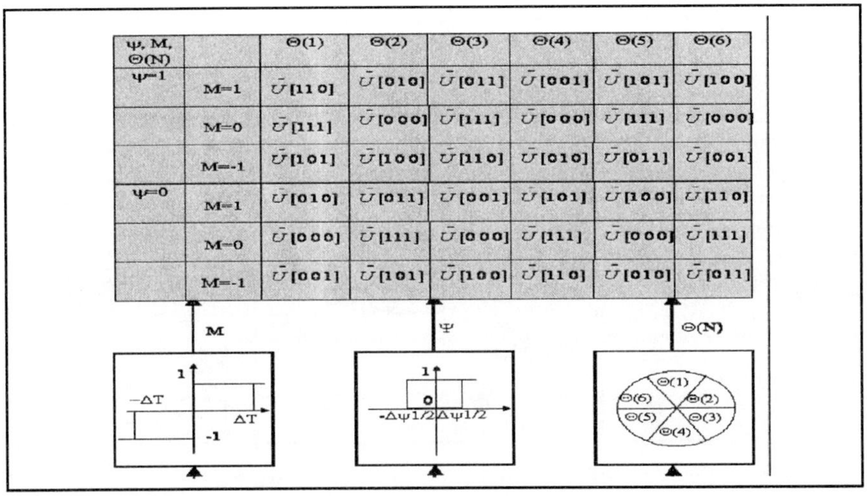

Fig. 4 Commutation table for optimizing voltage space vector

It can be expected that this approach results in more efficient operation when flow rates of industrial fluids are to be adjusted. Indeed, when a throttle is used in a system, reduction of the flow rate results in decreased efficiency. For example if the rate of fluids is changed from 60 to 40 m^3/h, efficiency of the pump can drop significantly. It should be noted that the power of the pump at 60 m^3/h is measured to be 30 kW, and only 25 kW at 40 m^3/h.

However, if the pump productivity is changed by adjusting the speed of rotation of its electric drive, efficiency remains essentially the same. For example, if the flow rate of a technological fluid is 40 m3/h, then power of the pump measured can be as low as 10 kW, which is 2.5 times less than what was required for a system with a throttle, where the remaining power of the motor was used to overcome resistance of a partially closed valve and; therefore, lead to unreasonably high losses of energy.

A control system presented in this paper has a simple architecture, which makes it suitable for compensating electromagnetic inertia of a motor in dynamic regimes when fast speed changes are required. This leads to improved response characteristics and facilitates better economical efficiency of the proposed control method. In addition, parameters of a technological process can be easily monitored.

3.3 Second Level of CSATP

On the second or middle level, we have modern microprocessors that can be used to collect, process, and analyze analog information from sensors and devices of the first level, calculate control efforts and apply those to the actuating mechanisms in accordance with our selected control criteria [6].

The second level CSATP is created using modern control and monitoring devices. An example of a system that includes sensors and actuators is presented in Fig. 5.

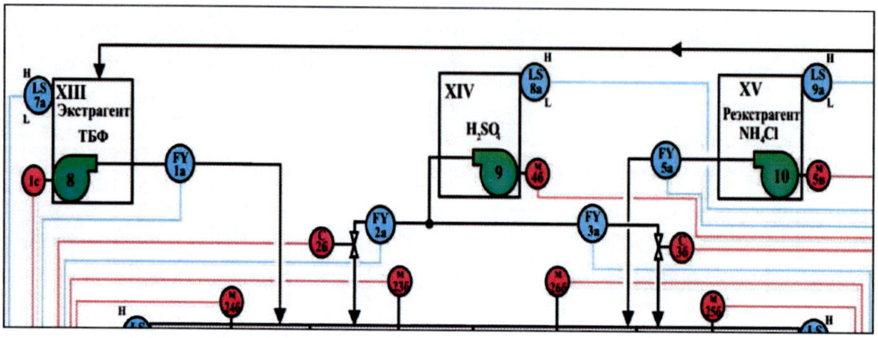

Fig. 5 System of process control and monitoring for tellurium extraction

The following tasks are included in the development of the second level of monitoring and control system:

- conversion of the signals received from sensors into a unified format and transfer of the signals to a control unit
- formulation of control laws and application of computed control efforts to actuating mechanisms
- formulation of algorithms for handling emergency situations, including emergency shut-off of equipment, emergency power-up, and engaging the reserves
- transfer of technological parameters to the third level of control.

As an implementation basis for generating control signals calculated by our algorithm, which uses appropriate mathematical models and takes into account various non-standard (emergency) situations, we have selected SIMATIC S7-300 microprocessor made by Siemens [4].

This family of microcontrollers has a modular structure, can be easily programmed, and operates without forced cooling. Its modular design, a possibility of creating distributed control structures, and user-friendly man-machine interface make this microcontroller suitable for solving a wide range of problems of automatic monitoring and control of technological processes. At the same time, it is also characterized by high flexibility and ease of service.

3.4 Top or the Third Level of CSATP

The top or third level of the automated system of control and management includes computers, monitors, local computational networks and other tools that form automated work stations (AWS) of the technological personnel [5].

An automated work station of a technologist is created in the form of a hybrid control model of a technological process that intelligently combines modern sensors, actuating devices and mechanisms, computational parameters, mathematical models and heuristic knowledge of the technologist to perform selection of optimal operational regimes. Hence, this complex can be called intellectual interface of a technologist.

This interface is equipped with modern means, which makes it suitable for performing continuous control of operation of technological equipment, to track the main dynamic characteristics of the physical and chemical processes, to display the state and operational regimes of equipment in the graphical form, and to extrapolate the course of a technological process and predict possible faults and failure situations [6].

This level of automation is characterized by the following features:

- Pre-launch preparation, power-up and shut-off of technological units and the complexes;
- Monitoring of the current values of technological parameters with indication of the signals that deviate from the acceptable ranges;
- Presenting information on a course of a technological process in the graphical form;
- Change of settings, emergency levels and other characteristics;
- Providing any other information pertaining to a course of a technological process.

3.5 Functioning of the System: Example of Flow Meters

Functioning of the system discussed in the previous sections can be demonstrated using a flow meter as an example. This type of device is widely used in technological automation processes.

The principle of operation of a differential-pressure flow meter is based on the hydrodynamic effect when the rate at which the reagents are used in the process can be related to the pressure difference created in a special narrowing tube. These devices are widely used in industrial applications for measuring single-phase and, within some practical limits, two-phase mixtures. They are known for their versatility and convenience of mass production.

When used in automatic control systems for tellurium extraction, the flow meters with narrowing tubes sense multiple important technological parameters that can be used to determine consumption characteristics of technological liquids or gases used in the process

The basic equations establishing a relationship between reagent consumption and differential pressure in these flow meters can be given in the following form [7]:

$$Q_m = \alpha\varepsilon \; F_0\sqrt{2\rho}(p_1 - p_2) \tag{1}$$

$$Q = \alpha \varepsilon \varepsilon \; F_0 \sqrt{\frac{2}{\rho}} (p_1 - p_2) \tag{2}$$

where:

- Q_m – mass consumption, kg/s;
- Q – volume consumption, m^3/s;
- α – coefficient of consumption of the narrowing device;
- ε – correction coefficient that is less than one;
- ρ – density of the measured substance;
- F_0 – aperture area of the narrowing device;
- p_1 – pressure at the reference point 1;
- p_2 – pressure at the reference point 2.

The above equations are applicable to both compressible and non-compressible reagents. In the latter case $\varepsilon = 1$, and equation (2) becomes a special case:

$$Q = \alpha F_0 \sqrt{2 \rho (p_1 - p_2)} \tag{3}$$

In order to find the actual consumption of reagents we need to know a Reynold's number correction coefficient calculated as the follows:

$$k_{Re} = (C + B(106/Re)0.75)/(C + B) \tag{4}$$

where:

$$C = (0.5959 + 0.0312m^{1.05} - 0.184m^4) / (1\text{-}m^2)^{1/2} \tag{5}$$

$$B = 0.0029\, m^{1.25}/(1\text{-}m^2)^{1/2} \tag{6}$$

Then the actual consumption of reagents can then be found from:

$$Q = Q^* \, k_{Re} \tag{7}$$

At the stage of advanced design we used a mathematical model of the hardware, which allows emulating operation of a realistic technological line on the basis of a logic microcontroller SIMATIC S7-300.

The third-level automation system can be integrated into a centralized system of analysis and monitoring of the state of a large-scale metallurgical process and allows optimum operation of the lead and zinc production of the entire metallurgical complex.

4 Experimental Results

Experimental verification can be obtained from a comparative analysis of the data collected from an extraction line where the proposed approach was implemented. The results are presented in Table 1 where the first number for each element

Table 1. Chemical analysis of products from tellurium extraction process

	Content of Metals (Impurities)				
	Tellurium	*Copper*	*Silver*	*Arsenic*	*Tin*
g/l	50-60 / 60-70	0,001 / 0,0005	0,005 / 0,0025	0,3-0,4 / 0,2-0,3	0,4-0,6 / 0,2-0,3

corresponds to traditional approach, and the second number reflects chemical composition of the end product from a modified production line.

5 Conclusion

In this work, we presented an engineering solution to the problems of first, second, and third levels of industrial automation. It confirms and justifies the necessity to create complex systems of automation with a three-level hierarchical control and monitoring of the technological processes and of the entire production cycle. This task is necessary to assure stabilization of the regime parameters of a technological process, to obtain objective information about the state of the process, and to increase its control efficiency.

The scientific formulation and engineering solution to the existing problems result in improvement and optimization of any technological process in the metallurgical field that leads to increased quality of the end product.

The approach described in this paper allows increasing the content of metals on the output of the process by means of regulation and control of its main parameters and also by injecting optimal amounts of main and auxiliary chemical components. In addition, smaller power consumption will result from these changes.

As a result of this research, we can show analysis and propose a unified concept of introducing modern means of automation that facilitate three-level hierarchy for monitoring and control of the product quality. This approach is based on creating complex automation system with three-level controls.

Ideas of these engineering solutions were tested in chemical and metallurgical branches of the shop of refinement and affinage, and the measurements obtained in the process indicate increased extraction of a valuable component and smaller fraction of impurities in the final product.

References

[1] Darovskih, V.D.: Prospectives of complex automation of technological systems, pp. 121–135. Kirgizstan, Frunze (1989)

[2] Pavlov, A.V., Abitova, G.A., Belgibayev, B.A., Ushakov, N.N.: About perfection of technology of tellurium extraction from alloys of lead manufacture. In: Complex use of mineral raw materials, vol. 6, pp. 36–40. SRZ Gylym, Almaty (2004)

[3] Belgibayev, B.A., Abitova, G.A.: Optimization of tellurium extraction process on the basis of hardware-software by SIEMENS. In: The Bulletin of National Academy of Sciences of Republic Kazakhstan, vol. 2, pp. 93–96. Publishing house "National Academy of Sciences of Republic Kazakhstan", Almaty (2004)

[4] Belgibayev, B.A., Abitova, G.A.: Hybrid model in management of hydrometallurgical processes. In: Proceedings of VKGU Publishing House, Ustkamenogorsk, pp. 3–5 (1997)
[5] Abitova, G.A., Tarasuk., I., Shakhmuhamedov, B.A.: Optimization of flow characteristics in technological pipelines. In: Proceedings of VKGU Publishing House, Ustkamenogorsk, pp. 6–8 (1997)
[6] Belgibayev, B.A., Abitova, G.A.: Calculation of parameters of a differential pressure flowmeter for hydrometallurgical manufacture. In: Proc. International Scientific-Practice Conference, pp. 123–125. Publishing house of KazGASA, Almaty (2003)
[7] Belgibayev, B.A., Shakhmuhamedov, B.A., Abitova, G.A.: The intellectual interface of a digital flowmeter. In: Proceedings of the International Scientific Practice Conference, p. 103. Publishing house of KazGNU of al-Farabi, Almaty (1999)

Chapter 23
Realizing the Benefits of Enterprise Architecture: An Actor-Network Theory Perspective

Anna Sidorova and Leon Kappelman

"A great architect is not made by way of a brain nearly so much as he is made by way of a cultivated, enriched heart." – Frank Lloyd Wright

Abstract. There is growing interest among IT practitioners and academics in Enterprise Architecture (EA) as an effective response to increasingly rapid business, economic, and technological change. EA has been proposed as a path towards better achieving and sustaining stronger business-IT alignment and integration, cost reductions, greater agility, reduced time to market, and other important objectives. Yet there is little theoretical basis to explain how EA work can lead to such achievements; moreover, the creation of a holistic and resilient EA remains an elusive goal for most enterprises. In this paper we use concepts from Actor-Network Theory to highlight some important socio-political and socio-technical aspects of EA work in the context of complex organization situations. Specifically, we focus on such challenges as actor identification in EA negotiations, the importance of soft skills, integration and reconciliation of multiple EA representations, discovering hidden interests and reflecting them in EA representations, dealing with misalignments of interests, as well as creating an environment for continuous EA, and thereby enterprise, improvement.

Keywords: Enterprise Architecture, Actor-Network Theory, Politics, IS Architecture, Technology Architecture, Socio-Technical, Business-IT Alignment, Strategy, Agility, Integration, Complexity, Information Systems, Soft Skills, Systems Analysis, System Design, IS Development, Analysis and Design, Enterprise Architect.

Anna Sidorova · Leon Kappelman
Information Technology and Decision Sciences Department
College of Business, University of North Texas
e-mail: {Anna.Sidorova,Leon.Kappelman}@unt.edu

1 Introduction

"An architect is the drawer of dreams" – Grace McGarvie

The increasing complexity of modern enterprises, as well as the growing heterogeneity of information systems and services used to support business operations, has lead to renewed attention towards Enterprise Architecture (EA) among information system (IS)[1] practitioners and researchers alike (Kappelman, 2010; Ross, et al., 2006; Ross 2003; Venkatesh, et al., 2007). EA has been proposed as a necessary condition for attaining and maintaining business-IS alignment (Sidorova & Kappelman, 2011). In addition, several technological and business trends point to the increasingly important role of the holistic EA approach, including enterprise-wide ERP adoption, cyber security, enterprise application integration, virtualization, data warehousing, business intelligence, service orientation in IS, IS and business process outsourcing including cloud computing, to name but a few. Increasing focus on business agility also makes it increasingly important to have a well-defined, yet flexible enterprise architecture. In spite of the recognized importance of EA work, the creation of a comprehensive and resilient EA remains an elusive goal for most enterprises.

In this paper we examine the process of enterprise architecture development and change through the radically relational lens of the Actor-Network Theory (ANT) (Callon, 1986; Latour, 2005, 1992; Law, 2000). Using ANT concepts, we conceptualize EA and EA processes and activities as flexible and constantly evolving. We further define the role of architectural representation in effectively determining both the present and future architectures of an enterprise, and discuss how such representations are created, used, and modified in the process of IS development and implementation. We then discuss the implications of this conceptualization of EA for EA practice and research.

2 EA Practice, Research, and Theory

"A doctor can bury his mistakes, but an architect can only advise his clients to plant vines." – Frank Lloyd Wright

The importance of Enterprise Architecture and its role in guiding managerial and technological decisions has long been acknowledged by business and IS professionals from industry and governmental institutions. The conceptual foundations of EA evolved from academic and practitioner, public and private, for-profit and not-for-profit, as well as federal, state, and local government efforts. The data modeling techniques and system analysis, design, and development methods developed and promulgated in the 1970s and 1980s by ideas like Ed Yourdon's

---

[1] The terms "information systems" and "information technology" and their respective acronyms (IS and IT) are used interchangeability in this paper when discussing the departments, people, processes, and technologies that process, manage, transmit, and store information for enterprises.

structured analysis and design methods (DeMarco, 1978; Yourdon, 1975), Peter Chen's (1976) entity-relationship diagrams, and Clive Finkelstein's Information Engineering (Finkelstein & Martin, 1981) laid some of the foundations. EA practice can be traced back at least to IBM's Business Systems Planning (BSP) systems development methodology developed in the 1970s. The development of an enterprise ontology by John Zachman was another important milestone in the evolution of EA theory and practice: Zachman's ontology of the enterprise and its architecture, used inside IBM in the early 1980s in conjunction with BSP, was first published externally in 1987 (Zachman, 1987; Zachman & Sowa, 1992) and to some extent continues to influence all EA concepts and practices.

Many other major developments have shaped EA practices. In 1992, the US Defense Department (DoD) initiated its Technical Architecture Framework for Information Management (TAFIM) project and developed the Command, Control, Communications, Computers, Intelligence, Surveillance, and Reconnaissance (C4ISR) Architecture Framework in the mid-1990s to promote interoperability across systems and services. The Open Group Architectural Framework (TOGAF) Version 1 released in 1995 was based on the TAFIM (Hagan, 2004). In 1996, responding to "best-practices" in IS studies conducted by the General Accounting Office (GAO)[2], the US Congress passed the Clinger-Cohen Act, which requires that every federal agency have a Chief Information Officer (CIO) responsible for all IS spending, equipment, and personnel as well as the Information Technology Architecture (ITA) for their agency. Since the ITA of an enterprise is a vital part and a reflection of the larger enterprise of which it is a part, for practical purposes ITA has been operationalized as EA in the US federal government. The DoD also developed the Joint Technical Architecture (JTA) in 1997 to facilitate the flow of information in support of warfare and C4ISR evolved into the DODAF (DoD Architecture Framework).

Responding to the need for guidance as federal agencies began to create their EAs, the CIO Council of the Office of Management and Budget (OMB) sponsored the development of the Federal Enterprise Architecture Framework (FEAF) in 1999 (CIO Council, 1999). OMB and the GAO published *A Practical Guide to the Enterprise Architecture* in 2001 to provide guidance on setting up an EA program and for developing and maintaining an EA (CIO Council, 2001). Over the years, many groups have emerged to offer various kinds and qualities of EA-related trainings and certifications, both Gartner and Forrester have EA research practices, and many vendors offer EA-related conferences, services, and products. A Society for Information Management (SIM) EA Working Group (SIMEAWG) was formed in October 2006.

In spite of the significant interest in EA, practitioners acknowledge that EA work is full of challenges, many of which are socio-political in nature. Moreover, notwithstanding enterprise architecture skills being ranked at the top of the "business domain" skills by CIOs (Collet, 2006), evidence suggests that business managers and even senior IT practitioners, treat EA work as belonging to the technical IS domain (Salmans & Kappelman, 2010). This is perhaps not surprising as many

---

[2] GAO has since changed its name to the General Accountability Office.

practitioners continue to focus on IS architecture, thus undermining the potential of EA to act as a bridge between business and IS.

Throughout the 1990s the term "enterprise architecture" appeared in a number of academic publications; however, such studies either adopted a black-box approach to EA (e.g., El Sawy, et al., 1999), or treated EA as a close synonym to Information Architecture (e.g., Miller, 1997). Academic interest in EA was reinvigorated in the 21st century with EA being proposed as a solution to achieving business-IT alignment and overcoming IT integration challenges. In her 2003 article "Creating a Strategic IT Architecture Competency: Learning in Stages" MIT's Jeanne Ross concluded that "the payback for enterprise IS architecture efforts is strategic alignment between IT and the business" (p. 43). Jerry Luftman's (2003; Luftman & Kempaiah, 2007) assessment of "IT-business strategic alignment maturity" included the degree to which "the enterprise architecture is integrated". Ross, with her MIT colleagues Peter Weill and David Robertson, released the book *Enterprise Architecture as Strategy* in 2006.

Yet in spite of the significant academic and practitioner interest in EA, there appears to be little consensus with regard to conceptualizations of EA. For example, while some treat EA as a description of the status quo, others subscribe to the view of EA as a set of standards and blueprints for the future enterprise and other still include both along with the transition plan between those present and future states. Similarly, some simply equate EA with IS or technology architecture, while others conceptualize EA as enterprise-wide requirements aimed at providing an all-encompassing model or approach for planning and running the business, capturing and providing management with all the knowledge about the enterprise, and serving as a shared "language" to align the ideas of strategy and with the reality of implementation (Kappelman, 2007). Furthermore, the focus among many practitioners and academics is on "doing EA" and so they tend to view EA as a process.

In this paper we adopt the conceptualization of EA as an inscription of aligned interests (Sidorova & Kappelman 2010, 2011), which is based on concepts from Actor-Network Theory. We elaborate on the process of developing EA as a negotiation process among heterogeneous actors both within and often outside the enterprise, and highlight the key challenges of EA development. In the next section we review some concepts from the Actor-Network Theory that are particularly useful for our discussion and elaborate an ANT-based conceptualization of EA.

3 Actor-Network View of Enterprises and Information Systems

"Architecture is politics." – Mitchell Kapor

Actor-Network Theory was originally proposed in the early 1980s to describe the creation of socio-technical networks of aligned interests (Callon & Latour, 1981) and was later extended to focus on the dynamics of relationships among such networks (e.g., Law, 2000). ANT was also recently further formalized and elaborated upon in the book *Reassembling the Social: An Introduction to Actor-Network-Theory* (Latour, 2005). Actor is the central element of the theory, and in its original conceptualization is defined as "any element which bends space around itself, makes other elements dependent upon itself and translates their will into the language of its own" (Callon & Latour, 1981, p. 286). Through such translation of interests the actor seeks to create networks of aligned interests, or actor-networks.

The creation of actor-networks by a focal actor through the process of translation is detailed in the study of scallops and fishermen (Callon, 1986). The translation process is defined from the point of view of a focal actor and its goal is to align the interests of other actors and actor-network with the interests of the focal actor. The translation processes is described as a multi-step process involving *problematisation, interessement,* and *enrollment* stages (Callon, 1986). Once the alignment of interests is achieved, it is often inscribed into technical artifacts (e.g., a computer application) or other elements that are difficult to change, such as legal contracts, or even such "mundane artifacts" as a car seat belt (Latour, 1992). The inscription process may, in turn require recruitment of yet additional actors (such as programmers or lawyers) and consequently may lead to the need to consider their interests.

The term "actor-network" reflects the fact that the resulting actor-networks are often perceived by external observers as individual actors and their coherency (the internal alignment of interests) is taken for granted, a phenomenon referred to as *punctualisation* (Monteiro, 2000). The heterogeneity of the elements of the actor-networks is only observed by the external actors when misalignment of interests occurs within the actor-network. The Actor-Network Theory takes a "radically relational" approach to defining actors, where "entities [...] achieve their significance by being in relation to other entities" (Law, 2000, p. 4). For example, the student registration system can only be defined as such when placed within a larger network of an educational institution. ANT also does not make an *a priori* distinction between human and non-human actors, thereby making it appropriate for examining the role of human entities as well as those that are comprised of social and technical elements (such as information systems or organizations) and purely technical ones (e.g., a server, building, or manufacturing robot).

The flexibility of ANT with regards to the level of analysis and its ability to include both the technical and social dimensions made it attractive for studying problems related to the development and use of information systems (Walsham, 1997). Among the early applications of ANT in IS research, Walsham and Sahay (1999) used ANT concepts for analyzing the case of GIS implementation in India. Recently ANT was used to examine a variety of IS-related phenomena; for example, to examine causes of failure of a large business process change initiative (Sarker, et al., 2006) and to examine issues related to standardization in IS (Hanseth, et al., 2006). ANT was also used for exploring a variety of organizational and business issues (e.g., Newton, 2002). In the next section we apply concepts of ANT to describe EA and its related processes.

4 The Architecture of Enterprises

"Our architecture reflects truly as a mirror." – Louis Sullivan

If the enterprise exists then the architecture of the enterprise exists whether or not it is known or written down. The same can be said of the architecture of a building, airplane, computer chip, and just about any other object. A modern enterprise, as well as the information systems within the enterprise, can be viewed as examples of complex actor networks. The process of enterprise creation and development can be viewed as a series of translations of interests of the various actors comprising the enterprise actor-network (Sidorova & Kappelman, 2011). Enterprises are often established as a result of a translation process between the interests of entrepreneur(s) and investor(s). The development of an enterprise proceeds with the enrollment of new actors, including employees, physical assets, customers, suppliers, production equipment, and information technologies. The enrollment of each of these actors is usually associated with the creation of artifacts in which the interests of the newly created or expanded networks are inscribed. For example, hiring an employee usually involves the creation of a contract and a job description. Such artifacts usually include references to the design of the enterprise, such as the legal and governance structure, the business model which implies the key entities of interest to and the core business processes of the enterprise, as well as references to technology, personnel, and often location requirements. As the enterprise grows, the enterprise actor-network grows to include vendors, customers, suppliers, employees, production technology, information technology, contracts, facilities, annual reports, SEC filings, and so on.

Thus, when viewed through the ANT lens, an enterprise is typically created through an organic process of multiple translations, as opposed to a planned undertaking where an enterprise is a realization of some pre-defined architectural plan. Consequently, the *architecture of an enterprise is not typically defined a priori, but rather emerges through the translation process and reflects the current state of alignment of the interests of various heterogeneous actors representing the enterprise actor-network.* Thus the process of creating and maintaining the enterprise and its architecture can be regarded as a process of managing the various translation processes that involve the enterprise actor-network. If the architecture of the enterprise is written down, then these architectural artifacts become critical to the communication and translation processes within the actor-network; and thus vital to the creation, management, and evolution of the enterprise. In this way, the role of an enterprise architect emerges largely as a strategic management role.

Why then is enterprise architecture usually discussed in the context of IS management, even by IS professionals (Salmans & Kappelman, 2010)? Perhaps it is because the processes of creating and maintaining information systems have long relied upon written architectural artifacts (e.g., Chen, 1976; DeMarco, 1978; Finkelstein & Martin, 1981; Yourdon, 1975; Zachman, 1987). Moreover, ISs are an essential and mission critical subsystem or component of the enterprise, much as the circulatory or nervous subsystems are to the human body, and significant human and financial resources are required to create and sustain those ISs. Thus, IS professionals have in effect played an increasingly important role in creating and maintaining the architecture of the enterprise (whether explicitly memorialized or not as architectural artifacts) because the IS artifacts themselves become the immutable mobiles (Latour, 1992) into which the aligned interests of the enterprise actor-network are inscribed. In fact, those information systems often themselves become actors in the enterprise actor network. Thus, one might conclude, that it is largely historical accident by which the responsibility for EA has "landed on the desk" of the IS department. However, because EA is by definition strategic, it is not likely to stay there (Ross, 2010). In the next section we illustrate how enterprise architecture can be shaped in the processes of IS development and implementation.

5 IS Development and Enterprise Architecture

"We shape our buildings; thereafter they shape us." – Winston Churchill

In order to illustrate the role of enterprise architecture in IS development (ISD), in either a build or buy situation, let us consider a typical procurement process which includes preparing a purchase requisition, preparing a purchase order based on the purchase requisition, sending the purchase order to the vendor, receiving the goods, and receiving and paying the invoice, see Figure 1 (Magal & Word, 2009). The process involves several actors including the buyer (the actor interested in purchasing the goods), the purchasing department, the warehouse, the vendor, the legal department, and the accounting department. While all the actors, perhaps with the exception of the vendor, are a part of the enterprise actor-network, they each have distinct interests. For example, it is in the interest of the buyer to get the goods as soon as possible, and he may have very little concern about the price the enterprise is paying, the vendor selection, or the record keeping associated with the procurement process. On the other hand, the accounting department is primarily concerned with ensuring proper record keeping and disbursement of funds. The interests of the vendor include selling as many goods for the highest possible price and collecting the money as soon as possible. The interests of other actors may include cost minimization, warranties in the purchase contract, and ensuring that enterprise funds are not spent inappropriately.

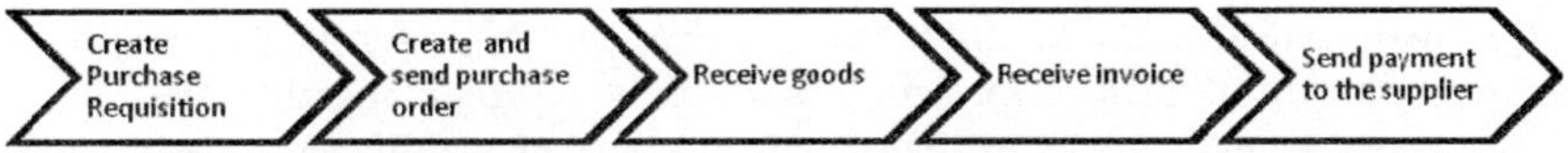

Fig. 1 A simplified view of a typical procurement process (Magal & Word, 2009)

For the enterprise actor-network to function efficiently and effectively, these seemingly contradictory interests need to be aligned, which is done by the key actors agreeing on a standard procurement process. For example, the interests of the buyer and the enterprise are aligned through the process of submitting and approving of the purchase requisition: if the buyer wants to receive his goods, he has to submit a purchase requisition. On the other hand, the enterprise (represented in this case by the purchasing department) has to approve a purchase requisition if it contains legitimate requests. The resulting procurement process becomes a part of the functioning enterprise and its enterprise architecture, regardless of whether it is inscribed into any architectural representations or not. It is however more likely that the process is followed (i.e., the agreement regarding the alignment of interests is enacted, thus making the functioning enterprise more true to its architecture) if it is inscribed in, and thereby memorialized and communicated by, artifacts such as procurement policies, purchase order forms, job descriptions, vendor lists, decision tables, and process maps. In ANT terminology, a procurement actor-network (AN) is created within the enterprise actor-network, which includes human actors and artifacts into which aligned interest are inscribed (see Figure 2). Such artifacts contain information about the enterprise's architecture as it relates to procurement, and thus constitute architectural representations, similar to a mix of drawings, models, bills of materials, and blueprints in building construction (Zachman, 1987).

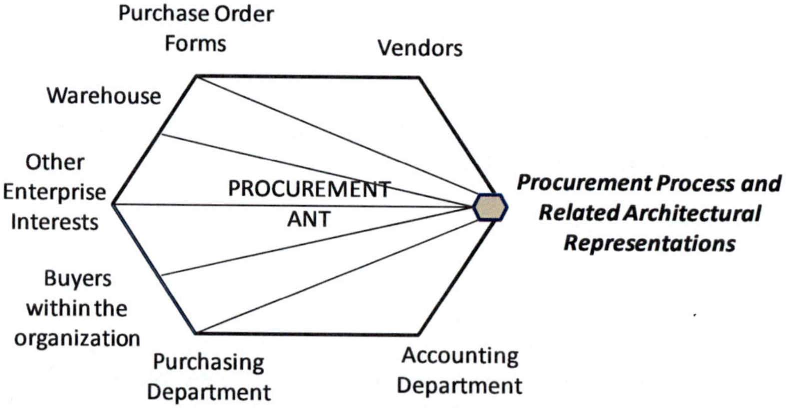

Fig. 2 Procurement process as an inscription of aligned interests of the procurement AN

Let us now consider that a decision has been made to automate the procurement process using information technologies. In EA and ANT terms, the original espoused purpose of such a project could be to further inscribe the existing procurement process into IS artifacts and thus further stabilize the de-facto (i.e., current or as-is) enterprise architecture. Alternatively, the purpose could be to improve the process thus bringing changes in the form of a future (i.e., target or to-be) enterprise architecture. Interestingly, regardless of the original goal, the development and implementation of the IS artifact (e.g., a computer-based information system) is likely to result in changes to the existing EA as it involves the enrollment of new actors, and thus requires a re-alignment of interests inside the enterprise to accommodate the interests of the new actors (see Figure 3). Initiation of such an automation project typically involves a team of "analysts" (e.g., systems analysts, designers, and architects). The project is also likely to follow some variant of the systems development life cycle utilizing some systems development methodology: This will involve a set of activities, although not necessarily in an entirely linear sequence, centered on (A) architecting (e.g., project initiation, planning, analysis of requirements, system design); (B) instantiation (e.g., coding, procuring, configuring, testing); and (C) deployment (e.g., system implementation, user and technician training, and transitioning the organization to the new system).

During project planning, the development team (whose interests include the successful completion of the project) is likely to align its interests with the project sponsor, who, we will assume for this discussion, represents the aligned interests of the enterprise as a whole. As a result, an actor-network is created, representing the aligned interest of the development team and the enterprise, and the agreement is inscribed into documents such as project charter, statement of work, project goals, and project plans (see Figure 3a). Such logical idea documents correspond to the upper rows in Zachman's enterprise ontology, whereas the later-developed physical architectural artifacts such as screen designs and data record specifications correspond to the lower rows (Zachman, 1987, 2002, 2010a, 2010b), as the project concept moves architecturally from idea to physical reality.

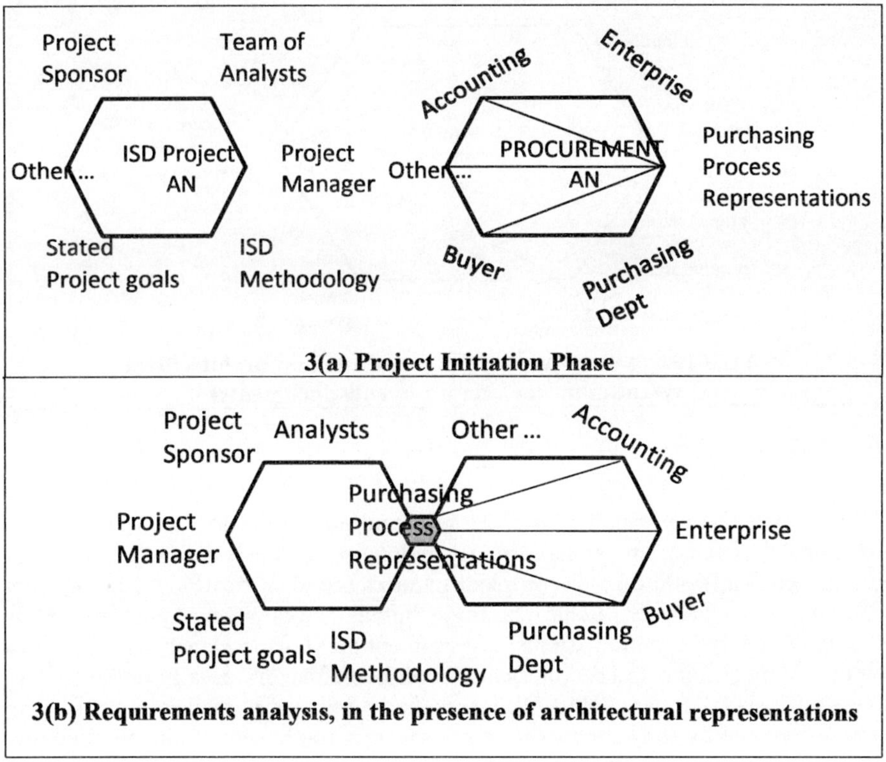

Fig. 3 The translation processes associated with IS development projects

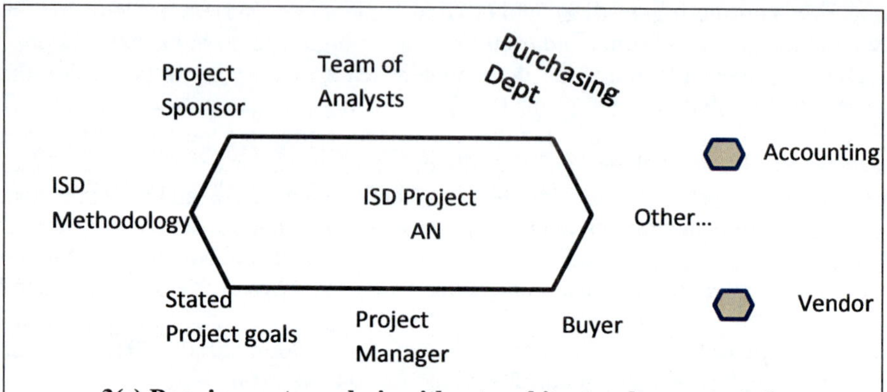

3(c) Requirements analysis without architectural representations

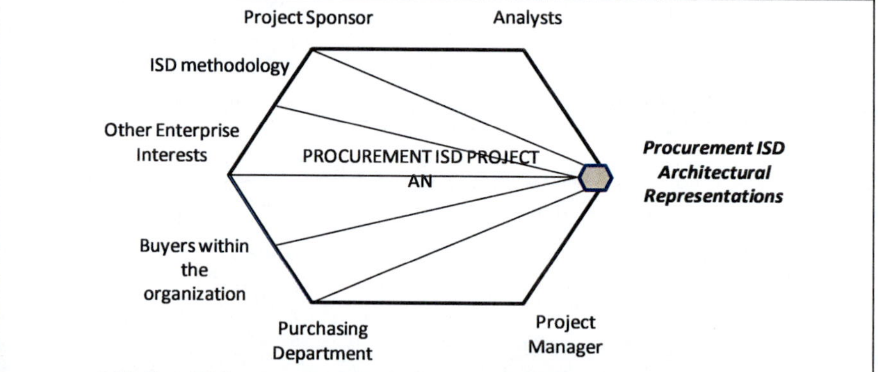

3(d) Inscribing new state of alignment into new architectural representations (i.e., requirements documents)

Fig. 3 *(continued)*

Analysis activities usually involve the recruitment of actors involved in the procurement process, and thus currently belonging to the existing procurement process actor-network. Such recruitment requires the identification of the relevant actors and their interests. Identification of human actors is usually referred to as stakeholder analysis in the systems development and business analysis literatures, whereas identification and recruitment of non-human actors usually involves document analysis (Brennan, 2009). Whereas the interests of human actors may have shifted since the existing procurement process was implemented, the architectural artifacts are likely to be relatively more objective and faithful representations of the alignment of interests embedded into the existing procurement process. Thus, if the goal of a system development process is to stabilize the de-facto enterprise architecture by automating the existing procurement process, existing architectural representations are likely to be particularly helpful. Recruiting the existing architectural representations, for example in the form of adopting existing job descriptions and process and data models, as a basis for the design of the new procurement information system, is likely to ensure easy enrollment (i.e., minimal resistance) on the part of other actors in the procurement AN (see Figure 3b). The existing architectural representations are also likely to be instrumental in the identification, interessement, and enrollment of human actors. In the absence of such representations, the IS development process is likely to include a lengthy rediscovery and a re-negotiation among the actors involved in the procurement process. Also, in the absence of such representations, important actors and their interests may be overlooked, leading to future misalignment(s) of interests (see Figure 3c).

Even if the new information system development and implementation activities require changes to the existing procurement process, and thus would require a realignment of interests within the procurement actor-network, the presence of architectural representations (artifacts) into which the current state of alignment is inscribed are also useful. Such representations can be "recruited" into the new procurement system actor-network and serve as the "voice" of the current processes and, as the new alignment emerges, serve to facilitate communication, negotiation, and finally the memorialization of the new process. Such recruitment may be easier (i.e., less political) for more abstract and logical representations, such as conceptual models, because relatively fewer modifications may be needed. Moreover, abstract architectural representations are likely to more faithfully represent the interests of the enterprise actor-network, and are likely to be most helpful in the process of IS development and implementation. More specific and physical representations are likely to represent the interest of specific human or technical actors and are likely to be less flexible with regard to enrollment of new actors. Finally, new architectural representations are created as these defining and architecting activities proceed to completion (see Figure 3d).

As a part of physical design (i.e., the creation of architectural artifacts regarding the lower rows of Zachman's enterprise ontology), decisions are made to use specific technologies. Ideally, from the perspective of the enterprise, the design and implementation of a technological solution should faithfully inscribe the alignment of interests achieved during the requirements analysis activities. Of course, in more iterative development situations (e.g., prototyping, agile methods) these architecting and instantiation activities occur more concurrently. In any event, however, each technology represents a complex actor network which includes multiple actors, such as software, hardware, vendors, programming languages, implementation guides, and so on (see Figure 4). Thus, these are the activities by which the interests of the technology actor-network are aligned with the interests of the enterprise procurement ISD project interests. This may require compromises on the part of the technology actor-network, as well as on the part of the enterprise procurement ISD actor-network. In other word, some user requirements may be sacrificed for the system implementation to be completed within project constraints such as time, money, or existing technologies.

The hypothetical system development process discussed here highlights several important aspects of the ANT conceptualization of EA and ISD. These are summarized below:

1. The current architecture of the enterprise, written or not, reflects the current state of alignment of interests in the enterprise actor-network.

2. Architectural representations (such as organizational charts, data flow diagrams, use cases, and process maps) and technical artifacts (such as an IS) are inscriptions of aligned interests, and thus serve to memorialize and stabilize the enterprise architecture at the time of their creation.
3. System development and implementation projects involve recruitment of new actors (e.g., humans, architectural artifacts, technical artifacts), and therefore result in changes to the EA. Such changes are in turn reflected in new architectural representations and finally technological artifacts.
4. The presence of architectural representations inscribing an alignment of interests within an enterprise makes it easier to ensure that all the aligned interests are taken into account during the software development and implementation processes.
5. During system development and implementation processes existing architectural representations need to be recruited into the new actor-network by means of updating the existing representations, such as process maps, data models, as well as existing technological artifacts. Failure to do so is likely to lead to future misalignment(s) of interests.

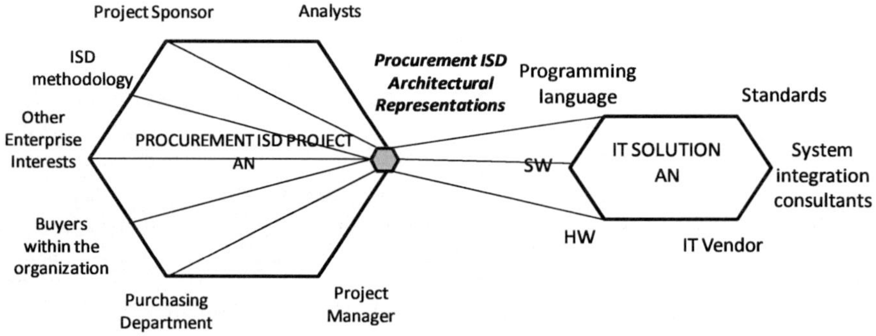

Fig. 4 New procurement system actor-network

6 The ANT View of EA: Implications and Conclusions for Research and Practice

"Architecture, of all the arts, is the one which acts the most slowly, but the most surely, on the soul" – Ernest Dimnet

The proposed conceptualization of EA as the reflection all enterprise interests in their current state of alignment illuminates the political and strategic nature of EA work. It also brings attention to the integration, transparency, actor-identification, and alignment challenges associated with EA. The identification of actors and interests is critical but also challenging. But since the as-is EA is the de-facto alignment of interests within the organization, one of the key aspects of EA work is to ensure that misalignment does not occur, or given that its occurrence is likely, that such conflicts are resolved. Since such conflicts may involve both technical and sociological actors, this points to the critical importance of both soft and technical skills for enterprise architects. Because enterprise architecture reflects the alignment of both human actors and technical artifacts, managing such alignment requires a combination of soft people skills, as well as technical skills. Even if the architectural decisions appear only to concern human actors and their interests, it is likely that realignment of such interests may require making changes to technical artifacts. And, even when the modification to the enterprise architecture may appear purely technical (e.g., switching to a different operating system or type of servers), such change is likely to involve interests of human actors, such as support staff and vendor preferences.

The need for integration of the various interest-inscribing artifacts constitutes another key EA challenge. Broadly, this integration challenge can be decomposed into the identification of all interests and the reconciliation of these interests. While the issue of interest identification is related to the actor identification challenge, identification and reconciliation of all inscribed interests can be a significant advance toward actualizing EA as a reflection of the shared vision of the human actors of the enterprise AN. As the first step, a comprehensive *taxonomy or typology of all such inscriptions* could be developed. Zachman's enterprise ontology (1987, 2002, 2010a, 2010b) can provide insight into the types of inscriptions and inter-relationships among them. In addition to identifying key classification principles, the typology should necessarily imply the hierarchical structure distinguishing among more or less influential inscriptions. The need for such a hierarchy brings the typology development from the primarily data management and knowledge management domains, into the realm of strategy and policy. EA, of course, includes both domains as both are part of the enterprise. The presence of a typology will allow for easier identification of all-important inscriptions, and will also serve as a guide for the resolution of conflicting interests.

Once all the inscriptions of interest are identified and classified, the integration and reconciliation of their content is required. Markup languages, text-mining technologies, and simulation and modeling tools offer a possibility for comparing different inscriptions and thus pave a way for their reconciliation. Clearly the need for different dialects and specialty vocabularies and models may be required to architect certain aspects of the enterprise, but alignment and integration can only be optimized if the ability to translate and reconcile exists. Thus there is a critical need for building an EA on a complete and comprehensive enterprise ontology and having tools capable of supporting not only model creation but also translation and reconciliation. While useful tools do exist, and are in general improving, in light of the vision of an adaptable, holistic, enterprise-wide, universal modeling, decision-making, simulation, and management EA repository, such capabilities do not exist commercially at this time (Simons, Kappelman, & Zachman, 2010)

In part the transparency challenge arises from the presence of covert interests. The need for the elicitation of such covert interests calls for the development of new architectural and requirements gathering approaches that do not assume that candor be present in such situations. Negotiation and mediation approaches from the conflict resolution literature may also be helpful. The other part of the transparency challenge is related to the need for EA information during negotiations of the enterprise with other actor-networks. Addressing this challenge will require creating appropriate interfaces that could provide limited access to the EA repository. Such interfaces should ensure that only necessary and sufficient EA information is presented in an appropriate format each time it is requested by an actor, including human and non-human actors. In fact, in the ideal situation, such interface should assist in assessing how enrollment of other actors into the enterprise actor-network will affect the alignment of interests inside the enterprise. Here, decision support and expert systems research may offer useful theoretical foundations. Research is needed also to examine the appropriate degree of accessibility to different parts of the EA repository in terms of appropriate practices regarding security, intellectual property, privacy, as well as competitive and other propriety matters.

Considering the aforementioned challenges of EA work would be significantly easier if complete alignment of all interests within the enterprise existed. Unfortunately, as the enterprise grows, the enrollment of numerous actors usually leads to multiple misalignments, and the risk of sub-optimal compromises. Such misalignments are often hidden due to low transparency of interests within the enterprise, and an attempt to create an integrated representation of all interests is bound to uncover such misalignments. As this situation is natural and expected, a certain level of misalignment needs to be tolerated within any enterprise. Therefore EA methodologies and tools should be able to accommodate and reveal it and provide decision tools for optimization in terms of trade offs such as those among the enterprise and its subsystems (e.g., departments, functions) and between long-term and short-term priorities. Research and practical guidance are needed to develop guidelines for the level of misalignment acceptable for different types of interests and actors within the enterprise. On the technical side, to facilitate awareness, understanding, and reconciliation of interests, tools need to be developed with tolerance for misalignments, as well as accommodation for the transitional states of the enterprise, its architecture, and its ANs.

In this paper we have used concepts from the Actor-Network Theory to re-examine the meaning of enterprise and of EA through the lens of interest negotiations and actor-network creation. Such re-examination led us to an idealized definition of EA as an integrated and transparent representation of all interests within the enterprise and their current state of alignment. Thus EA cannot only serve as a negotiation interface between the various actors in the enterprise, but also between that enterprise's actors and external actors (such as vendors, suppliers, or customers). Such a view of EA opens several additional directions for EA research.

First, research is needed to devise approaches for the identification of all significant interests and resolving potential misalignments. Because it is impractical that all interests within the enterprise are included perfectly in the various ANs and thus the EA, criteria for interest inclusion need to be developed, as well as guidelines of the acceptable level of misalignment among such interests. Strategic planning literature, as well as literature on negotiations is likely to provide a source of relevant theoretical frameworks. Second, on a more technical note, research is necessary to develop appropriate capabilities and interfaces to enable digital EA artifacts to serve as an important tool for communication, simulation, and negotiation among internal and external actors. This would include the development of appropriate modeling and storage capabilities, and the user and technology interfaces that would provide internal actors representing the enterprise AN (or any part of it) with access to the EA repository. Moreover, tools need to be developed and tested which would allow for checking the consistency of all interests inscribed within the EA and identifying potential misalignments. Still other capabilities are necessary to check how the proposed alterations to the enterprise actor-network fit into the existing network of interests. Such validation would allow for *a priori* identification of sources of resistance to change initiatives and facilitate making appropriate managerial and strategic choices.

From the practitioner point of view, the ANT view and definition of EA highlights the important and often overlooked political aspect of doing EA. Such a definition should raise an interest in EA among C-level executives and strategists. The definition also highlights the important challenges of EA, which in the absence of necessary tools, including intellectual and conceptual ones, may discourage some business managers from embarking on EA initiatives. We believe this is an ill-advised option given the facts that:

1. the creation of value producing processes and practices best precedes tool procurement (i.e., a fool with a tool is still a fool and likewise automating poor processes);
2. EA practices and programs are still in the early stages and the playing field is still pretty flat and there are many opportunities to create advantage through EA work;
3. maximizing EA's benefits typically involves a significant degree of learning and culture change which takes time (Senge, 1990); and
4. it is of critical importance for public and private management and policy makers in general to have a much more holistic view of their enterprises in light of the plethora of enterprise catastrophes due to the failure of management to see risks, dependencies, and misalignments (e.g., GM, FNMA, AIG, Bear Sterns, Lehman Brothers, Landsbanki, Allied Irish Bank, Fortis, Northern Rock, and RBS to name but a few).

We hope, however, that the benefits of EA and ANT for understanding the interests of the enterprise in negotiations with fast-changing internal and external environments, combined with the benefits of EA in managing change and complexity, outweigh the perceived risks, and that this article will inspire more organizations to embrace the challenge of EA development. Of course, "No one has to change. Survival is optional" (W. Edwards Deming)[3].

[3] For more information about Dr. Deming visit http://deming.org/

References

Brennan, K. (ed.): A Guide to Business Analysis Body of Knowledge, 2nd edn. International Institute of Business Analysis (March 31, 2009)

Callon, M.: Some elements of a sociology of translation: Domestication of the scallops and the fishermen. In: Law, J. (ed.) Power, Action and Belief: A New Sociology of Knowledge, pp. 197–225. Routledge, London (1986)

Callon, M., Latour, B.: Unscrewing the big leviathan: How actors ma-cro-structure reality and how sociologists help them to do so. In: Knorr-Cetina, K.D., Cicourel, A.V. (eds.) Advances in Social Theory and Methodology: Towards an Integration of Micro and Macro-Sociologies, pp. 277–303. Routledge, London (1981)

Chen, P.P.: The Entity Relationship Model - Toward a Unified View of Data. ACM Transactions on Database Systems 1, 1 (1976)

CIO Council. Federal Enterprise Architecture Framework (FEAF). United States Office of Management and Budget (September 1999)

CIO Council. A Practical Guide to Federal Enterprise Architecture, Version 1. 0, Chief Information Officer Council of OMB and the US General Accountability Office (February 2001), http://www.cio.gov/documents/bpeaguide.pdf (January 1, 2011)

Collett, S.: Hot Skills, Cold Skills. Computerworld (July 17, 2006), http://computerworld.com/action/article.do?command=viewArticleBasic&articleId=112360 (January 1, 2011)

DeMarco, T.: Structured Analysis and System Specification. Yourdon Press, New York (1978)

El Sawy, O., Malhotra, A., Gosain, S., Young, K.: IT-intensive value innovation in the electronic economy: Insights from Marshall Industries. MIS Quarterly 23(3), 305–335 (1999)

Finkelstein, C., Martin, J.: Information Engineering, vol. 1,2. Prentice Hall, Englewood Cliffs (1981)

Hagan, P.J. (ed.): Guide to the (Evolving) Enterprise Architecture Body of Knowledge. MITRE Corporation, McLean, VA (2004), http://www.mitre.org/work/tech_papers/tech_papers_04/04_0104/04_0104.pdf

Hanseth, O., Jacucci, E., Grisot, M., Aanestad, M.: Reflexive Standardi-zation: Side Effects and Complexity in Standard Making. MIS Quarterly 30 (Special Issue), 563–581 (2006)

Kappelman, L.A.: Bridging the Chasm. Architecture and Governance Magazine 3(2), 28, 35–36 (2007); Also in Kappelman, pp. 35–36 (2010)

Kappelman, L.A. (ed.): The SIM Guide to Enterprise Architecture. CRC Press, NY (2010)

Latour, B.: Where are the missing masses? The sociology of some mundane artifacts. In: Bijker, W.E., Law, J. (eds.) Shaping Technology/Building Society, pp. 225–258. MIT Press, Cambridge (1992)

Latour, B.: Reassembling the Social: An Introduction to Actor-network-theory. Oxford University Press, New York (2005)

Law, J.: Networks, Relations, Cyborgs: On the Social Study of Technology. Centre for Science Studies, Lancaster University, Lancaster LA1 4YN, UK (2000), http://www.lancs.ac.uk/fass/sociology/papers/law-networks-relations-cyborgs.pdf (January 1, 2011)

Luftman, J.: IT-Business Strategic Alignment Maturity Assessment. Society for Information Management research report, Chicago (October 7, 2003), http://simnet.org

Luftman, J., Kempaiah, R.: An Update on Business-IT Alignment: "A Line" Has Been Drawn. MIS Quarterly Executive 6(3), 165–177 (2007)

Magal, S.R., Word, J.: Essentials of business processes and Infor-mation System. John Wiley and Sons (2009)

Miller, D.: Enterprise client/server planning. Information Systems Management 14(2), 7–15 (1997)

Monteiro, E.: Monsters: From systems to actor-networks. In: Braa, K., Sorenson, C., Dahlbom, B. (eds.) Planet Internet, pp. 239–249. Studentlitteratur, Lund (2000)

Newton, T.: Creating the new ecological order? Elias and actor-network theory. Academy of Management Review 27(4), 523–540 (2002)

Ross, J.W., Weill, P., Robertson, D.: Enterprise architecture as strat-egy creating a foundation for business execution. Harvard Business School Press, Boston (2006)

Ross, J.W.: Foreword. In: Kappelman, pp. xli-xlii (2010)

Ross, J.W.: Creating a Strategic IT Architecture Competency: Learning in Stages. MISQ Executive 2(1) (2003)

Salmans, B., Kappelman, L.: The State of EA: Progress Not Perfection. In: Kappelman, pp. 165–217 (2010)

Sarker, S., Sarker, S., Sidorova, A.: Understanding Business Process Change Failure: An Actor-Network Perspective. Journal of Management Information Systems 23(1), 51–86 (2003)

Senge, P.: The Fifth Discipline. Doubleday, NY (1990)

Sidorova, A., Kappelman, L.: Enterprise Architecture as Politics: An Actor-Network Theory Perspective. In: Kappelman, 70–88 (2010)

Sidorova, A., Kappelman, L.: Better Business-IT Alignment through Enterprise Architecture: An Actor-Network Theory Perspective. Journal of Enterprise Architecture, 39–47 (February 2011)

Simons, G., Kappelman, L., Zachman, J.: Enterprise Architecture as Language. In: Aiguer, M., Bretaudeau, F., Krob, D. (eds.) Complex Systems Design and Management. Springer, Berlin (2010); Also in Kappelman pp. 127-146 (2010)

Venkatesh, V., Bala, H., Venkatraman, S., Bates, J.: Enterprise Architec-ture Maturity: The Story of the Veterans Health Administration. MISQ Executive 6(2), 79–90 (2007)

Walsham, G.: Actor-network theory and IS research: Current status and future prospects. In: Lee, A.S., Liebenau, J., DeGross, J.I. (eds.) Information Systems and Qualitative Research, pp. 466–480. Chapman and Hall, London (1997)

Walsham, G., Sahay, S.: GIS for district-level administration in In-dia: Problems and opportunities. MIS Quarterly 23(1), 39–66 (1999)

Yourdon, E.: Techniques of Program Structure and Design. Prentice Hall, Englewood Cliffs (1975)

Zachman, J.A.: A Framework for Information Systems Architecture. IBM Systems Journal 26(3), 276–292 (1987)

Zachman, J.A.: John Zachman's concise definition of the Zachman Framework. In: Kappelman, pp. 61–65 (2010a)

Zachman, J.A.: Architecture is Architecture is Architecture. In: Kappelman, pp. 37–45 (2010b)

Zachman, J.A., Sowa: Extending and Formalizing the Framework for Information Systems Architecture. IBM Systems Journal 31(3), 590–616 (1992)

Chapter 24

Introducing the European Space Agency Architectural Framework for Space-Based Systems of Systems Engineering

Daniele Gianni, Niklas Lindman*, Joachim Fuchs, and Robert Suzic

Abstract. System of Systems (SoS) engineering introduces a higher level of complexity compared to conventional systems engineering, for the number and geographical distribution of resources, and for interoperability and agreements issues, for example. In domains such as defence or Information Technology, architecting methodologies have been introduced to address engineering needs deriving from this increased level of complexity. From ongoing and future European space programmes, *ESA* has identified new engineering needs that cannot be addressed with existing methodologies. These needs concern the use of the methodology to support decision making, the representation of European regulation and policies, and the representation of space-specific domain concepts. In this paper, we introduce the *European Space Agency Architectural Framework* (*ESA-AF*), an architecting methodology that aims to address these new engineering needs by improving on existing architecting methodologies. In addition, *ESA-AF* introduces exploitation tools for user-friendly interactive visualisation of SoS architectural models and for textual reporting of model data, enabling non technical users to exploit these models. We also briefly present example applications of *ESA-AF* in support of SoS engineering activities for the *Galileo* navigation, the *Global Monitoring for Environment and Security (GMES)*, and the *Space Situational Awareness (SSA)* programmes.

Keywords: SoS, ESA-AF, enterprise architecting, architectural framework, MODAF, TOGAF, space.

Daniele Gianni · Niklas Lindman · Joachim Fuchs · Robert Suzic
European Space Agency

Niklas Lindman
ESA/ESTEC, Postbus 299, NL-2200 AG, Noordwijk, The Netherlands
e-mail: niklas.lindman@esa.int

* Corresponding author.

1 Introduction

The space systems capabilities of global access and remote sensing have contributed to the pervasive use of space systems to improve the quality of living by supporting the implementations of advanced services such as climate monitoring, disaster management or global telecommunication. New services are often demanded to further improve our living; however, available funding is limited, and therefore optimising the reuse of existing resources becomes necessary. Many of these resources were originally designed for purposes different from the provision of the new services. Moreover, these resources are often managed and owned by independent organisations. A number of new issues arise when aiming to integrate these resources in new configurations—namely Systems of Systems (SoS) configurations [1]—for the provision of a new service. For example, these new issues can concern interoperability—incompatibility of physical interfaces or different data interpretation; agreements—resource use and expected performance; SoS stability and evolution—contrasting objectives of the organisations involved in a SoS configuration, resulting in different life-cycles and life-spans of the associated systems; number of resource—SoS can be composed by a very large number of resources. These new issues can considerably increase the complexity of the engineering activities, and therefore they motivate the introduction of new methodologies to support design consistency, decision making effectiveness, and managing the increased complexity, in general. In domains such as defence or Information Technology (IT), architecting methodologies have been introduced to support the decision making of systems and programme managers in SoS engineering activities. However, the European space context is considerably different from the typical contexts in these domains, for the institutional role of the *European Space Agency (ESA)* and the unique characteristics of the space domain. As a result, the same methodologies do not entirely suit the new engineering needs that *ESA* has identified from the European space context. These needs, which inherently derive from organisational necessities for cost reduction, concern the use of the architecting methodology, the representation of European regulations and policies, and the representation of space-specific domain concepts. In this paper we introduce the *European Space Architectural Framework (ESA-AF)*, an architectural methodology aiming to address these needs while leveraging on the established methodologies *TOGAF* [2] and *MODAF* [3]. Specifically, *ESA-AF* introduces concepts for the representation of interface specifications, data policies, security requirements, and financial regulations, for example. Moreover, *ESA-AF* defines new analysis methods to guide the model definition and support issues identification and solution. *ESA-AF* consists of a software infrastructure and a set of business rules for the definition, maintenance and use of the architecting methodology.

The paper is organized as follows. The *European Space Context* section outlines the organisational and technical background motivating the introduction of *ESA-AF*. *The Needs for an ESA Architectural Framework* section presents the needs arising from the European space context. Following, the *ESA-AF* section describes the methodology, including the technical requirements and the framework structure. The *Example Application* section presents preliminary use cases of

ESA-AF in support of the SoS engineering activities for the *Galileo* navigation [4], *the Global Monitoring for Environment and Security (GMES)* [5], and the *Space Situational Awareness (SSA)* [6] programmes.

2 European Space Context

The European characteristics of wide diversity of national cultures, regulations, and socio-economical directives introduce new political and technical challenges when aiming to implement increased capabilities and optimise resources at European level. Space systems can be financially very demanding and the most ambitious types of programmes (e.g. Global Navigation Satellite System or Space Situational Awareness) can be delivered only through joint efforts, integrating available capabilities and reusing existing resources in SoS configurations. However, these configurations differ considerably from conventional systems as they are *"large scale integrated systems that are heterogeneous and independently operable on their own, but are networked together for a common goal"* [1]. In addition SoS configurations present new properties (e.g. operational and managerial independence of the composing systems, evolutionary development, and geographic distribution [1]) that increases the complexity of the engineering activities compared to conventional systems engineering activities. To support SoS engineering activities, architectural models of SoS are an essential mean to support decision making as these models can be used to concisely provide comprehensive and consistent views on relevant facets of a SoS.

Within this context, *ESA* has the mission of promoting the cooperation among European states in space research and technology, for peaceful purposes, including the elaboration and implementation of activities in the space, the coordination of European and national space programmes, and the elaboration and implementation of industrial policies for space programmes [7]. To achieve its mission, *ESA* cooperates with European institutions to define space programmes and initiates new technologies and methodologies in support of systems design, development, integration and operation. Considering the increasing demand for new services with a space component and the respectively limited funding, the orchestration of European industry and national roles becomes a key issue for the effective implementation and provisioning of advanced space-based services. The European space context inherently shows the properties of the above defined SoS, where however the autonomy properties are further exacerbated for the inherent national influence and regulations constraining the degree of freedom of individual systems.

3 The Needs for an ESA Architectural Framework

In non space domains, architecting methodologies have been introduced to support system and programme managers in SoS engineering activities. For example, in the defense domain, *DoDAF* [8], *MODAF* or *NAF* [9], are the most prominent standard methodologies for SoS representation. Differently, *TOGAF* addresses more conventional IT architectures by defining an architecture planning and design process. However, these methodologies are not tailored to the unique characteristics of European space-based SoS. In particular, the above architectural methods do not satisfy the following extra needs that *ESA* has identified:

Usage need: In a European space programme, the decision making often involves a large variety and number of actors. These include technical actors (e.g. end-user and systems engineer), management actors (e.g. programme manager and procurement manager), and political actors (e.g. politician defining the strategic goals and program board evaluating the programme feasibility). The architecting methodology must thus consider this variety and number of actors by enabling the actors to communicate at several abstraction/concreteness levels, using the most suitable means and tailoring the communication to European cultural differences. Addressing this need is key to maximise the effectiveness and alignment of technical and strategic decisions while ensuring the support of political institutions.

Regulation need: In a European space programme, several EU member nations are typically involved in the various programme phases, depending on the national interests and on the contribution to the European space programme. This multinational involvement requires that national regulations (e.g. national design practices, data and security policies, governance procedures) are considered to effectively contribute to the successful programme development.

Domain need: The space domain presents different characteristics from other domains such as defence or IT. For example, space capabilities can be characterised by space-specific types and large set of parameters that are often unique within the entire domain. Similarly, the number of systems implementing a capability can be very limited if not unique. Furthermore, some interface control specifications might require a detailed level of representation to support the integration with legacy systems and promote the development of new and compliant ones. In addition, aspects like procurement policies also affect the space context in several ways. For example, ESA is subjected to geographical return criteria. In addition, ESA aims to harmonise technologies among European partners and to identify synergies at European level.

The above needs cannot be entirely addressed by existing architectural methodologies. For example, *DoDAF* and *MODAF* are tailored to the military domain and do not address issues such as data policies. In addition, these methodologies do not adequately support the representation of programmatic and procurement activities, which are central in the European space context. Similarly, *TOGAF* focuses on the modelling process. Moreover, by initiating a new methodology, *ESA* can also establish the infrastructure for the methodology modification and evolution, coordinating and receiving feedback from the European space industry for a long term standardization plan. These and similar considerations have motivated the introduction of the *ESA-AF* methodology described below.

4 ESA Architectural Framework (ESA-AF)

ESA-AF introduces an architecting methodology for European space-based SoS engineering, aiming to address the above mentioned engineering needs. *ESA-AF* is based on the standard methodologies *TOGAF* and *MODAF*, tailoring and extending these methodologies to satisfy the above needs. We achieve this by deriving a set of technical requirements and by structuring *ESA-AF* in governance, modelling and exploitation levels.

4.1 Technical Requirements

The engineering needs have driven the identification of *ESA-AF* technical requirements. For example, from the *usage need*, we derived the technical requirements:

- *ESA-AF* shall improve the logical and technical consistency of existing architecting methodologies;
- *ESA-AF* shall reduce the complexity of existing architecting methodology, without affecting their effectiveness in European space-based SoS engineering activities;
- *ESA-AF* shall support its users in the exploitation of architectural data, by providing user-friendly visualisation of models.

Similarly, from the *regulation need*, we derived the requirements:

- *ESA-AF* shall introduce specific concepts for security policies, data policies, and financial regulations;
- *ESA-AF* shall enable the representation of European member nations.

Follows the *domain need*, from which we derived the requirements:

- *ESA-AF* shall enable the accurate representation of space domain concepts and their relationships;
- *ESA-AF* shall introduce a multi-resolution modelling approach where needed (e.g. in critical Signal-In-Space interfaces).

Finally, to support the possible long term standardisation, we have identified the requirements: *ESA-AF* shall be very adaptable to support the needs of ongoing and future European space programmes and to implement feedback from the European space industry; *ESA-AF* shall introduce software infrastructure to support the methodology evolution; *ESA-AF* shall provide accurate documentation for the use of the software infrastructure. All the technical requirements have been implemented by *ESA-AF*, which structure is illustrated below.

4.2 Framework Structure

The framework structure is organised in governance, modelling and exploitation levels, as shown in Fig. 1. Each level involves the participation a number of

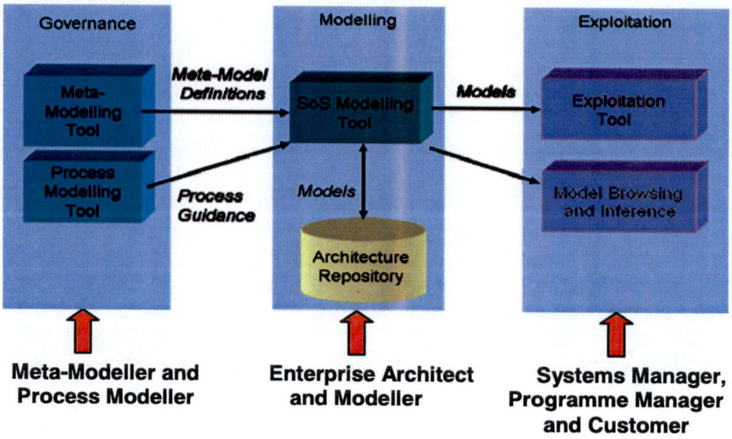

Fig. 1 ESA-AF Structure

professional figures and supports a framework phase. Specifically, the governance level supports the maintenance, the modelling level supports the use, and the exploitation level supports the model data use.

4.2.1 ESA-AF Governance

The *governance level* defines the software infrastructure and the business processes to support the methodology evolution, involving the meta-modeller and the process modeler.

The *software infrastructure* consists of data files representing *ESA-AF* definition data and the programs for the manipulation of these files. The data files are *ESA-AF* interactive glossary and *ESA-AF* meta-model implementation. The interactive glossary is a navigable set of web pages containing the definition of all the terms used in *ESA-AF*. *ESA-AF* meta-model implementation is the Eclipse-based implementation of *ESA-AF* meta-model, which is a reviewed and extended version of *MODAF* 1.2.003 meta-model (M3). Specifically, *ESA-AF* meta-model improves M3 logical consistency, simplifies M3 by removing unneeded concepts and views in the European space context, and introduces new concepts and viewpoints. *ESA-AF* meta-model consists of the following viewpoints: strategic, operational, systems, availability, technology, standards, programme, agreements, risk, financial, services, data policy and security. The viewpoints strategic, operational, systems, availability, technology, standards, programme are inherited from *MODAF*. These viewpoints were subjected only to minor reviews, such as the introduction of concepts and tagged values, to contribute to address the usage need. The remaining viewpoints have been specifically designed to address the regulation and domain needs. More specifically, the *Financial viewpoint*, which includes the *Cost and Funding views*, can be used to represent financial resources and the links of these resources to programmes, contributing nations, institutions and systems. The *Data Policy (DP) viewpoint* can be used to represent policies for any sort of information, including paper documents, scientific data, or project data. This viewpoint includes the *DP Definition*, *DP Validity*, *DP Provisioning* and *DP Use views* for the representation of the policy properties. The *Security viewpoint* can be used to represent factors that can affect the security, including properties of information items and security regulations. This viewpoint includes the *Information Asset*, *Security Requirements* and *Security Solution views* for the representation of the respective security concerns. The programs are customised versions of the Eclipse Process Framework (EPF) and Eclipse Meta-modelling Framework (EMF) [10]. The *ESA-AF* EMF also includes a generator for the *ESA-AF* plug-in to be deployed within the software infrastructure of the modelling level.

The *business processes* define the procedures that meta-modeller and process modeller to update and modify the interactive glossary and the *ESA-AF* meta-model implementation. Currently, the processes are described in the *ESA-AF* documentation, which also refers to the Eclipse user guide and the documentation of the modelling software infrastructure.

Within this level, the *meta-modeller* maintains the modelling structure underlying any ESA-AF architectural model. Similarly, the *process modeller* maintains the business procedures describing and regulating the use of the methodology.

4.2.2 ESA-AF Modelling

The *modelling level* defines the software infrastructure and the business processes that the modeller and the enterprise architect can use to represent European space-based SoS.

The *software infrastructure* defines the capabilities for the digital representation of European space-based SoS. These capabilities include the modelling tool Magic Draw (MD) [11], the *ESA-AF* plug-in, and the MD Teamwork Server. MD is a standard and popular tool for modelling, including UML. The *ESA-AF* MD Plug-in is the MD-based implementation of the *ESA-AF* meta-model. MD Teamwork server is a configuration management system for the reuse and sharing of individual model blocks across programmes and *ESA-AF* users. In addition, MD Teamwork also enables modellers and enterprise architect to concurrently and synchronously extend, review and visualise stored models. The business processes define the procedures for the improvement, analysis and evaluation of an *ESA-AF* enterprise model. These processes mostly conform to the *TOGAF* standard. However, improvements are introduced to guide the modeller in the SoS representation and to coordinate the modeller and enterprise architect for the identification of possible issues in European space-based SoS engineering, including design consistency and completeness [12].

Within this level, the *modeller* can represent European-based SoS architectures using *ESA-AF*. Similarly, the *enterprise architect* contributes to the architecture representation by supporting the modeller with domain knowledge.

4.2.3 ESA-AF Exploitation

The *ESA-AF* Exploitation defines the software infrastructure and the business processes that the modeller and the enterprise architect can use to extract information for decision making of the systems managers, programme managers and customers.

The *software infrastructure* consists of a configurable suite of tools for graphical and interactive visualisation and for tabular reporting generation. A visualisation tool provides enterprise architects and programme managers with a graphical and interactive overview of the SoS architecture, including relationships between service providers and service consumers, agreements regulating service use, systems involved in the operational and maintenance chains. The visualisation tool is configurable and can host new plug-ins offering improved functionalities, such as data flow interactive visualisation, data policy reporting or potential security risk detection. Similarly, a report generator tool can provide PDF documents of the main SoS properties, such as services involved in a scenario or owners of the systems implementing a set of services.

The *business processes* define a set model inference patterns that can guide the modeller to extract information from the architectural model by identifying specific systems properties [12]. Using these patterns, the modeller can identify the open interfaces (i.e. those system interfaces used across a stakeholder boundary) or support the mitigation of failure risks of the SoS or determine whether there are mismatch between agreement conditions and service levels, for example.

Within this level, the *modeller* can operate the visualisation tool, generate PDF reports and extract architectural information using the business processes. The *enterprise architect* can guide the graphical visualisation and enquire the modeller for more detailed data. The *system* and *programme managers* can visualise relevant parts of the SoS architecture and gain an insight into the development of the engineering activities while identifying strategic issues. Similarly, the *customer* can visualise the progress of the developments and the entire scope of the programme. An example part of the *ESA-AF* exploitation is illustrated in the following section.

5 Example Applications

In our modelling support for space programmes, we have been applying *ESA-AF* to support SoS engineering for the *Galileo*, *GMES*, and *SSA* programmes.

5.1 Galileo

Galileo is the upcoming European Global Navigation Satellite System. *Galileo* is per se a complex system; however, *Galileo* will also form SoS configurations with systems of third-parties, enabling these parties to implement and provide advanced space-based services. Currently, the *Galileo*-COSPAS/SARSAT configuration is planned to be formed for the provision of global Search-and-Rescue (SAR) service [13]. In support to the engineering activities for this integration, we have developed a preliminary architectural model of this configuration, specifically for the *Galileo*-based SAR scenario shown in Fig. 2. The scenario is initiated by a human rescuee who activates a *Galileo*-enabled SAR device. Using *Galileo* direct link, the device transmits the distress signal to a local user terminal centre, which subsequently forwards the request to a local mission coordination centre. Using the *Galileo* reverse link, the mission centre acknowledges the reception of the distress signal to the human rescue's device. Next, the mission centre plans the SAR mission, which is assigned to a rescue team. The team continues to rely on Galileo signal for the determination of the current position, while reaching the distress location.

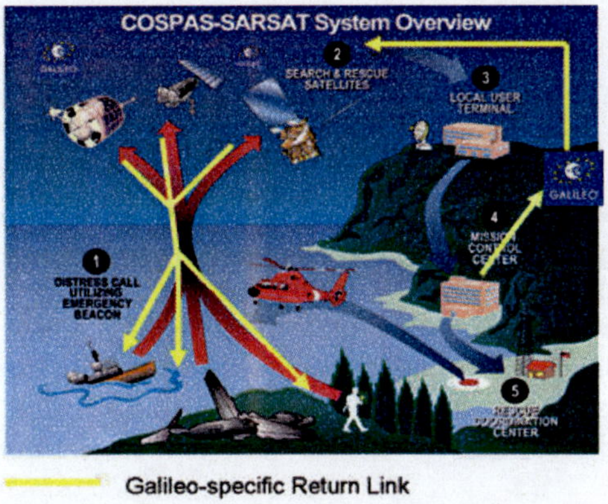

Fig. 2 Galileo-based Search and Rescue Operational Scenario

Critical to the provisioning of the SAR service is the identification and definition of all the open interfaces (i.e. the interfaces between independently managed or owned systems) in the *Galileo*/COSPAS-SARSAT configuration [13]. The model purpose was to identify these interfaces, determining the responsibility for their definition and providing formal specification for the most critical interfaces [14]. The modelling began with the characterisation of the operational roles playing in the above scenario. Next, we have identified the systems that will play these roles, and basing on this we have determined the interaction occurring across organisation boundaries. For each of these interactions, the organisation responsible for the definition of the respective interface was identified, and the main critical interfaces formally defined through *ESA-AF*, including the one with the *Galileo* receiver [15].

5.2 Global Monitoring for Environment and Security (GMES)

GMES programme aims to establish a European capacity for Earth observation to provide services in Land Monitoring, Marine Environment Monitoring, Atmosphere Monitoring, Emergency Management, Security, and Climate Change [5]. GMES SoS will collect data coming from a disparate variety of national and

European assets. Critical to the SoS design and integration activities are the identification of the GMES governance dependency chains and the service performance evaluation in terms of various performance metrics [16].

Using *ESA-AF*, we have represented part of the *GMES* SoS architecture and supported the evaluation of the latency time for the Oil Spill Detection and Monitoring service and of the response time for the Oil Spill Forecasting service, producing a graphical overview of the architectural model by means of *ESA-AF* Exploitation. Fig. 3 shows the graphical output for the Oil Spill Detection and Monitoring service. Using *ESA-AF* Exploitation, ESA programme managers can display all the systems and enterprises involved in the operation and governance of the SoS for this service. Through *ESA-AF* interactive visualisation, programme managers can also identify missing agreements between these enterprises for the use of the needed resources. Moreover, the managers can visualise the operational chain, thus gaining further information for the negotiation of these agreements. The same information could have been retrieved manually by the modeller, accessing directly the architectural model, and theoretically obtaining the same outcome. Practically, programme managers would have been exposed to a considerably higher quantity of data, with continuous switching from a diagram to another one. This manual process can easily be error prone and certainly diverts the attention of managers from the essential information needed. Furthermore, the effectiveness of this process would have largely relied on the modeller browsing and presentation abilities, which can easily become less effective as the enterprise model grows.

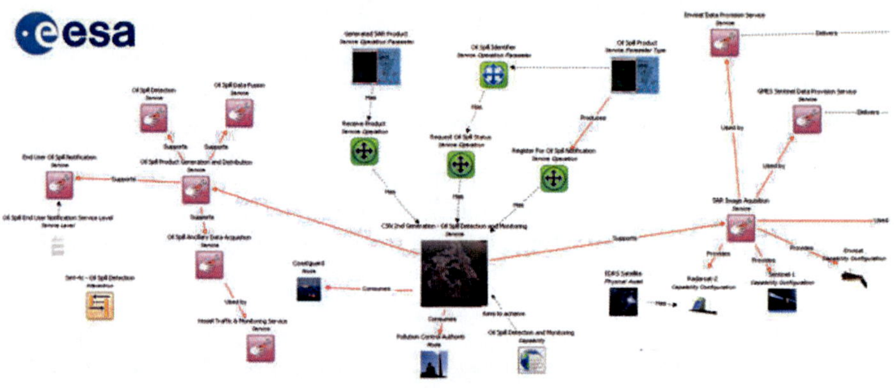

Fig. 3 Overview of an operational chain for the Oil Spill Detection and Monitoring service [16]

5.3 Space Situational Awareness (SSA)

The European *SSA* programme aims to establish a European system of warning about dangerous situations in the outer space [6]. In the programme preparatory phase, SoS functional and physical architectures are designed by evaluating possible integrations of existing assets located over Europe. Critical to ensure the reuse of the largest set of existing assets is that observation data are guaranteed to be disseminated and used only according to the data policy requirements of the individual European institutions owning or operating the assets. To gain the trust of these institutions and to ensure that their requirements are actually satisfied, we are applying the *ESA-AF* methodology to guide the functional architecture design and physical architecture verification activities for the *SSA* SoS. Currently, our model addresses the definition of an example data policy for Space Surveillance and Tracking (SST) data. This data policy specifies many attributes, including authorised data recipients and communication requirements. For example, authorised data recipients are only European military organisations and communication requirements include the use of physically secure connections. The model also includes the operational architecture, which involves *SSA* roles (e.g. *SSA* portal, *SSA* SST front-end, *SSA* SST back-end) and European partners (e.g. European MoDs, National Space Agencies, National Data Hubs and Data Collectors). In addition, the model defines the possible interactions among these roles and the information flow. Using *ESA-AF*, we have been able to specify values of the functional architectural parameters (e.g. protocol minimum encryption time, protocol repudiability time, or definition of interaction between two roles) basing on the example SST data policy. In the model, we have also outlined a prototypal representation of the *SSA* SoS physical architecture, including the mapping with operational roles on actual systems and capability configurations. Using *ESA-AF*, we will be able to ensure that the physical architecture meets the data policy requirements by verifying the congruence between the data policy specification and the properties of the systems storing or communicating the data [17,18].

6 Conclusion

In Systems of Systems (SoS) engineering activities, the *European Space Agency (ESA)* has identified needs concerning different types of use for architecting methodologies, representation of European regulations and policies, and representation of space-specific concepts. These needs cannot be completely addressed using existing architecting methodologies, and therefore a new methodology must be introduced to effectively support SoS engineering for ongoing and future European space programmes. In this paper, we have introduced the *European Space Agency Architectural Framework (ESA-AF)*, an architecting methodology that improves on existing methodologies by introducing methods and structure to address the identified needs. Specifically, *ESA-AF* aims to enable SoS engineering actors to communicate at several levels by defining information exploitation methods such as inference patterns for architectural information extraction and software tools for user-friendly architectural visualisation. In addition, *ESA-AF* introduces new concepts, viewpoints and views for the representation of European and space-specific characteristics. *ESA-AF* also provides an infrastructure for the methodology evolution to support the implementation of feedback from European space industry and possible long term standardisation. We have shown three example applications of *ESA-AF* in support of SoS engineering activities for the *Galileo*, *GMES*, and *SSA* programmes.

References

1. Jamshidi, M.: Systems of Systems Engineering: Innovation for the 21st Century. Wiley (2009)
2. The Open Group, TOGAF 9™ Enterprise Edition (2011)
3. MODAF Guidance, UK MoD (2009)
4. ESA, What is Galileo, `http://www.esa.int/esaNA/GGGMX650NDC_galileo_0.html`
5. GMES Project web page, `http://www.gmes.info`
6. Bobrinsky, N., Del Monte, L.: The Space Situational Awareness Programme of the European Space Agency. Cosmic Research 48(5), 392–398 (2010)
7. ESA, Convention for the Establishment of a European Space Agency and ESA Council, ESA Publishing Division, SP-1271 (October 2003)
8. DoDAF V2.02: Introduction, Overview, and Concepts - Manager's Guide, vol. 1
9. NAF, Nato Architectural Framework v.3 (2010)
10. Budinsky, F., Steinberg, D., Merks, E., Ellersick, R., Grose, T.J.: Eclipse Modeling Framework. Addison Wesley (2004)
11. Magic Draw, `http://www.nomagic.com`
12. Gianni, D., Bowen-Lewis, J.: Inference Patterns for ESA-AF Models. ESA Technical Report (July 2010)
13. Lisi, M.: Engineering a Service Oriented System: the Galileo Approach. In: Proceedings of the 4th International Workshop on System & Concurrent Engineering for Space Applications (SECESA 2010), Lausanne, Switzerland (October 2010)
14. Gianni, D., Lewis-Bowen, J., Lindman, N., Fuchs, J.: Modelling Methodologies in Support of Complex Systems of Systems Design and Integration: Example Applications. In: Proceedings of the 4th International Workshop on System & Concurrent Engineering for Space Applications (SECESA 2010), Lausanne, Switzerland (October 2010)
15. Gianni, D., Fuchs, J., De Simone, P., Lindman, N., Lisi, M.: A Model-based Approach to Signal-In-Space Specifications for Designing Galileo Receivers. InsideGNSS 6(1), 32–39 (2011)
16. Vega, SoSDEN Project Final Report (June 2010)
17. Gianni, D., Lindman, N., Moulin, S., Fuchs, J.: SSA-DPM: A Model-based Methodology for the Definition and Verification of European Space Situational Awareness Data Policy. In: Proceedings of the 1st European Space Surveillance Conference (June 2010)
18. Gianni, D., Lindman, N., Fuchs, J., Suzic, R., Fischer, D.: A Model-based Approach to Support Systems of Systems Security Engineering for Data Policies. INCOSE Insight, Special Feature on Systems of Systems and Self-Organizing Security 14(2), 18–22 (2011)

Chapter 25
Continuous and Iterative Feature of Interactions between the Constituents of a System from an Industrial Point of View

Patrick Farfal

Abstract. Commonly used definitions of a system rightly emphasize the dynamic interactions between its constituents. In fact this characteristic property is ambivalent and liable to a double interpretation: one, scientific, usual (flow exchanges: data, energy…), the other industrial, additional (evolutionary and iterative feature of the interactions).

Important changes, significant delays, failures, can be attributed to a poor characterization of interactions, although inherent to the concept of system.

From actual cases, the paper intends to develop the industrial interpretation, by highlighting poorly characterized interactions, which may result in increased risks, or, even, lately identified interactions, which are possible sources of failures.

The presentation is illustrated with several examples from military and space fields.

Keywords: dynamic interaction, characterization, iteration, industry.

0 Introduction

A system is essentially made of components in dynamic interaction. Generally, the large number of interactions is highlighted: that number characterizes the amount of complexity, which makes the system more or less difficult to master.

So as to control those interactions, among other purposes, during design, realization, validation phases, in order to mitigate risks, engineers have developed (for a long time), theorized (recently) about methods and tools of Systems Engineering.

Patrick Farfal
PatSys, Conseil et formation en Systèmes
25 rue Jean Leclaire 75017 Paris - France
Phone/Fax:+33 (0)1 42 52 89 60 - Mob.:+33 (0)6 72 14 82 40
e-mail: `pfarfal.patsys@sfr.fr`

From the history of large military and space systems it can be seen that the interactions between the components of a system are essentially evolutional; that evolution is ongoing all along the system life: this is *one of the meanings* of the concept of "dynamic interaction".

First, the design and development of a complex system is an iterative process, through the natural work of pluridisciplinary teams ("system loops" concept) and design and manufacturing reviews.

Moreover, possible changes in the usage of a system during its operational life reinforce that evolutional feature.

Beyond those right and normal cases, the possible lack of characterization of some interaction, which may lead to a late statement of non-compliance, also gives a new meaning to the dynamic feature of the interactions. Precisely, in spite of the existence of widely proven and used methods and tools of Systems Engineering, the fact remains that significant changes or delays, or failures, can be attributed to a poor characterization of interactions.

Several actual cases make it possible to develop that industrial interpretation of "dynamic interaction": examples from military and space fields illustrate new, or late, characterizations of interactions, which result in increased risks, or even late identification of interactions, which may lead to failures.

1 Nature of Interactions in a System

A definition of the system was proposed by Joel de Rosnay (ref. [11]) in 1974 and is still generally accepted (it is us who underline): "A system is a set of elements in dynamic <u>interaction</u>, organized to achieve a purpose." That rather old definition is taken up again by recent authors: "A system is an integrated set of constituent parts that are <u>combined</u> in an operational or support environment to accomplish a defined objective." (Federal Aviation Administration, 2006); a very close definition is given by the INCOSE (ref. [3], 2004); another one, hardly different, is given by Martin Wilke, François Fournier (ref. [12], 2010), referring to ISO/IEC Standards (ref. [4], 2008): "a combination of <u>interacting</u> elements organized to achieve one or more stated purposes".

Those concise definitions have the virtue of condensing the essence of the system in a brief expression the words of which all matter, and especially "interaction", "combined", "interacting". However, as the author himself admits in ref. [1], the definition is too general to be really useful; that comment is taken up by Gérard Donnadieu and Michel Karsky (ref. [2]): systems definitions "are poorly operative for action [...]. The richness of the system concept can only be revealed by its utilization".

The ambivalence of the words "dynamic interaction", one meaning of which is mentioned in the introduction, perfectly illustrates the previous comment.

First, the word "dynamic" plays an essential role: a system with only static interaction between components would be dead (even in a building, a cathedral, essentially static, stones or concrete beams do warp). This is the scientific and traditional interpretation: the structures of subassemblies of an aircraft, space launch vehicle, warp in relation to each other (accelerations, vibrations, bendings), more or less accordingly to the flight phases; on-board electronic pieces of equipment exchange data which by nature vary in time, etc.

Second, another interpretation of the word "dynamic" is brought out when we consider:

- the concept of "system loops": a complex system is pluridisciplinary by definition; each expert team iterates their calculations, sizings, processes, as many times as needed, while periodically validating the consistence of their input and output data with other teams; this is really a *dynamic* feature: the evolutionary character of the interactions arises all along the design-realization-validation cycle (see fig.1)
- the changes decided during the operational life of a system, which may bring about new interactions to be characterized and need the activation of "system loops" again (to be compliant with new requirements from the end user or customer, e.g. facing new or more severe environments)
- the reasons of difficulties (changes, delays), or failures, when "system loops" have not been properly operated or not activated at all.

The last point means that in spite of numerous iterations of "system loops", some interactions may escape a pertinent characterization. And, which is worse, some "system loops" may have been never activated.

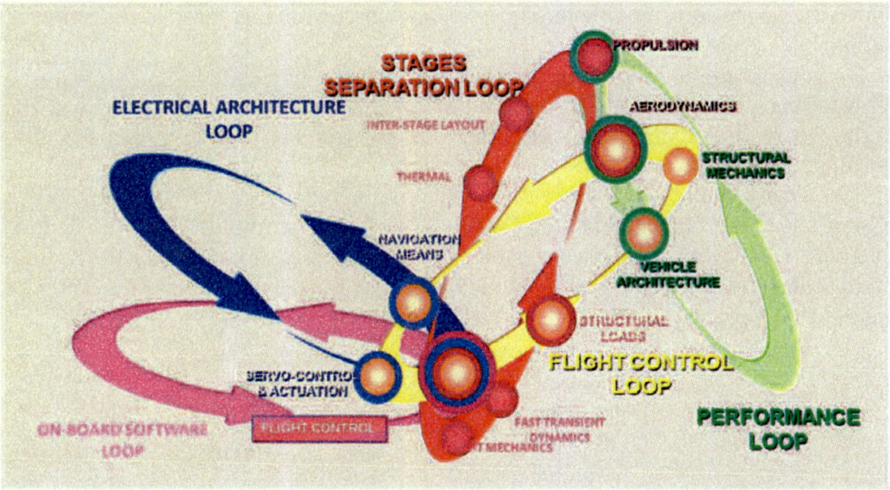

Fig. 1 "System loops" (ref. [8])

To sum up, the analysis of the words « dynamic interaction » leads to distinguish (not disjunctively) between two interpretations of the typical feature of systems, made of components in dynamic interaction:

- on one hand, the components exchange fluxes of data, matter, energy (scientific, classical and natural interpretation; that aspect is widely developed by authors such as Ludwig von Bertalanffy: ref. [1]) ;
- one the other hand, the exchanges between the components, while keeping their nature, may be differently characterized in the beginning and at the end of the program (changing and iterative feature); in some extreme cases, they

may be uncharacterized at all in the beginning and be discovered very lately (industrial interpretation; that point of view is the one considered by authors in Systems Engineering: ref. [3], [4], [9], [12]).

As written in the introduction, the latter point of view is the subject of the present paper.

The paper will first deal with an example resulting from the evolution of the operational use of a system, then will mostly focus on examples of late characterization and late identification of system interactions.

2 New Interactions Resulting from Changes in the Operational Use

2.1 Example of Electromagnetic Transmission under the Fairing of a Launch Vehicle

A launch vehicle carries one or two payloads (or more), generally satellites, to be put into some orbit. Those payloads are located on the top of the launch vehicle, above the upper stage and the equipment bay which contains most of the avionics equipment.

So as to ensure a good aerodynamic behavior, the payloads are protected by a nose fairing which prevents from aerodynamic efforts for the first minutes of the flight. As soon as those efforts are tolerable, at relatively high altitudes, the nose fairing is jettisoned, so as to lighten the vehicle.

The fairing is made of electrically conducting material, generally carbon. When it is in place, the fairing constitutes with the top of the equipment bay an electromagnetic cavity. In case of two payloads, two cavities are found under the fairing, more or less electromagnetically coupled (through slots, cables across the walls…) with each other and with the equipment bay which is a third cavity (fig. 2):

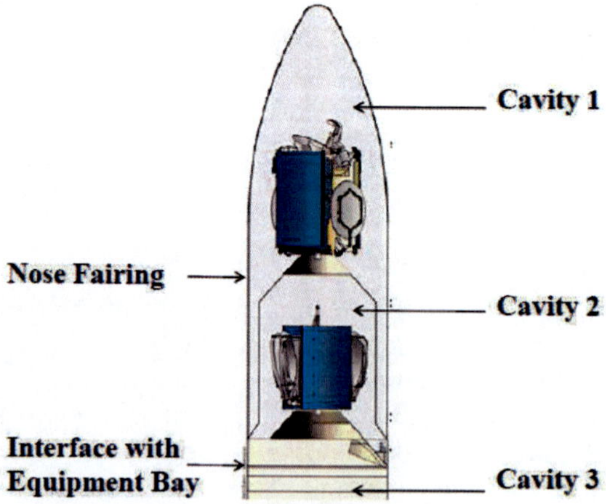

Fig. 2 Electromagnetic cavities under the nose fairing

Payloads are equipped, among others, with telemetry systems. At the beginning of a program, it can be a wise statement to forbid any radiofrequency transmission from the payloads when the fairing is in place: such a transmission, if allowed, might a priori result in resonances inside the cavities and coupling to the equipment bay, which might cause electromagnetic interferences to the electronic pieces of equipment and electrical harness.

On the other hand, when the mechanical and electrical definition of the vehicle is definitely known and the prime contractor has at his disposal powerful electromagnetic calculation codes and methods adapted to the high frequency fields radiated by the payloads, the radiofrequency transmission under the fairing can be considered: the advantage for the customer (operator of the satellite) is the possibility to power on the radiofrequency systems of the satellites before jettisoning, which simplifies its implementation, especially just after being put into orbit.

Of course, since that way of implementing the payloads inside the launch vehicles generates a new electromagnetic environment to the equipment bay (and also to various pieces of equipment, electrical and pyrotechnic devices, under the fairing), new characterizations have to be done: assessment of the electromagnetic field intensity inside the cavities under the fairing, assessment of the field intensity inside the equipment bay, comparison between that additional environment and the electromagnetic susceptibility of the electronic boxes, appraisal of the safety margins and final statement.

So, the interaction between the payloads and the on-board equipment of the vehicle may vary over time.

3 Late Characterization of Interactions

The question here is an interface between two components of the system or between the system and its environment, characterized from the beginning, but poorly.

3.1 Example of Trident D5 (ref. [5], [6])

The *Trident D5* – or *Trident 2* – missile is the latest ICBM (InterContinental Ballistic Missile) of the US sea-based nuclear deterrence. It succeeded the *Trident C4*, deployed in October 1979, which took over from the *Poseidon C3* submarine missile, deployed in 1971, taking over from *Polaris*, deployed in 1960.

Those missiles are operationally submarine-launched: experimental tests consist of first launches from the ground, then underwater launches *(Performance Evaluation Missile – PEM)* from a submarine.

The operational life of such a missile is divided into several phases: storage in a tube of the submarine, implementation for launch, ejection from its launch tube by the pressure of expanding gas, submarine ascent up to the emergence, ignition of the first stage at emergence, and controlled flight (three stages in row) until the delivery of several payloads.

Among those phases, the submarine ascent is short but fundamental: the missile must reach the sea surface with a correct attitude and limited angular velocity, so as to be controllable by the autopilot from the very ignition of the first stage. During the ascent, due to a complex fluidic process, a water jet – or water dart – goes up behind and with the missile, and may violently hit the nozzle used for flight control, located inside the aft skirt of the missile.

The *Trident D5* experienced 19 ground tests from 1987 to 1989, 15 of them succeeded.

On March 21, 1989, the *PEM-1* (*Performance Evaluation Missile n° 1*) flight of *Trident D5*, first submerged flight of the *D5*, from the *Tennessee* submarine, ended in a failure: after an apparently nominal submarine phase, the first stage of the missile properly ignited after emergence, but the missile looped like a pinwheel and eventually sank less than ten seconds after emergence, after tearing apart.

The failure was attributed to a high speed water jet, sucked up into the nozzle as the missile is ejected from the submarine. The analysis resulted in strengthening the nozzle in view of the second flight (*PEM-2*), which succeeded, on August 2, 1989.

The *PEM-3*, fired on August 16, failed.

The total of resulting changes (redesign, additional development and tests), among which four important changes in the missile bottom: shock absorber, attenuating device, shield preventing water from getting into the nozzle, modification of the igniter, came to more than 100 M$.

Submarine tests resumed on December 4, 13 & 15, 1989, then January 15 & 16, 1990; all succeeded. The *Trident D5* missile was operationally deployed in 1990.

The question of water jet was well known, for *D5* it was not a matter of discovery of interaction, but of late characterization of the interaction between the submarine environment and the missile.

3.2 Example of Underestimated Thermal Environment

A number of examples of inadequately characterized interactions can be found, hopefully identified as such before flight and less spectacular.

An example is given by the change of the upper stage propulsion technology on a launch vehicle, leading, for propulsion performance reasons, to the usage of cryotechnic technology, with liquid oxygen and hydrogen. It must be pointed out that, whatever the launch vehicle, the equipment bay, which shelters most of the avionics, is located in the vicinity of the upper stage (see fig. 3), in that case close to the oxygen and hydrogen tanks; hydrogen liquefaction temperature is 20 K.

Fig. 3 Vicinity of cryogenic stage and equipment bay

It happened that, only a few months before the maiden flight of a launch vehicle, it was necessary to install heaters inside the batteries, due to the late statement of the actual thermal environment in the vicinity of the cryotechnic stage and malfunction of the batteries. Therefore the pieces of equipment had to be shipped from the launch pad to the manufacturer, in order to design and implement the change, get an opinion on the need for a possible additional qualification; the onboard harness had to be modified to drive heating power…

And yet, qualification tests for the pieces of equipment located in the vicinity of the cryotechnic tanks had been revisited at the very beginning of the program…

4 Late Identification of Interactions

4.1 Changes Decided without Care of Possible Impacts on the Rest of the System

An utmost situation (actual!) is the change in the definition of a piece of equipment, decided by the development manager, without taking care of the possible impacts on the rest of the system (subassembly, subsystem); that change may be decided for acceptable reasons (one of the requirements cannot be met) or not (it is more simple, or faster, "to do so"); the latter situation can be met! The person is said not to have a "system approach". That case is normally trapped in project reviews, but inevitably results in redesign, additional delays and extra costs.

In the same way, some changes may be decided to improve some performances, but might have consequences unfortunately unexpected by people who made the decision. For example, metallic parts of stages structures can be replaced by carbon fiber ones, which lightens the vehicle. Carbon fiber has a rather good electrical conductivity, but much lower than a metal; so, due to a lower shield effectiveness (lower electromagnetic attenuation), the harness and pieces of equipment located inside the carbon fiber structure may experience more severe electromagnetic environments, which may have consequences on their behavior.

Once again, that case is normally trapped during project reviews, but may lead to additional verifications, or even changes in definition (metallization of the inside face of the carbon structure).

4.2 Ariane 501(ref. [7])

The failure of the first *Ariane 5* flight is widely dealt with in all the courses about System Architecture or Systems Engineering; it perfectly illustrates the risk associated with the reuse of a piece of equipment (hardware *and* software) which had been fully satisfactory on the previous program.

Ariane 5 is equipped with totally redundant electrical and electronic subsystems (avionics boxes) for navigation, guidance, flight control and vehicle management; that redundancy prevents from mission failures in case of random breakdowns in some piece of equipment.

Among those avionics boxes, the launch vehicle uses an Inertial Measurement Unit (*Système de Référence Inertielle - SRI)*, also redundant, already fully proven on *Ariane 4*.

The horizontal velocity of *Ariane 5*, at lift-off, is noticeably higher than *Ariane 4* velocity; in the maiden flight *501*, that feature resulted in a data overflow in the *SRI*, inside a program only used on ground; the fact was indicated by a diagnosis information (dedicated bit pattern of the *SRI* 2 computer breakdown), which was understood by the main on-board computer as a functional flight datum, while redundant *SRI 1* had already declared to be broken down. On the basis of those data, the main on-board computer elaborated steering commands inconsistent with tolerable aerodynamic loads, which resulted in the flight termination.

Among other failure reasons (software functionality remaining activated in flight without necessity, no functional data sent by the *SRI*, breakdown linked to a software exception affecting both redundant *SRI*s), the lack of ground verification of *SRI*s behavior facing *Ariane 5* trajectories led to be unaware of the interaction which brought about the failure (non-compliance of the piece of equipment with the application).

4.3 Mars Climate Orbiter (ref. [10])

The mission of *Mars Climate orbiter (MCO)*, first interplanetary satellite, was to study the atmosphere and climate of Mars; it had to serve as a communication relay to another American probe, *Mars Polar Lander*, which was to be launched a few months later. Both were part of *Mars Surveyor 98* program.

MCO was launched on December 1998; on September 1999, the ground noted the final loss of communication with *MCO* while the probe was in the vicinity of Mars (the probe did not reappear after its occultation by the planet in the beginning of its capture by Mars atmosphere). The most probable hypothesis was the destruction of the probe due to atmospheric turbulences and frictions.

MCO cost was 125 M$, the total of the program came to 328 M$.

The failure responsibility was attributed to a defect in NASA Systems Engineering process.

The failure was due to a navigation error before and when putting *MCO* into orbit around Mars. A first correction revised from 193 to 140 km the expected flying over altitude; 140 km altitude was deemed acceptable, the lowest limit was 85 km. In fact, the probe entered Mars atmosphere at 57 km only: the flying over air line was much lower than the theoretical one, non-compliant with a safe flying over.

The investigation team established that the on-board software had underestimated the probe thrusts by a ratio of 4, 45 (*pound-force/newton*).

The probe sent navigation measurements to the ground system in metric units; the ground system made calculations of some parameters in Anglo-Saxon units (*Imperial measurement units*), namely *pound.second*, sent without conversion to the navigation module which expected them in metric units (*newton.second*).

Besides an illustration of the "domino effect", that failure points out that an informatory interaction was forgotten: the Systems Engineering process had not specified the system of units.

5 Conclusion

The interactions between subsystems, subassemblies and equipment of a system, the interaction between a system and its environment, are, by nature, inherent to the concept of system.

Those interactions are, by nature again, dynamic, in the sense that the system components continuously exchange data, matter, energy, while they are operating. The characterization of those exchanges, by the processing of material and immaterial interfaces, is of course an integral part of the system design.

The examples shown, most of which take on a spectacular feature, point out that interactions are evolving along the development of a system, since their characterization is a resolutely iterative process - therefore dynamic again -; and in some extreme cases, interactions are identified very (too) lately.

Such cases come from the high level of complexity of military and space systems, including the number of people involved. As written in ref. [9], partly taken up in ref. [3], "a system is an integrated set of constituent parts that are combined in an operational or support environment to accomplish a defined objective. These integrated parts include <u>people</u>, hardware, software, firmware, information, procedures, facilities, service, and other support facets. People from different disciplines and product areas have different perspectives on what makes up a system."

Methods and tools of Systems Engineering were established in order to master at the best those interactions along the development of the system. One of the concerns of a program manager the day before a flight, whether maiden or not, is that the Systems Engineering process has forgotten no interaction and characterized well enough the interactions in actual conditions.

Systems Engineering has limits, which in fact are fortunately exceptional, even if those limits may feature spectacular consequences.

Advances in system design and development mastering are achieved by drawing lessons learnt from those exceptions.

References

[1] Von Bertalanffy, L.: General System theory: Foundations, Development, Applications. George Braziller, New York (1968); revised edition 1976

[2] Donnadieu, G., Karsky, M.: La Systémique, penser et agir dans la complexité, pp. 29–30 (2004); Liaisons (ed.)

[3] INCOSE SE Handbook, version 2a, p. 10, 11 (June 2004)

[4] ISO/IEC Standards 15388-2008, Systems and software engineering - System life cycle processes, ch. 4, 2nd edn. (February 01, 2008)

[5] Kolcum, E.H.: Aviation Week & Space Technology. Navy Assesses Failure of First Trident 2 Underwater Launch (March 17, 1989)

[6] Kolcum, E.H.: Aviation Week & Space Technology. Three Successful Launches Verify Design Fixes to Trident 2 C5 ICBM (January 8, 1990)

[7] Lions, J.-L.: ARIANE 5 - Flight 501 Failure - Report by the Inquiry Board; the Chairman of the Board: Prof. J. L. LIONS (June 23, 1966)

[8] Morey, J.-C.: ABOUT Gilles, Engineering and Validation on Space Transportation Avionics Systems. In: Complex and Safe Systems Engineering, Arcachon, France, June 21-22 (2004)

[9] National Airspace System (Federal Aviation Administration). In: System Engineering Manual, Version 3.1, ch. 2, pp. 2–2 (June 06, 2006)

[10] Report on Project Management in NASA by the Mars Climate Orbiter Mishap Investigation Board (March 13, 2000)

[11] De Rosnay, J.: Le Macroscope, p. 92, 100–101 (1974); Ed. Points

[12] Wilke, M., Fournier, F.: Astrium Space Systems Engineering Training (dedicated training) (2010)

Author Index